Also by George H. Daniels:

American Science in the Age of Jackson, 1968
Darwinism Comes to America (editor), 1968

SCIENCE IN
AMERICAN SOCIETY

SCIENCE IN AMERICAN SOCIETY
A Social History

GEORGE H. DANIELS

Alfred A. Knopf New York
1971

THIS IS A BORZOI BOOK
PUBLISHED BY ALFRED A. KNOPF, INC.

ISBN: 0-394-44386-1

Library of Congress Catalog Card Number: 79-118708

Manufactured in the United States of America

First Edition

FOR GEORGIALEE AND DONI

Contents

Preface

While it is commonplace to assert that science is the most vital force in American civilization, and while scholars know that, throughout our history, it has played a role at least comparable to literature and the arts, no effort to tell the complete story of science in America has ever been made. This failure is not due simply to the ignorance of American historians. In part, it has been due to the powerful prejudice held by historians of science against studying their subject in national cultures. Science is an international enterprise, so they have argued, and it makes no sense to speak of "American Science," or "English Science," or "Russian Science." They generally attribute the origin of what they regard as such an error to the Marxists, who were, indeed, the first scholars to see clearly that science was intimately bound up with culture. But Marxist over-enthusiasm should not obscure the basic truth of their position. Although "Science" in the sense of "Truth," as some insist upon using the term, *is* independent of national cultures or ideological origins, science, understood as an activity of men looking toward the comprehension of nature, differs markedly from society to society, and it is very much dependent upon the national ideological framework. In the past few years, many scholars have overcome their prejudice, and a number of good books and articles have been written about particular periods in the history of American science and about particular aspects of American science.

A second prejudice has also militated against an understanding of the role of science in this society. Traditionally, historians of science have seen their role as more celebratory than strictly historical. That is, they have adhered to the belief that, when studying science, they should study its "advances"—its successes. Throughout most of its history, American science has not been particularly successful in terms of adding to the theoretical structure of world science. Therefore, the scholar who chose to study American science before the twentieth century, unless he chose Benjamin Franklin or Willard Gibbs as his subject, could be accused, by definition, of misdirecting his efforts. Was not Darwin an Englishman, Helmholtz a German, Volta an Italian, and Ampère a Frenchman? How could one presume to study science in the nineteenth century without studying these people? American science was, after all, "derivative" and this, to the celebratory school, meant that it was not worthy of serious attention; except, perhaps, in an effort to explain its failures.[1] The more sophisticated history of scientific ideas school, represented at its best by Thomas Kuhn, A. Rupert Hall, and Alexandre Koyré, likewise denigrated the study of science in relation to culture, but on the grounds that scientific ideas are essentially autonomous.

The celebratory school has all but vanished among professional historians, and the history-of-scientific-ideas school, although still strong, has been losing its exclusive hold on the field in the past few years. A growing number of scholars have turned their attention to American science as a subject worthy of study in itself, simply because it is an expression of the culture of the American people, and because understanding the American approach to science helps one to understand Americans. A historian needs no other justification for devoting some attention to a subject.

Insofar as possible, I have used the works of those scholars who have begun to pay serious attention to American science in writing this book. The story is still so incomplete, however, that even for those subjects I chose to deal with, it was not possible to write completely from secondary sources. Each chapter is a hybrid, based partly on the works of others and partly on original research including, in most cases, archival and manuscript work. I have not presumed to tell the complete story of science in American society; I am painfully aware that a great many subjects of equal importance to the ones I have chosen have been omitted. But I do not believe that an author should have to apologize for selectivity, provided that he make his purpose clear. My purpose has been simply to provide an outline for the study of science in American society, focusing upon topics that seem important to me. I would be pleased if my book did no more than stimulate others to study important topics that I have neglected.

In every chapter except the last, I have sought to give a reasonably complete interpretation of a single theme in the history of American science. But the subject of science in the past thirty years is so vast that this approach is not possible. Therefore, in the form of an Epilogue, I have simply attempted to survey some of the major changes that have occurred in the pursuit of science in America during the present century and, insofar as possible, to account for their origins and suggest implications of the changes. To do more would require a separate book and many more years of study.

I have tried, throughout the text, to give credit when I borrowed the works of others. But there are three scholars whose works have been so useful that they deserve a special mention. The first of these is Theodore Hornberger, whose articles on Colonial science written during the 1930's and 40's are a mine of information which should be neglected by no one interested in the subject. Even when I disagreed with him, his works saved me an enormous amount of time. Brooke Hindle's *The Pursuit of Science in Revolutionary America* is a model for period studies and has been useful in several chapters. Finally, A. Hunter Dupree's *Science in the Federal Government* is a work without which this one could not have been written. It, too, is a model of its type.

I was aided in the present endeavor by a grant from the American Philosophical Society for work in Early American science, by another from the National Science Foundation for work in the late nineteenth and twentieth centuries, and by Northwestern University in providing leave time. Librarians at the Massachusetts Historical Society, the Pennsylvania Historical Society, the American Philosophical Society, and Harvard University were particularly helpful. My friend, Professor Robert Dykstra of the University of Iowa, as usual read every word and offered valuable suggestions for improvement, as did Professor A. Hunter Dupree of Brown University. Professors E. William Monter and Clarence Ver Steeg of Northwestern and Peter Carroll of the University of Minnesota read chapters that impinged upon their specialties, and also offered helpful advice. Finally, Angus Cameron of Alfred A. Knopf has provided everything from intelligent advice and encouragement as the work went along, to indulgence for my tardiness in completing the manuscript.

SCIENCE IN
AMERICAN SOCIETY

ℭ NOTES

Footnote numbers in the text are for bibliographical notes likely to be of interest primarily to scholars. These notes are placed at the back of book and may safely be ignored by the general reader.

I

~~~~~~

Medieval Science and the
New World

C After his first American trip, the European naturalist and
world traveler Alexander von Humboldt, commented on the
early impact of the New World on Old World thought. Its importance, he said, extended far beyond that of merely offering new
curiosities of nature for man to delight in; it had a particular impact on knowledge about physical geography, the varieties of the
human species, and the migrations of nations:

> It is impossible to read the narratives of the early Spanish
> travellers, especially that of Acosta, without perceiving the in-
> fluence which the aspect of a great continent, the study of
> extraordinary appearances of nature, and intercourse with men
> of different races must have exercised on the progress of knowl-
> edge in Europe. The germ of a great number of physical truths
> is found in the works of the sixteenth century; and that germ
> would have fructified had it not been crushed by ignorance and
> superstition.

Von Humboldt, friend of Thomas Jefferson and the very model
of an Enlightened philosopher, could not help but moralize upon
the "ignorance and superstition" that had so long delayed the
knowledge of the great physical truths that he thought were implicit in the discovery of America. In Enlightenment fashion, he
tended to dwell more upon the failures to take advantage of the
new opportunities than he did upon the actual impact of the New
World—which impact was indeed important, touching as it did
virtually every area of Renaissance thought. New discoveries out-
moded old knowledge faster than the new could be assimilated,

preconceptions ranging from cosmology to schemes of classification were abandoned, and men were introduced, at the same time, to boundless new possibilities and apparently insuperable problems. Out of it all came the wholly different world view known as the "Enlightenment," important aspects of which were conditioned by the existence of a previously unknown world.

Modern physical science would probably have arisen in Western Europe during the sixteenth and seventeenth centuries whether America had been discovered or not, since the necessary groundwork for it had been laid during late medieval times and the frame of mind necessary to develop science had already arisen. The same "this-worldly" orientation that led Columbus to seek a more convenient way to the treasures of the East provoked others to enter trade and commerce on a grand scale, and still others to seek to manipulate nature for the material benefit of man. Moreover, the change from feudal economy to merchant capitalism and manufacture had presented European science with many problems, the solutions to which were important components of the scientific revolution. More efficient methods of mining were demanded to supply larger quantities of metals for machinery and artillery. New sources of power were urgently needed. Among the most imperative needs were the development of more adequate means of marine transportation to settle and exploit the new colonies successfully, and swifter modes of communication to make possible a vigorous trade with distant parts of the world. This required studies in the determination of latitude, longitude, and exact time, and in the cause of tides. There was no direct route from the countinghouse to the chemistry laboratory—as some extreme economic determinists would have us believe—yet in a milder form, the connection between countinghouse and laboratory was a reality. Only a society like that which had developed in Western Europe by the fifteenth century could have produced either; and the greatest scientific discoveries of that era did reflect the needs of those rapidly changing times.

The groundwork for the scientific exploitation of the New World had been well laid with the revival of learning in the fourteenth and fifteenth centuries. The curiosity of Renaissance men had led to an enthusiasm for the collection of "all things naturall or artificial," including wild animals in zoos, botanical specimens in herbaria, fossils, minerals, gems, shells, human artifacts, and human customs. One recent writer has seen the beginnings of ethnology in a 1499 publication of a collection of human customs by Polydore Virgil.[1]

But although the original impetus existed before the period of exploration, enthusiasm would not have been enough to produce a science. Furthermore, the exact form taken by the scientific revolution would have been dramatically different had it not been for the fact of America, for many questions that were central to the formation of the scientific world view of the period simply would never have arisen. New plants, new animals, new stars, even new human beings and new types of human societies were found, and all these challenged traditional ideas and assumptions. They also brought new opportunities; in a certain sense, before the explorations opened up the Western Hemisphere and brought greater knowledge of the outlying areas of the Old World, the data that had been collected could form only the raw materials of science. With half of the human societies in the world unknown, for example, a science of ethnology was impossible; with more than half the plants and animals of the world unknown, even as rudimentary a task as natural-history classification could not be accomplished with even approximate faithfulness to nature.

The New World's impact, to reduce the matter to its essence, was in terms of the problems it raised, and the questions that it consequently suggested. The first great problem posed by the New World was the very fact of its existence—a fact regarded as highly improbable if not downright impossible in fifteenth-century cosmology. It is well known, for example, that Columbus was so committed to his belief that another land mass could not exist that he was never able to recognize his own discovery of it. All his learning and all his experience dictated that it could not be so; until the day of his death, he persisted in the belief that he had reached the Orient.

It is not surprising that Columbus should have been so deluded, for the world of the educated European on the eve of the Columbian voyages extended roughly from Iceland south to the Guinea coast of Africa (70°N. to 12°N.) and from Shanghai west to Cape Blanc, West Africa (120°E. to 17°W.). The land mass thus described, representing about one-fourth of the earth's surface, was the *Oikoumene*—the limits of the habitable earth—an area only slightly larger than Ptolemy's world. The very few pre-Columbian geographers who knew about Greenland—which had well over 5,000 settlers by the middle of the thirteenth century—considered it to be a peninsula of Europe and grossly misrepresented its position on the globe.

Medieval cosmologists placed various barriers about the *Oikoumene* in order to explain why it represented the limits of the habitable world: to the north of Europe, it was thought to be solidly frozen; to the south of the Canary Islands, off the coast of Africa, began the dread Tenebrous Sea, where a mixture of the boiling waters of the tropics and the freezing ones flowing from the poles met to produce a thick fog of vapors. These vapors, mixed with sands carried out of the African desert by the winds, formed an impenetrable mass. Beyond were the most terrible creatures the mind of man could imagine. And beyond the monsters was ocean—empty, uninhabitable, and probably boiling.

The barriers placed around the habitable world in medieval thought represented an important characteristic of the entire medieval world view, which could be described as essentially a closed one. To medieval man the universe had a roof on top, his towns had walls around them, his churches had "hedges" about them, and all roads ended definitely somewhere. The concept that the habitable earth had firm limits was, therefore, simply one aspect—although an important one—of a cloistered and introspective world view that pervaded all of life. It is no wonder that men were so reluctant to give it up, nor is it surprising that the act of giving it up implied profound reorientations in thought.

Although throughout the medieval period there had been some who argued that there might be land in other quarters of the globe, and isolated thinkers such as Albertus Magnus (c. 1200–80) and Roger Bacon (1214–94) even argued for the possibility of an inhabited equatorial belt and populated antipodes, majority opinion was on the side of Aristotle, who held that there was no other *Oikoumene* outside his own. The generally accepted size of the *Oikoumene* was that given by Eratosthenes: about 9,000 miles from east to west and 4,500 miles from north to south. Since the generally accepted figure for the circumference of the earth was 18,000 miles, the estimate of Posidonius, medieval thinkers in general postulated a great watery void for the major portion of the earth. Columbus, like others of his time, accepted the concept of the watery void; accepting also the erroneous measure of the earth and the exaggerated size of Asia, he concluded, therefore, that the distance from Spain to India was only about 3,500 miles. Despite the availability of better advice, the Spanish crown accepted Columbus's reasoning, agreeing, as an act of desperation in its rivalry with Portugal, to support the Genoese sailor's venture. When on his momentous voyage Columbus sighted what appeared to be continental land, the accumulated wisdom of the ages, calculation and his own fervent hopes dictated that it must be Asia. Never mind that the naked natives did not at all resemble those described by Marco Polo, Mandeville, or others he had read; the fabulous civilization of the Orient must be nearby.[2]

There were many and cogent reasons in medieval cosmology why the unknown portion of the earth must be presumed to be void—that is, without land and certainly without inhabitants. Even though few educated men doubted the fact of the earth's roundness, their conception, nevertheless, was of a very different earth from ours. It was a round earth, hanging immobile in the heavens with the sun, planets, and a rotating sphere of "fixed stars" traveling about it. And even though it was round, there was a "top" and a "bottom" to it. Most medieval thinkers seriously believed, therefore, that everything on the "other side" of the earth would be upside down—just as Lactantius and the Church fathers had said. Those who dared to believe in the antipodes were ridiculed with images of people marching with their feet up and their head down, of trees growing downward, and of rain falling from the earth to the heavens. As able a thinker as the Jesuit priest José de Acosta had to grant that, on grounds of reason alone, such images were justified. The only explanation Acosta could offer was that he and others who lived in Peru "finde not our selves to bee hanging in the aire, our heades downward, and our feete on high"[3]—certainly a convincing argument, albeit not a very intellectually satisfying one.

Medieval men had to struggle with a great deal more than a conceptual difficulty, however; the physical possibility of land existing at the antipodes could not easily be reconciled with Aristotelian physics. According to Aristotle, earth was heavier than water and the masses of each must precisely balance each other or the earth would not remain immobile in its appointed position in the center of the universe. Secondly, while planetary influences could be invoked to explain why one land mass, although heavier than water, nevertheless stayed above the water, the addition of a second land mass in another quarter of a station-

ary globe would hopelessly complicate matters. According to
Aristotle, in fact, in principle the sphere of water should cover
the sum total of the globe; it should form a regular sphere outside
earth, the heavier element. The Bible, however, explained that
God commanded the sea to withdraw in order to allow a portion
of the earth to appear. Putting these unimpeachable sources together, a medieval thinker could only conclude that all land not
submerged was something exceptional; the land is therefore of an
insular nature. From this point of common agreement, two rival
physical hypotheses were invoked to explain the uncovering of
the Island of Earth: (1) The center of gravity of the earth sphere
did not coincide precisely with its center of magnitude; since this
was so, a portion of the earth—a bulge—protruded above the
ocean. (2) A star attracted the waters, making an immense mountain of water on one side of the globe and uncovering the Island
of Earth on the other. The proponents of both these hypotheses
agreed that land occupied approximately one-fourth of the surface, and both virtually ruled out the existence of land in other
hemispheres.[4]

The second hypothesis was the preferred one, for it had
much greater explanatory power. Various biblical events, such as
Noah's flood, could also be physically explained on the same assumption that it was only the virtue of the heavenly bodies which
defied the predisposition of water to find its own level. If once the
stars relaxed their hold on the huge reservoir of water, piled up at
the south, then naturally a universal deluge would result. God, it
was thought, in the time of Noah had been so incensed by the sins
of men that he had temporarily suspended the planetary influences, allowing the waters to flow back over the land mass to the
north. This was a perfectly legitimate explanation that fit in quite
well with the accepted physics, but, of course, the flood could not
have been "universal" on this theory if there existed any land to
the South. Aristotle, the Bible, and the common sense of the age
were as one in rejecting the other *Oikoumene*. The discovery that
there was indeed an enormous land mass on "the other side of the
world" therefore reopened physical questions of such a fundamental nature as the size of the earth, its motion, and its constitution—all of which Aristotle had presumably closed; and it also
reopened the religious questions that had been so easily resolved
in terms of Aristotle's physics.

Given the importance of the fifteenth-century explorations,
it is understandable that the first major advance in modern

science occurred in astronomy, a discipline which has historically been closely related to geography and to navigation. Under Prince Henry the Navigator (1394–1460), of Portugal, an observatory had been erected at the extreme southwestern point of Europe and numerous expeditions had been sent along the coast of Africa to chart the coastline and establish contact with the natives. Prince Henry's twin ambitions—to extend the Christian faith and to circumnavigate the African continent—had resulted in giving a tremendous impetus to astronomy and cartography in Portugal.

The impact of the Portuguese discoveries and, later, those of their Spanish rivals, is evident even in the work of Copernicus, the man who inaugurated the revolution in astronomy. In his *De Revolutionibus*, Copernicus devoted one chapter to the subject "How Earth, with the Water on it, forms one Sphere," arguing on the basis of the recent geographical discoveries that the globe was predominantly made of earth, and that both water and earth were required to make the globe a sphere. His reasons for this argument are evident: earth breaks up less easily than water when moved; motion of a solid globe is more plausible than of a liquid one. Since Copernicus held that the earth moved naturally in circles because it was a sphere, he needed to show that both earth and water were essential to the composition of the sphere, in order that both would participate in the sphere's natural motion. The explorations conveniently provided him with the evidence he needed to force a reconsideration of the nature of the sphere.

Major arguments against the mobility of the earth had been that at one time the region of water would be above that of earth, and that men in Europe would occasionally be upside down. But geometrical argument, Copernicus said, demanded that the mainland of America be diametrically opposite the Ganges basin in India; since both areas were now known to be inhabited, the old argument for a stable globe was therefore invalid.[5]

The very knowledge of Copernicus's facts did not, of course, automatically imply a belief in the motion of the earth. It is well known that even the highly intelligent did not immediately flock to Copernicus's standard. The Spanish priest José de Acosta, for example, also argued that the earth "imbracing and ioyning with the water makes one globe or round bowle," but clung fast to the immobility of the earth. His concern extended no further than explaining how the earth could be kept together. Yet, while one should not make too much of the relevance of America to Copernicus's argument, he should also beware of

making too little of it. Even though the new land did not imply
motion, the least that can be said is that, as Copernicus recognized, it did remove the major objections to the possibility of
motion, and it therefore seriously weakened the position of the
supporters of Ptolemy. Where once they had been able to point
to the absurdity of a heliocentric theory, opponents now had no
recourse but to debate it on its merits. Surely, the suspicion that
Copernicus was right must have entered the minds of many who
made the Atlantic crossing and stepped onto land on the other
side of the world. At least one early Maryland resident reported,
half jokingly to be sure:

> We had a blowing and dangerous passage of it, and for some
> days after I arrived, I was an absolute Copernicus, it being one
> main point of my moral creed, to believe the world had a pair
> of long legs, and walked with the burthen of Creation upon her
> back.[6]

It was the Spanish priest José de Acosta who, more clearly
than any other early writer demonstrated that the function of the
New World was to reopen questions that the science of the Old
World had thought to be closed. The book written by Acosta while
on a Jesuit mission in Peru is usually known by its Spanish title,
Historia Natural y Moral de las Indias (Seville, 1590). It was
published in Latin in 1588; by 1604, it had been translated into
Spanish, Italian, French, Dutch, German, and English. It became
even more widely known when large portions of it were reproduced in *Purchas His Pilgrimes,* a compilation of travel literature
published by Samuel Purchas, an English priest and geographer,
in 1625.

Acosta's intent, as stated explicitly in the book, was to
correct those parts of natural philosophy which the New World
had demonstrated to be in error. He had had the inadequacy of
some of the older concepts thrust upon him on the voyage itself,
he said. Having read

> What Poets and Philosophers write of the burning Zone, I
> perswaded my selfe, that coming to the Equinoctiall, I should
> not indure the violent heate, but it fell out otherwise; for when
> I passed . . . I felt so great cold, as I was forced to go into the
> sunne to warme me; what could I else do then, but laugh at
> Aristotle's Meteors and his Philosophie, seeing that in the place
> and at that season, when as all should be scorched with heat,
> according to his rules, I, and all my companions were a colde.

Acosta's shock at the discovery, even at that late date, was entirely natural, for the strength of the old theory was that it not
only had centuries of tradition on its side, but was also a prediction based on the best available generalizations concerning the
causes of the distribution of temperature. Others regularly reg-
istered their surprise at the same discovery. Pigafetta, diarist of
the Magellan voyage, for example, entered under date of October
3 the following remark:

> We had opposing winds, with brief calm spells, and it rained
> all the way to the equator; and the weather was damp for
> sixty days, against the opinion of the ancients.

As Pigafetta's comment indicates, the fact that it was "against
the opinion of the ancients" was not lost either to the travelers or
their audience. Queen Elizabeth I is said to have charged Raleigh
with atheism when he ridiculed Aristotle's notions about the
Torrid Zone.

Clearly, said Acosta, the older, simple explanation that it
was hottest when the sun was most directly overhead—however
well it may have fit the facts in the limited part of the world
known to Aristotle—was false. Along with this factor, Acosta
found it necessary to discuss such matters as the prevailing
winds, the altitude and proximity to water. These considerations, he concluded, not only helped one understand why the
equatorial zone was habitable, to the confounding of Aristotle
and Pliny, but sometimes actually more comfortable than the
regions outside the Tropics.

Acosta was also fascinated by the problem of the trade
winds—those regular northeasterly and southeasterly winds
which blow from the subtropical belts of high pressure toward
the equatorial belt of low pressure. Since their explanation
depends upon the motion of the earth, however, Acosta was
only able to state the problem posed by their regularity, not
to solve it.

His interest in weather phenomena and climate was
typical of most scientifically inclined colonial Americans. Early
Spanish explorers, including Acosta, Gómara, Hernández, and
Oviedo, who wrote the first systematic treatise on the natural
history of the New World in 1526, contributed not only criticism but information which enriched the physical and natural
sciences. Their works are filled with observations on the dis-

tribution of temperature on the earth, variation of climate in
latitudes of mountains, the limits of perpetual snow in each
latitude, the relation between the sea and continental areas,
composition of the atmosphere, and general information on
the conditions of life. Oviedo, as Acosta did later, speculated on
the habitability of the Torrid Zone, concluding that it was caused
by rains, the many rivers, marshes and other sources of water.
He also noted that there were many tall mountains and cool
breezes, and thought that perhaps the equality of night and
day provided a longer cooling period than was customary in
Europe during the summer. Oviedo also wondered about the
difference in tidal activity in the Atlantic and Pacific oceans,
and in his account of the origin of coal deposits and their relation to gold, suggested an evolutionary theory of land formation.[7] Three British-American Colonials, John Mitchell, John
Bartram, and Benjamin Franklin, were later led into some of
their boldest hypotheses by their observations of the characteristic differences of American weather.

Acosta is only the best of many sixteenth-century examples
of the impact that the physical fact of the New World had on
Western European thinkers. He illustrates the truth of the
assessment made many years earlier (1530) by Jean Fernel,
physician to the King of France:

> This age need not, in any respect, despise itself and sigh for
> the knowledge of the ancients. . . . Our age today is doing
> things of which antiquity did not dream. . . . Ocean has been
> crossed by the prowess of our navigators, and new islands
> found. The far recesses of India lie revealed. The continent of
> the West, the so-called New World, unknown to our fore-
> fathers, has in great part become known. In all this, and in
> what pertains to astronomy, Plato, Aristotle, and the old phi-
> losophers made progress, and Ptolemy added a great deal more.
> Yet, were one of them to return today, he would find geography
> changed past recognition. A new globe has been given us by
> the navigators of our time.[8]

The possibility that life might exist on the "other side
of the world" was one that was even more frightful to medieval
thinkers than was the existence of land. Pierre d'Ailly in his
Tractatus de Imagine Mundi (c. 1410) makes one reason for
the unacceptability of this belief clear. Some people, he observed,
say that the zone between the Tropic of Capricorn and the
Antarctic Circle is as temperate and habitable as our own.

They explain that there can be no communication between those living in the two temperate regions, however, because of the impossibility of crossing the intervening Torrid Zone and the Tropics. But this opinion borders on heresy, D'Ailly said, for the population of the region beyond the Torrid Zone would be ignorant of the teaching of Christ and the Apostles, clearly contrary to the sacred affirmation that "their sound went into all the earth and their words into the ends of the world."

The thought that there could be human beings on opposite sides of the equator, and that there nevertheless could be no communication between them, as D'Ailly and many others both before and after him recognized, raised possibilities too frightful to even be considered seriously. Plainly, given the theory of the equatorial heat, belief in an antipodal race was equivalent to belief in the polygenetic origin of mankind— that is to say, there must have been two Gardens of Eden. "If Aristotle will forgive the expression," as Dante said, "one may well call those [people] asses who can believe any such thing as this."[9]

Faced with the undeniable fact, not only of land but of inhabitants, Europeans reacted in two different ways. On the one hand the discovery of the New World gave credence to the marvelous. The genuine marvels found in it led some men to believe, for example, in Pliny's assertions concerning the monstrous races of men. Aldrovandi depicted a black African race with their lower lip hanging to the breast, believed in Patagonian giants and in the prevalence of hermaphrodites in Virginia. Eusebius told of natives in the vicinity of California whose ears drooped almost to the ground.[10] The sailor David Ingram, who left Hawkins's expedition in 1568 to go overland from the Gulf of Mexico to Cape Breton—a distance of 2,000 miles by Indian trails—provided a fantastic tale of marvelous adventures and marvelous beasts which stirred the English imagination for some time after his return. Even as skeptical a thinker as Raleigh helped spread stories of monstrous races of men.

Although such stories were given currency for a while, the more significant reaction—and really, I think, the dominant one—was the direct opposite of this. It soon became clear that the new continents, despite their novelties, were not es-

sentially different from the old—neither in physical structure
nor in inhabitants. Voyagers did continue to search for monsters for a very long time, but as voyage after voyage ended
in finding no dragons, no sirens, no monstrous races, the belief
in them declined. A characteristic reaction is that recorded by
Pigafetta, diarist on the Magellan voyage around the world.
They were told, he reported, that not far from Molucca there
was an island where there were men of only one cubit in height
but with such big ears that they lie upon one and cover themselves with the other. "But owr men wolde not sayle thyther,"
he explained, "bothe bycause the wynde and course of the
sea was agaynste theym, and also for that they gave no credite
to his reporte."[11]

Such skepticism became more and more a characteristic
reaction, especially with the ripening of humanism in Renaissance times. Often it was the mariners themselves who laid
to rest the monstrous stories. Maximilian Transilvane, secretary to the Emperor Charles V, told of his persistent questioning of returning seamen. He had discussed their experiences
and observations at length, he said, in order to discover whether
there was any truth in the ancient stories:

> All which gave the selfe same information, and this with such
> faythfulnesse and sinceritie, that not only they are judged of
> all men to have declared the trewth in all thynges, but have
> thereby also given us certeyne knowledge that all that hath
> hiterto byn sayde or written of owlde autours as touchyinge
> these thynges, are false and fabulous.

Writing in 1522, just after the Magellan voyage around the
world, Maximilian wondered who would any longer believe
that there were men with only one leg or with feet whose
shadows covered their bodies, or who were only one cubit
tall. All such creatures would be monsters rather than men,
and surely if they existed either the Spanish or the Portuguese
would have reported seeing them. The inevitable conclusion, he
thought, was that antiquity, having no such knowledge of the
world as did men of the sixteenth century, had been exces-
sively credulous.[12]

In such manner, it gradually came to be realized that
all the marvels in the New World were strictly natural. There
were marvels enough of this variety. Francis Bacon, for example, was astonished by rivers larger than any known; he
thought the Amazon particularly impressive. "They have such

pouring rivers," he said, "as the rivers of Asia and Africk are but brooks to them." He also claimed for the Andes—falsely, as it turned out—an altitude in excess of anything the Old World knew. The influence of the voyages on Bacon is indicated by the fact that much of the imagery in his writing is borrowed from the explorers. He expressed himself, for example, as aspiring to be the Columbus of a new intellectual world, to sail through the Pillars of Hercules (a symbol of the limits of the old knowledge) into the Atlantic Ocean in search of new and more useful knowledge. At one point he explicitly stated that "by the distant voyages and travels which have become frequent in our times, many things in nature have been laid open and discovered which may let in new light upon philosophy." Sir Thomas Browne, another close student of the voyages, noted that discoveries in the New World had upset theories about which were the highest mountains. The "enlarged Geography of aftertimes" had shown that heaven-kissing Olympus was almost insignificant as compared with the "Andes in Peru." Having relinquished the authority of the ancients in some things, said Browne, "it may not be presumptuous to examine them in others." Surely, he concluded, on the experience of "our enlarged navigations," it would be most unreasonable to consider the ancients infallible. It is clear from such statements as these that in the minds of both Bacon and Browne, the impulse to dominate the world geographically was intimately related to the growing impulse to dominate nature.[13]

The natural products of the New World were themselves regarded as marvelous in nature. The virtues of tobacco as a cure-all for every disease were highly touted not only by Portuguese traders who had a vested interest and had encouraged a gullible public to accept the weed as a medical cure-all, recommended in the treatment of fifty-nine diseases, but also by serious-minded physicians who no doubt did feel that they had made a discovery that would prove remarkably beneficent to man. Guaicum (lignum vitae) was regarded as a sovereign remedy for syphilis, the potato as an aphrodisiac of wondrous potency—"They comfort, nourish, and strengthen the bodie, procure bodily lust, and that with greedinesse," said Gerald in his *Herball of 1597*—and sassafras was characterized by Davenant as being "dearer than unicorn's horn."

As these illustrations indicate, the discovery of the New

World raised great hopes in the minds of medical men and the
sick of Europe. Although time was to prove that, except for
quinine and ipecac, the Americas were to make no important
contributions to the pharmacopeia, the delusion is characteristic
of most of the medical writings produced in the sixteenth century. The first book on New World botany was, in fact, the
Spanish physician Nicholas Monardes's *Dos Libros, el uno trata
de todas las Cosas que sirven al uso de Medicina* (Seville,
1565); Monardes devoted an entire chapter to tobacco and
"his great vertues." Far more revealing of the point of view is
the title of the 1577 English translation of Monardes by John
Frampton: *Joyful Newes out of the New Found World Wherein
is Declared the Rare and Singular Virtues of Diverse and Sundry
Herbs, Oils, Plants, and Stones, with Their Applications as
well for Physic or Chirurgery, The Said Being Well Applied
Bringeth Such Present Remedy for All Disease, as May Seem
Altogether Incredible.* The same point of view underlies some
of the work of the early English travelers to the colonies, such
as John Josselyn.

The stimulus given to natural history by the New World was
really incalculable. It was, I think, probably responsible for the
new interest in nature so evident at the time. The idea that the
study of nature for its own sake is a worthwhile enterprise
was not common in late medieval thought; its development
was contemporaneous with the explorations. Oviedo expressed
the new point of view very well when he prefaced his account
of New World insects with the following explanation:

> Although some of the things I am about to describe are filthy
> and not so clean and agreeable as those already described,
> nevertheless they are worthy of being noted, so one can see
> the many things in nature.

Along with the new desire to see "the many things in nature," the
broadening of perspective and the knowledge of new forms,
unaccounted for in the old system, gave a powerful impetus
to the search for new classificatory systems. This search became a major concern of the seventeenth and eighteenth centuries. One can, in fact, say that the beginnings of modern
classificatory systems were in large measure a response to the
explorations.

As long as man's knowledge of nature is limited to what

he can observe in his immediate vicinity, he has no need for a sophisticated way of classifying the objects he encounters. Aristotle had used only two classificatory terms, *species* and *genus,* and with the limited number of forms known to him this was adequate. In late medieval times, botanical objects were typically divided into herbs, bushes, and trees; the only point of interest being individual forms and their medicinal virtues. Conrad Gesner's arrangement of animals, published about 1550, is illustrative of the very best medieval practice. The zoological world was divided into six large classes: viviparous quadrupeds, oviparous quadrupeds, birds, fishes, reptiles, and insects. Under these class headings, the arrangement was alphabetical and each animal was discussed under eight headings: (1) its name in different languages, (2) habitat and origin and description of external and internal parts, (3) "the natural function of the body," (4) the qualities of the soul, (5) its use to man in general, (6) utility as an article of food, (7) utility for medical purposes, and (8) poetical and philosophical speculations about the animal, anecdotes and resemblances to be found in different authors. In such an arrangement, of course, there appeared no notion of connections in nature, no idea of relationships among different animals; all of this was obscured by the alphabetical arrangement.

The new data at first seemed to be a bit too much for the sixteenth-century naturalists to handle. There was an element of the incomplete and the miscellaneous about their works —so much so that they struck the historian of science Lynn Thorndike as inferior to a thirteenth-century botanist like Rufinus in observation of nature and descriptive ability, and to Albertus Magnus in intellectual range and grasp and in critical and logical power. The only virtue Thorndike could see was that there were simply more of the naturalists in the later period.[14]

Surely this is too harsh a judgment; and, besides, it is based only upon Thorndike's study of those who stayed at home, whose knowledge of the novelties was at best based upon the isolated specimens that made their way back to Europe, and at worst, on hearsay. None of the men he studied *do* have either the descriptive power or the critical ability of Oviedo, for example, who spent thirty years in the West Indies, or, better yet, of José de Acosta, who spent seventeen years traveling between Peru and Mexico. Even Thomas Harriot, by far the

intellectual inferior of either of the Spanish writers, surpassed
in descriptive ability any of the writers discussed by Thorndike.

But aside from a few individuals like these, there is no
doubt that the tendency was at first to present the new data in
the old patterns and framework, and in consequence they were
often lost sight of. Certainly as isolated facts they not only
failed to illuminate, but often bred confusion. It is, however, only
natural that this should have been a first reaction. Most of the
early explorers believed that practically the same plants and
animals existed everywhere, with only minor variations. Animals were thought to have all spread to the East and the West
from Mount Ararat after the subsidence of the Deluge; consequently they were all either those familiar to a European
or those of which he had read as belonging to the East. It
was the same with plants. The botanist-physicians of the early
Renaissance remained well satisfied if an herb could be identified
in the pages of Dioscorides, cultivated, and its therapeutic
qualities roughly assessed. Every effort was made to force a
new plant into some description found in Dioscorides or Pliny.
Peter Ryff, for example, tried to identify guaicum, a rare South
American wood valued in the treatment of syphilis, with the
ebony of Dioscorides.[15]

Columbus, six days after landing on San Salvador, wrote,
"All the trees are as unlike ours as day is to night, and so are the
fruits and also the plants"; and on his second voyage, he sent
home a considerable collection of New World vegetables. But he,
of course, remained convinced that they were Oriental species.
The observations of Chanca, his physician on the second voyage,
also illustrate the tendency to identify new forms with the old.
He thought he saw trees bearing tragacanth, nutmeg, ginger,
aloes, cinnamon, mastic, and pepper—all of which were species
occurring in the Orient and only one of which was actually found
in the New World.

The tendency to identify new forms with old ones persisted for quite a long time. If a bird or animal appeared larger
than some similar European species, it was because of the vastness
of the strange country and the difficulties of gaining a living there,
whether by man or beast. As late as 1762, for example, one finds
the Frenchman Nicolas Denys making the same mistake of thinking the American forms, especially of the birds he described, to
be similar to European species. Eighteenth-century writers reg-

ularly listed the bullfinch, the nightingale, and the hedge sparrow among the birds of the New World.[16]

The idea that there could be something essentially new—something unknown to either the Jews or the Greeks—was alien to the thought of the time, and its discovery was often painful. Perhaps the first seeds of doubt in the infallibility of the ancients was planted by explorers of the sixteenth century, who brought back plants from their travels which even with great imagination could not be found in the works of either Dioscorides or Pliny. As Garcia da Orta, a Portuguese physician living in Goa, remarked in his book on the *Medical Simples of India* (1563), "It now seems that the ancients may not have exhausted this subject." A few years earlier (1539), Antonio Musa Brasalova, physician to Pope Paul III, suggested that healing herbs could be found outside the pages of Dioscorides. The suggestion brought an official reproof from the Pope.[17] By 1583, however, Rembert Dodoens could write in his preface, "We have preferred to describe herbs under their common names rather than rashly to ascribe ancient appellations to them." The proliferation of plant names and the consequent confusion bred by the heaps of unrelated data being collected were necessary preludes to the more accurate classification systems that began to be produced during the following century.

Even after the discovery that the species were new, it is not surprising that the sixteenth-century naturalists should have been unable to cope immediately with the situation. Information was coming in far too rapidly for assimilation—so rapidly, in fact, that for the first time the inadequacy of the old systems became obvious. In the years between 1571 and 1577, Francisco Hernández described 1,200 species of American plants. Explorers often brought back specimens with them; Cortez, for example, brought vanilla from Mexico, balsam from Peru, and other species new to Europeans. García de Orta, Ciesa de Léon, López de Gómara, and Cabeza de Vaca each added in a major way to the accumulation of natural history knowledge. From his 1585 voyage to Virginia, Thomas Harriot brought back descriptions, specimens or drawings of 28 species of mammals, 86 species of birds and a larger number of new plants. Among the strange mammals never before seen in Europe that Harriot brought back were the opossum and raccoon, the American gray squirrel, the black bear, the otter, the cony, and the skunk—which, he said, he needed no eyes to observe. Among the birds, the American mockingbird, the bald eagle, and the turkey (which the Spanish had previously brought to Europe) were especially interesting to European zoologists. How important the New World contributions proved to be is illustrated by the fact that as late as 1766 Linnaeus could describe no more than 210 species of mammals from all over the world; of these, 78 came from the observations of American naturalists. Of the 790 species of birds that he catalogued, the descriptions of at least one-third had been contributed by early American naturalists.

Late-sixteenth-century naturalists began the experimentation with systematic arrangement that reached its culmination with Linnaeus nearly two centuries later. Aldrovandi, for example, objected to the alphabetical arrangement on the grounds that it separated related objects and brought together dissimilar things. After Gesner, the last and greatest of the medieval botanists, most naturalists did try to group things that seemed related. Rondelet tried to distinguish fish by differences in habitat, food, taste, odor, and the arrangement of their parts. Cesalpino, head of the botanical garden at Pisa, after many years of trying to classify the new flora brought by the explorers and assign them the old Greek, Arabic, or Latin names, was finally convinced of the pressing need for some arrangement other than the alphabetical. He attempted something approaching a scientific classification by genera, which he felt would have the additional advantage of brevity, since it would not be necessary to repeat the common characteristics of each plant. He used such distinctions as that between trees whose seeds are external, trees whose seeds are in the heart of the fruit, and plants which bear solitary seeds under the same flower or seed receptacle. Apparently seedless plants were a separate classification.[18]

Such efforts were, of course, not entirely successful; and with by far the major part of the species of the world yet undiscovered, they were probably premature. Yet their efforts were important beginnings that Linnaeus and the later botanists built upon; and the efforts were, in themselves, an impressive testimony to the power of novelty to stimulate fresh thinking.

With the discovery that species of the New World did not, in fact, belong to the Orient, and that forms previously unknown did exist, there arose one of the most difficult intellectual problems of the sixteenth century. The explorers for the most part being practical men of affairs, were generally blind to the implications of what they saw, but careful comparisons of the flora and fauna of the Old World and the New increasingly disturbed men of learning and sensitivity.

Among the earliest of such individuals was Gonzalo Fernández de Oviedo, who came to Santo Domingo in 1514 as inspector of mining and smelting of gold for King Ferdinand; later he became governor of the island. His *Sumario de la natural historia de las Indias* (1526), produced as a report for the Emperor Charles V, although written entirely from memory after the author's return to Spain, contains both clear and sharp descriptions and a critical attitude toward what he had seen. His comments on the New World "tiger" (Jaguar, *Felis onca*) are illustrative. Are there any genuine tigers in the New World, he asked, or are certain beasts called tigers on the supposition that tigers are to be found everywhere? He was not able to determine positively, he said, but at any rate he did not believe that these were tigers, for despite the similarity in shape, they were very slow beasts, while Pliny and other writers had emphasized the swiftness and agility of tigers. It was, however, possible that the differences could be ascribed to geographical or astrological differences:

> It is true from what one can see of the marvels of the world and the great differences among animals, that these differences are greater in some places than in others, according to the locality or the constellations under which these animals have been bred. We see that the plants which are poisonous in some areas are healthful and useful in others; and that birds in one province are of good flavor while in other places they are not prized or eaten. Some men are black, while in other lands they are very white. Still they are all men. Granted all this, it may be true that tigers in one place may be swift, and that in your Majesty's Indies under discussion here they may be slow and awkward. Likewise the men of some kingdoms are courageous and bold while in others they are cowardly and timid.

Whatever one decided about the tiger, he concluded, the particular animals in the Indies could not have been learned about from the ancients, for in their exact form they existed in no land which had been discovered before his own time. "There is no mention made of these lands in Ptolemy's Cosmography, nor in any other work," he said, "nor were they known until Christopher Columbus showed them to us." He was also quick to point out that Pliny, in his *Natural History,* had failed to list the shark, the dogfish, and the spotted dogfish among the animals producing living young.

Such reflections as these, based on observation and comparative study, were of enormous importance in the development of the natural sciences. Later in the century, an even more perceptive naturalist, and one even more prone to speculate on what he saw, turned to a consideration of the same question. This was José de Acosta, the Jesuit Priest whose observations on the climate have already been noted. Acosta, in traveling regularly from one educational institution to another throughout a large part of South and Central America, was able to acquire a profound knowledge of New World flora and fauna that was unique for his time. Acosta wrote of the animals, insects, birds, fishes, plants, the mines and minerals, geological formations, precious stones, the soil, and the customs of the Indians. He was the first observer to notice the large fossil bones in South America, which he thought to be the remains of antediluvian giants—an interpretation that was current for almost two centuries after he wrote.

Like many after him, Acosta could not help but wonder how the first inhabitants came to South America. It was not likely, he decided, that a second Noah's ark might have transported men into the Indies; moreover, he refused to take seriously "the mightie power of God," as a special causal factor, but would entertain "only of that which is conformable unto reason, and the order and disposition of humane things." He was certain that ancestors of the Indians did not understand the art of navigation; therefore, if men had arrived by sea, it must have been by chance. This was possible, Acosta conceded, but such a conclusion led to a difficulty that troubled him greatly. How could the wild beasts be accounted for? Certainly men would not have brought them on a long sea voyage and kept them through the long and tempestuous period that must be imagined for the trip. Taking the biblical account of the flood as his beginning point, it was necessary for Acosta to explain the peopling of the New World as proceeding out of the Mountains of Ararat, where the Ark had landed. The conclusion which Acosta inevitably reached from this chain of reasoning was that the "whole earth is united and ioyned in some part," or at the very least the hemispheres approached quite close to one another. The fact that there were similar animals in the Old World and the New, if nothing else, led to this conclusion.

The question Acosta raised has a long history of perplexing people. The origin of the Indians, in particular, was an obvious problem to every European who saw them. "The wonder that far exceedeth all others," wrote Marc Lescarhot in 1606, "is that in one and the selfsame kind of creature, I mean in Man, are found more variety than in other things created." The first contacts with the Indians by Europeans were soon followed by full reports on their arts, manners and beliefs; and speculations about them always accompanied such reports. It early occurred to scholars that studies of the Indians could illuminate the history of man in general. Thus, William Gilbert pointed out that the savages of the West Indies then lived the same life as early man. Later, the Jesuit Joseph François Lafitan (1670–1740) spent five years in eastern Canada trying to find parallels with classical antiquity.

Opinions varied widely about the precise origin of the Indians. They were thought by some to be descendents of the Tartars, by others to be descended from the lost tribes of Israel; others suggested that they had descended from Moors of Africa, who must have come over by sea where the two continents were nearest, and still others thought them the offspring of Welsh explorers. All such speculations were based on superficial resemblance of one or, at most, a very few characteristics. For example, certain menstrual and dietary taboos convinced one eighteenth-century writer that they were descendents of Jews; the hide-covered boats used by an upper Missouri tribe convinced a still later writer that they were descendents of the Welsh.[19]

In the nineteenth century some prominent scientists, seeing that the resemblances were merely superficial, argued that Indians were a separate creation. In the seventeenth century, however, this was one option that was rarely taken and never fully developed. Francis Bacon did suggest that the people of the West Indies were probably "newer or younger" than people of the Old World, but whether he recognized the full implications of his statement is unclear. There was general agreement that Indians were children of Adam originally and therefore "children of wrath." To men who believed this, it was important to connect the Indians in some way with Old World people. Until well into the seventeenth century, in fact, it was common to deny any

essential color difference between Indians and Europeans. Captain John Smith was only one among many who believed that at
birth the Indian's skin was naturally white, but altered by several
dyeings of roots and barks into a cinnamon brown.[20] The existence of a black race and a white race had Biblical sanction, but
an additional one was difficult for Europeans to accept within the
old framework. Thus, their European origin, and the belief that
they would sooner or later be identified with some existing European stock, was taken for granted. But their later descent and
how they came to America, as Roger Williams put it, "seems
as hard to find as to find the well-head of some fresh stream,
which running many miles, out of the country to the salt ocean,
doth meet with many mixing streams by the way." Williams, in
his *Key into the Languages of America,* found many affinities
in their language both to Hebrew and Greek.

Any way one answered the question of the Indans, important religious problems were raised. Acosta's solution was perhaps as good as any. He argued against the most popular of the
current conceptions—that they were descendents of the Jews—
but he declined to speculate further on their ancestry than to
conclude that they had probably crossed a land bridge, or at most
a narrow strait, to the West, and that they were descendents of
Adam and Eve. He was not bothered, as some earlier commentators had been, by the apparent failure of the biblical promise
that the Gospel would be preached throughout the world. He had
no apparent need to grasp at such a slender thread as did the
English King James, who, upon being informed that the Virginia
Indians believed in immortality, concluded that the Gospel *must*
previously have been known in those parts and that only this
vestige of light remained. There were many nations to whom the
Gospel had *not yet* been preached, Acosta granted, but he and
others like him were at that very moment busy fulfilling the
prophecy.

But at this point in his speculations, Acosta was by no
means through with worrying. He understood very well that the
real problem was not with the people or with the similar animals,
but with the animals which had no Old World counterpart. An
entire chapter of his *History* is devoted to the question "How it
should be possible that at the Indies there should be anie sortes
of beasts, whereof the like are no where else." His statement of
the problem is, as usual, perceptive:

For if the Creator hath made them there, wee may not then alleadge nor flie to Noahs Arke, neither was it then necessary to save all sorts of birds and beasts, if others were to be created anew. Moreover, wee could not affirme that the creation of the world was made and finished in sixe days, if there were yet other new kinds to make, and specially perfit beasts, and no lesse excellent than those that are knowen unto us. If we say then that all these kindes of creatures were preserved in the Arke by Noah, it followes that those beasts, of whose kindes we finde not any but at the Indies, have passed thither from this continent, as we have saide of other beasts that are knowen unto us. This supposed, I demand how it is possible that none of their kinde shoulde remaine heere? And how they are found there, being as it were travelers and strangers.*

Acosta concluded his almost definitive statement of the problem of the distribution of species by remarking that "Truly tis a question that hath long held me in suspense." He would not resort to what he considered the shabby device of declaring that all the differences between Old and New World animals were "accidental" and reduce them to the same species as those of Europe. Faced with this same problem later, Sir Walter Raleigh attempted to solve it by postulating an almost infinite variability within species. Color, magnitude, and other "accidental" characteristics made so little difference for Raleigh that he assigned, for example, the dogfish of England and the shark of the South Ocean to the same species. Raleigh was primarily interested in explaining how all the species on earth could have found space in Noah's ark, and his solution to the problem remained a popular one. A slightly less radical deduction was that by Lambert Daneau, who in his *Christian Physics* of 1575 reduced the terrestrial animals to about thirty genera, assuming that the currently existing species had developed from the original generic pairs on board the Ark. For Acosta, however, who had a great deal more first-hand experience

---

*Compare the following similar statement by Sir Thomas Browne: "Another secret not contained in the Scripture, which is more hard to comprehend . . . and that is . . . how America abounded with Beast of prey and various Animals, yet contained not in it that necessary Creature, a Horse, is very strange. By what passage those, not only Birds, but dangerous and unwelcome Beasts, came over; how there be creatures there, which are not found in this Triple Continent; all of which must needs be strange unto us, that hold but one Ark, and that the Creatures began their progress from the Mountains of Ararat." Quoted in Loren Eisley: *Darwin's Century* (Garden City, N.Y., 1958), p. 3.

with New World species, the solution was not satisfactory. The
beasts of the Indies were so diverse, he said, "as it is to call an
egge a chestnut" to try to reduce them to the species of Europe.

Acosta's only answer to the grand problem he posed was
to suggest that after leaving the Ark "by a naturall instinct and
the providence of heaven" diverse kinds had dispersed themselves
into diverse regions, "where they found themselves so well, as they
woulde not parte; or if they departed, they did not perserve themselves, but in processe of time, perished wholy, as we do see it
chaunce in many things." Acosta did know enough about the rest
of the world to realize that it was not a problem peculiar to the
Indies, but general to different parts of Asia, Africa, and Europe.

In offering his tentative answer to the problem, Acosta had,
of course, come very close to the solution to the mystery, which
lies in evolutionary radiation and organic change. His recognition
of the peculiarly intimate relationship between animal species
and particular environments makes him almost unique for the
sixteenth century. It would be gratuitous to term Acosta a "forerunner of evolution" simply because of his recognition that
animals taken out of the environment suited for them would
probably die out—essential parts of the evolutionary concept are
absent from the mind of this thoughtful sixteenth-century scholar.
It is, however, not beside the point to observe that it was the very
phenomena that Acosta was observing and the very thoughts that
he had about them which, in a later world view, *did* result in an
evolutionary theory.

Acosta, whose speculations went no further than the search
for a way to explain how all the diverse animals in the
world could have come out of Noah's ark, was apparently satisfied with his hypothesis, and there is no indication in the subsequent history of this pious man that he was ever led to question
the concept of the Ark itself. Others were not so fortunate as he
in being able to assimilate the new knowledge while leaving their
religious convictions undisturbed. For example, there is good
reason to believe—although there is no positive evidence—that
Thomas Harriot, the first English scientist in the New World, was
led to his later skepticism by what he had seen on his trip. For
Harriot, too, noted that there were differences between the creatures of the New World and the Old, and he later became well-
known for his "disbelief." As his biographer, Anthony Wood,
wrote: "Notwithstanding his great skill in mathematics, he

[Harriot] had strange thoughts of Scripture, and always under-valued the old story of the creation of the world."

While the connection between Harriot's observations in the New World and his religious opinions is a matter of speculation, the evidence is more conclusive in the case of Robert Hooke. Drawing heavily upon the accounts of Acosta, Garciloso de la Vega, and other New World naturalists, Hooke reached several heterodox conclusions in his posthumously published "Discourse on Earthquakes." Basing his thoughts on an essentially correct explanation of the formation of fossils, Hooke concluded:

> That a great part of the Surface of the Earth hath been Since the Creation transformed and made of another Nature; namely, many Parts which have been Sea are now Land, and divers other Parts are now Sea which were once firm Land; Mountains have been turned into Plains, and Plains into Mountains, and the like.

Noah's flood, he thought, could not explain these changes in the earth's surface, for it had not lasted long enough. He preferred, instead, to emphasize the gradual action of water and of wind, the natural subsiding of heavy bodies and the rising of light ones, and volcanic action.

Hooke also accepted the conclusion that Acosta had tried so desperately to avoid: there were extinct animals, he insisted, and there were new ones existing that had not been made at the first Creation.

The implications of the New World had their clearest early statement in Hooke, although his conclusions were not generally accepted for well over a century. It is evident, however, that later naturalists remained aware of a possible conflict between the new discoveries and common religious views. For example, it had generally been held that Adam "undoubtedly could call each herb by its proper name," as a sixteenth-century botanist put it. Compare this view with the following exchange running from 1737 to 1738 between John Custis, an American botanist, and his English correspondent, Peter Collinson.[21] Custis, speaking of the American strawberry bush (*Enonymous Americanus L. Gray*) concludes that:

> This tree was not extant when the scripture was writ, which accounts mustard seeds the least of seeds; but there is no comparison between these seeds; the strawberry tree seeds are so small that I am not sure whether I saw any or not when rubbed to pieces . . .

A few months later, Collinson answered:

> As to the strawberry seed it is indeed small but the Evangelists
> were inspired men in Religious Matters, yett I presume were
> no great naturalists they represented things of that nature,
> according to their knowledge in those matters and suitable
> to the Ideas of their Country Men.

Custis, apparently not satisfied with Collinson's attempt to save
the Creation, presses the issue in his next letter, again saying
that he thinks the strawberry seeds were unknown to the evangelist. Collinson, still unwilling to yield the Creation, is willing to
concede that the seeds may have been *unknown* to the evangelist,
but he insists that they were extant. There must have been seeds
even smaller than the strawberry then, he insists, but "the Jews
were poor naturalists," except for Moses and Solomon, and they
spoke of things within their own knowledge.

   Both of these individuals, as far as is known, were pious,
churchgoing men; yet their faith was of a different kind from that
of previous generations. In their exchange of letters one sees the
more liberal theology, the interpretation of revelation in the light
of nature, that was common among Enlightenment thinkers. If
one multiplies the incident of the strawberry seed several thousand
times, he will begin to understand the profound intellectual
change set in motion by the medieval navigator who set out to
find the Orient.

   The explorers, the observers and the naturalists of the
sixteenth century had indeed found a new world of nature: one
that some found it difficult to live with, and that others used to
advantage in battering down the old cosmology and natural history. Aristotelean physics and meteorology, Ptolemaic geography
and astronomy, the orthodox views of the Creation, the Noachian
flood, and conventional notions about the distribution of animal
life on the globe were all—at least implicitly—called into question
by the new discoveries. The New World, therefore, played an
indispensable role in one of the most noticeable—and certainly
the most important—of the changes that took place in the two
centuries after 1450; that is, the change in attitude toward the
ancients. In 1450 men attempted no more than comprehension of
what the ancients had discovered, certain that this was the most
that could be known. By 1536, Petrus Ramus was publicly defending the thesis that everything Aristotle had taught was false.
Ramus's assertion was perhaps premature, but forty years later,

bright university students like Francis Bacon were saying that the study of Aristotle was a great waste of time. Ancient learning was increasingly old-fashioned, and every new discovery made it more evidently so. As Von Humboldt suggested, many of the leads provided by America did not come to fruition; but nevertheless they were there, and they remained to trouble men and to stir them into thought in efforts to formulate more satisfactory systems to contain their knowledge. Disconcerting and bothersome the new discoveries often were; but more than this, they were liberating and they were challenging.

II

~~~~~~

The Scientific Colonization
of America

The first impulse, upon the discovery of a new world, rich in novelties, is to find out exactly what is in it—to take a vast inventory, as it were, and then to classify and appraise the organic and inorganic environments. As one historian has pointed out, this is standard procedure whenever *any* new area is opened up to Western science, and it forms a necessary part of the process by which scientific culture is transmitted.* This kind of scientific activity, which was the characteristic type in America for the first century and a half, is generally undertaken as an appendage to geographical discovery and economic exploitation. Maps must be drawn, lists of poisonous and medicinal plants must be compiled, building materials must be identified, and native sources of food must be found. The indigenous population must be studied and in some measure understood, if only for the purpose of controlling it. These are minimal necessities if any colonization is to be attempted. Beyond them, there is the tempting possibility that sources of valuable minerals or other exportable products will be found. Although the impulse, in the beginning, is not strictly scientific, much work of scientific value is accomplished as a by-product of this effort to dominate and to exploit the natural environment.

In the sixteenth century, all the major colonizing powers

---

* George Basalla: "The Spread of Western Science," *Science*, Vol. 156 (May 5, 1967), pp. 611–22. Basalla, however, denies the primary role of economic exploitation.

made some efforts along these lines. The scientific information and the specimens that filtered back to the European centers of learning were furnished by the explorers themselves, by priests involved in the missionary activity associated with exploration, by physicians, and by a random assortment of adventurers, promoters, artists, and others, who managed to get themselves attached to expeditions. Frequently the military commanders themselves acted as amateur scientific observers; on occasion, they made genuine contributions. Thus Jacques Cartier, a French military commander who was well known as a cartographer, was also a skilled scientific observer. His journal, containing notes on places, peoples, animals, minerals, and other natural history data, provided the first written observations on the plants of western Canada. Most of the information on the flora in the book of André Thevet (1557–8) came from Cartier, for Thevet had never been to Canada. Cartier also compiled an Indian vocabulary and was a careful student of Indian customs and religious beliefs.[1] Of even more scientific importance was Samuel de Champlain, also a French military commander, who contributed substantially to the knowledge of the flora, fauna, and people of French Canada.

But despite the examples of Cartier and Champlain, the French embarked on no grand effort before the eighteenth century. Primarily interested in furs and fishing, and having little interest in colonization, the French gave little official encouragement to study beyond the necessary mapmaking. Some information did find its way back to France in the *Jesuit Relations*, submitted every year from 1632 to 1673, which often contained geographical, botanical, and ethnographical comments. Aside from these, Lescarbot's *Histoire de la Nouvelle France* (Paris, 1612) and Father Hennepin's *Description de la Louisiane* (1683) were the only notable contributions to the scientific exploration of the region, and they came quite late.

In the sixteenth century, Spain provides the model of a nation carrying out a scientific inventory; Spanish America was the subject of a much more intensive scientific survey than either French or British America. This was so despite the fact that the *Conquistadores*, generally more interested in empire and in gold than in natural history, contained none in their numbers who was the intellectual equal of Cartier or Champlain; and despite the fact that the military geographical surveys were generally kept in manuscript, for fear that the English and French would profit by knowledge of the terrain in Spanish territories. Most of the scientific work was done in the sixteenth century by priests who followed the Spanish empire from Peru to California, as well as by those who remained behind to man the universities and missionary centers in South America, Mexico, and the southwestern part of what is now the United States. José de Acosta, whose work in South America was discussed in the preceding chapter, had more broad-ranging interests than most of his fellow priests, but he was only one among many who made some contribution to understanding the New World. It is also important to note that the Spanish, unlike the French, often sent trained naturalists who were commissioned to explore and report on the natural history of the New World.

Motivated by a keen sense of the historical significance of the conquest—which in Spain was considered the greatest event since the birth of Christ—and also by practical problems of administration, the Spanish were vitally interested in systematically studying the New World. Thus in 1570, Philip II, the unusually curious king who took a deep interest in the scientific work of Spanish expeditions after he came to the throne, sent his personal physician, Francisco Hernández, to Mexico to study the land and its people. After touring New Spain with a group composed of two or three artists, an equal number of scribes, an interpreter and two or three herbalists, or gatherers of plants, Hernández returned to Spain with sixteen volumes of drawings and descriptions as well as a fund of information gathered from the Aztecs. Among the scientific results of Hernández's trip were the first book on North American medicinals and the descriptions of 1,200 species of American plants.[2]

Positive governmental policy, initiated early in the sixteenth century by the Council of the Indies, promoted the compilation of a mammoth natural history which would reveal the natural wealth of the Spanish possessions, assess the agricultural possibilities, and assist in the problem of dealing with the Indians. Thus Fernández de Oviedo's *Historia General y Natural*, termed by one scholar the "First American Encyclopedia," was begun in 1532 on the recommendation of the Council of the Indies when the author received a royal commission to undertake the work.[3] Oviedo, whose work has been discussed in the preceding chapter, took Pliny's *Natural History* as his model and set out to compile everything that was known or suspected about Spanish America—

the natives, the soil, the indigenous plants and animals, Spanish innovations—in short, everything that was necessary for efficiently administering the territory. For example, he described more than seventy trees native to the New World, giving the appearance, habitat, and rate of growth of each, specifying whether they were poisonous or medicinal and whether the timber was suitable for building or cabinetwork. Oviedo remained on salary all the years in which he was involved in the work, and all colonial governors were ordered to assist him in any way he might require.

The natural history assessment was put on a continuing basis in 1569 when Juan de Ovando, president of the Council of the Indies in Spain, had a detailed questionnaire drawn up and required every governor in America to submit specific data on the history, people, products, climate, and geography of the territory he administered. Begun as a brief inquiry, the questionnaire soon grew to fifty items and eventually became a printed volume of 350 separate questions. Printed instructions accompanying the questionnaire specified in detail how to complete it. These "Relaciones Geográficas," perhaps because they constituted such a wealth of data as to be unmanageable with the tools available to sixteenth-century men of science, were never consolidated into a systematic whole, and remain to this day in the archives unpublished.[4]

Anthropological studies, in which Spanish priests pioneered, were closely related to Spain's genuine concern for working out a just Indian policy and settling the question of the real nature of the Indians. The Spanish apparently took their obligation to Christianize the natives, which had been written into the Treaty of Tordesillas, quite seriously. Consequently, it was the friars, looking for souls to save, who first mastered the Indian languages and began to study Indian customs, history, and religion. The missionaries needed particularly to know the names and attributes of the Indian gods, the sacrifices made to them, and some indication of the mentality of the Indians in order to lead them away from their pagan rites toward Christianity. Thus the founder of American anthropology was Friar Ramón Pané, who accompanied Columbus on his second voyage for the express purpose of reporting on the Indians and their ways, and who was the first European to learn an Indian language.

The Crown encouraged ecclesiastics throughout the sixteenth century to study the Indians, and numerous volumes on their culture and many more on their language were prepared. Much of this work is remarkably tolerant and broad-minded, given the general image of the sixteenth-century Spanish priest. Thus Bartolomé de las Casas was notorious for his denunciations of Spanish cruelty to the Indians, and throughout his work he insisted that the Indians must not be measured by a Spanish yard-stick but must be understood within the framework of their own culture—a culture in which he found much to admire. The Franciscan, Diego de Landa, who did not share Las Casas's enthusiasm for the Indians, nevertheless felt a strong desire to set down everything he could find out about Maya culture, and he did admire certain parts of it—particularly their calendar, food, architecture, some of their moral ideas, and the beauty of their women.

The greatest of the early anthropologists, however, was the Franciscan Friar, Bernardino de Sahagún, whose *General History of the Things of New Spain* was a remarkable treatise on the Aztecs of Mexico. Sahagún began to collect material on the Aztecs in 1547. Ten years later his Provincial ordered him to prepare a history of Indian culture on the assumption that it would aid in the then-faltering missionary effort. Between 1558 and 1560 he lived in a native village, where he systematically questioned a dozen of the oldest and most knowledgeable inhabitants he could find, using a carefully prepared list of culture elements as the basis for his investigation. His subjects also drew many pictures to explain their history. Sahagún then moved to another village to check his data by using a fresh set of informants during 1560 and 1561. The next three years were occupied in organizing his data and only then did he begin writing. The result, completed in 1569, was the best anthropological study that had appeared to date; a methodically arranged mass of carefully verified information, on the gods worshipped by the Indians, their fiestas, their ideas on immortality and death ceremonies, their belief in astrology and witch doctors, their rhetoric and philosophy, lords, governments, merchants and mechanical arts, vices and virtues, their beliefs about animals, birds, fish, herbs, trees, fruits and flowers, and their view of the conquest of Mexico by the Spaniards.[5]

Thus the Spanish, before the English arrived in America, had virtually completed their rough inventory of the far-flung empire. That so much of the material they collected lay unused in the archives of the Council of the Indies can be attributed in part to secrecy requirements imposed by considerations of national rivalry, and in part to the inability of natural historians to assimilate such masses of new data. That which did find its way into print was, in most cases, quickly translated, and it served as a stimulus both to further work by other scientists and to colonization efforts by rival powers.

In the English possessions, the pattern very quickly became different from either the French or the Spanish. Colonization went hand in hand with exploration and sometimes, in fact, preceded it. Indicating the unusual pattern is the fact that it was the least populated sections of the coastal regions that were the first to be accurately mapped. Again reflecting the early settlement is the fact that with the single exception of Thomas Harriot's work, the early scientific reports were written by settlers having little or no scientific training.

Thomas Harriot, an accomplished natural philosopher, then twenty-five years old, had been specially selected by Sir Walter Raleigh to make a statistical survey of the land, to act as historian and geographer, and to bring back a detailed account of the mineral resources and the plant and animal life of the region and learn as much as he could of the "heathen." Harriot arrived at Wingandacoa, Virginia (now North Carolina) in August 1585, and, after a stay of nearly a year, published *A Briefe and True Report of the New Found Land of Virginia* (London, 1588). An edition of 1590 contained engravings of the water colors made by Harriot's companion, John White. Harriot contributed descriptions, specimens or drawings of twenty-eight species of mammals, eighty-six of birds, and an even larger number of new plants. The best known of these was the tobacco plant. Although it had earlier been cultivated by the Spaniards as a garden plant, and some medical use had been made of tobacco, Harriot became the first great enthusiast for smoking. He reported "many rare and wonderful experiments of the virtues" of the plant and praised its presumed great medicinal value.

After this promising beginning, the English apparently lost interest in systematic scientific exploration of their colonial possessions. Harriot had no immediate successors. Long before any other serious natural history work had been done in the British colonies, there had appeared Le Page du Pratz's study of Louisiana, that of Canada by Charlevoix, and that of California by

Venegas—each of which built upon previous work and long stood
as the authoritative study.

A certain amount of scientific information was, however,
common among the better-educated pioneers who settled in the
English colonies. Like other Englishmen of their time, they were
interested in the newness of nature about them, and, realizing
that a knowledge of botany, zoology, mineralogy, climatology,
and medicine might mean the difference between success and
failure of their ventures, they often brought along the necessary
books to supply whatever deficiencies they had.

Francis Higginson, a clergyman, writing of "New-England's
Plantation" in 1629, showed very clearly the pressing necessity
for early interest in natural history:

> And because the life and wel-fare of every creature heere
> below, and the commodiousnesse of the countrey whereat such
> creatures live, doth by the most wise ordering of God's
> Providence, depend next unto himselfe, upon the temperature
> and disposition of the four elements, earth, water, aire, and
> fire (for as of the mixture of all these, all sublunary things
> are composed; so by the more or lesse enjoyment of the
> wholesome temper and convenient use of these, consisteth
> the onely well-being both of man and beast in a more or less
> comfortable measure in all countreys under the heavens)
> therefore I will indeavour to shew you what New-England is by
> the consideration of each of these apart . . .[6]

It was only natural, given the background of information
available to the early settlers, that their earliest writings on the
natural history of their new home should be particularly marked
by an interest in the marvelous. The modern reader is conscious
of a feeling of awe-struck wonder at the novelties and mysteries,
real and imagined, that the settlers found in America. The writers
dealt, therefore, more with the prodigies and wonders of nature
than they did with the normal sequence of events, and the scientific value of their work suffered accordingly. Nevertheless, even
a work like John Josselyn's *New England's Rarities* (1672) contained evidence of a genuine scientific interest in explaining the
occurrences of nature. That work contains, for example, the first
known suggestion that the "Gullies" of the White Mountains
were cut by running water—over one hundred and fifty years
before the fact was generally accepted.

Credulous and haphazard as he was in his method of recording data, Josselyn at that time probably knew more of the

botany of New England than anyone else, and he did at least have a rudimentary idea of classification in mind, for he divided his plants as follows:

    I. Such plants as are common with us in England.
    II. Such plants as are proper to the country.
    III. Such plants as are proper to the country and have no name.
    IV. Such plants as have sprung up since the English planted and kept cattle in New England.[7]

His treatment of the animals, birds, and fishes, however, is far less sound than his discussion of plants.

Josselyn's work, like many of the seventeenth-century productions, was written in part as a promotional tract; it could therefore be counted on to romanticize the American environment to a degree. This same aim had inspired William Wood's *New England's Prospect* (London, 1634), Captain John Smith's earlier account of Virginia and of New England, and, in fact, Higginson's 1629 account. Further north, Pierre Boucher's 1664 account had been written in order to convince the King of France of the importance of colonization. Predictably, there always occurred a certain degree of exaggeration in such works, as William Wood's assertion that easily accessible wells in New England produced water almost as good as "good Beere" and that "any man will choose it before bad Beere, wheay, or Buttermilk." The bullfrog in Virginia which, according to Robert Beverly, was "big enough to feed four Frenchmen" was the product of a similar motivation. Higginson, after gravely observing that it "becometh not a preacher of truth to be a writer of falshod in any degree," reported that "Joseph's encrease in Egypt is out-stript here with us," that the deer brought forth three or four young ones at once, and that many marvelous cures had been effected by the air of New England. On the last subject, he concluded with words that should warm any New Englander's heart: ". . . a sup of New England's aire is better than a whole draught of old England's ale."

The promotional nature of early accounts is nowhere better demonstrated than in the fact that an account of Virginia by John Strachey, containing chapters on the Indians and their customs, lay buried in the archives of the Virginia Company "lest it prejudice prospective settlers for the colony." It was first published by the Hakluyt Society, in 1849.[8]

Because of the haphazard nature in which information was collected and also because of the inexperience and mixed motives

of those collecting it, before about 1660, knowledge in Europe
about American nature—especially North American nature—
remained scarce and unsystematic. One scholar has estimated
that probably no more than 150 species of plants were introduced
into England during the seventeenth century.[9] The next century
was an entirely different story; partly because the actual struggle for survival was no longer a prime concern, at least in the
seaboard colonies, and partly because of developments in
Europe.

The late seventeenth and early eighteenth centuries were
the age of scientific societies in western Europe. There were a
great many founded during this time and they played an indispensable role in nurturing science in the colonies. The relationship
was a reciprocal one, for the seemingly endless stream of plants,
cuttings, seeds, minerals, and other specimens sent by correspondents from the New World provided both a powerful force
in the founding of the societies, and a continuing stimulus to
research. Promoting natural history work in the colonies was,
in fact, a main *raison d'être* of most of the societies.

The most famous of these—and the one most relevant for
the present story—was the Royal Society of London for the Promotion of Natural Knowledge, an informal group of scholars
meeting before 1662 which was given corporate being by Charles
II in that year. In one of the earliest recorded meetings of the
founders, December 5, 1660, an exploratory group of 112 of the
Original Fellows agreed to continue its meetings for the "promoting of Experimental learning." At the same meeting, the founders
each agreed to contribute one shilling weekly to defray the costs
of experiments and other corporate expenses, and during the
following year a variety of experiments was performed at their
meetings, plans were laid to test the scientific reliability of a number of classical works formerly given wide credence (such as
Virgil's *Georgics*), committees were appointed "to consider of
proper Questions to be enquired of in the remotest parts of the
world," and scientific correspondence of a worldwide scope was
actively sought after. The Royal Society's correspondence was so
ordered, wrote Thomas Sprat in 1667, "that in a short time, there
will scarce a ship come up the *Thames*, that does not make some
return of *Experiments*, as well as of *Merchandize*."[10]

Forwarding of specimens, or at least descriptions of botanical, zoological, or mineralogical forms was encouraged in volumes
of earnest letters; as was the keeping of meteorological, medical,

and astronomical records, making reports of seismic phenomena, and dissecting strange animals. The savants never missed an opportunity to improve their knowledge of useful facts about the colonies—natural history facts or otherwise. For example, when John Winthrop, Jr., Governor of Connecticut and son of the first governor of Massachusetts Bay, visited England in 1662, he read papers on "The Manner of Making Tarr and Pitch in New England," "The Manner of Building Ships in New England," and he prepared his "Dissertation upon Indian Corn," which was read later. Even his voyage home was an opportunity the savants of the Royal Society tried to take advantage of—although with less success. In March 1663, while preparing to sail for America, he was prevailed upon to undertake with balls, leads, and a valve-fitted cylinder a careful examination of the Atlantic Ocean. The apparatus, however, "failed of a perfect triall," Winthrop reported, for the motion of the waves "unhookes the lead."[11]

The motivations of those promoting colonial science were partly utilitarian, partly scientific—the legacy of Bacon had so tended to merge these that they were largely indistinguishable. Early English science, like the earlier Spanish survey, was quite generally linked to economic exploitation of one sort or another. Men like Robert Boyle were in the habit of ransacking Hakluyt, Purchas, and other compilers of travel literature for information that might be of possible economic value. In the archives of the Royal Society are scores of sheets of notes that Boyle made from such literature. When he isolated an item that interested him, he would try to test its value by having Englishmen who were leaving Europe make careful inquiries on the subject and send reports back to him. He would then make abstracts, consolidate the material from various correspondents, and report to the Royal Society. Like so many others of the period, Boyle also carried on an extensive correspondence with men on the spot. Faced with the perennial problem of how to work with men largely untutored in science, Boyle finally worked out a systematic list of queries—characterized by one historian as an "Intelligent Man's Key to Intelligent Questions to Ask."[12] This list of queries, under the title "General Heads for a Natural History of a Country, great or small, imparted . . . by Mr. Boyle," was published in the *Philosophical Transactions* for 1666.

An example of this practical interest is Boyle's curiosity about foreign methods of farming. He requested a report on:

What the Nature of the Soyle is, whether Clays, Sandy, etc.
or good Mould; and what Grains, Fruits and other Vegetables,
do the most naturally agree with it: As also, by what particular
Arts and Industries the Inhabitants Improve the Advantages,
and remedy the Inconveniences of their Soyle: What hidden
qualities the Soyle may have (as that of Ireland, against
Venemous Beasts, etc.)

As Boyle's queries indicate, there was a great deal of interest
in agricultural innovations in England at that time. As early as
1582, the elder Richard Hakluyt, England's greatest propagandist for colonization, was urging the importation of new plants
into England. He recorded a list of plants and other things not
native to England but in his own time already adding richly to
its life. Vigilance in aiding in further imports might, he said to
one correspondent, "do more good to the poore . . . then ever any
subject did in this realme by building of Almeshouses and by
giving of lands and goods to the relief of the poore."[13] By Boyle's
time, agricultural innovations were a major preoccupation of
many English intellectuals. Samuel Hartlib, one of John Winthrop, Jr.'s, regular correspondents, had similar interests. The
Civil Wars had brought about the economic ruin of many land-
lords, and so "spirited farming," rather than the traditional conservative methods, began to be thought of more and more as a
restorative of vanished fortunes. Many thoughtful men of the
period of the Restoration observed that an earlier importation, the
potato plant, had become of vital importance in the economy of
Ireland; they knew of the success of Lord "Turnip" Townshend
with his importation, and they showed great interest in other
crops that might be domesticated in England.

This practical motivation remained important throughout
the colonial period, and, indeed, persisted well into the national
period. Shortly after American independence, for example, John
Fraser was in South Carolina seeking plants and grasses that
would be useful in England. When he returned home with a new
grass for meadowland or pasture, Fraser wrote a book in which
he acknowledged the indispensable assistance given him by
Thomas Walter, a local resident. Walter was without books or
herbaria, said Fraser, but he "made his descriptions with an
accuracy that is allowed to be by no means inferior to the most
eminent botanists in Europe."

Other nations were also interested in exploiting the wonders
of New World flora. Charlevoix, for example, who went down the

Mississippi between October 1721 and February 1722, in the volume on plants of his six-volume *Histoire* dealt chiefly with useful, harmful, or medicinal plants] He gave particular notice to the different species of forest trees, always with the preoccupation of marking their eventual utilization when they were introduced into France. His greatest enthusiasm was prompted by the discovery of a species of holly (*Ilex vomitoria*) which provides an emetic.

Le Page du Pratz, who was sent to Louisiana in 1718, spent seventeen years collecting and dispatching to France all the seeds, seedlings, and plants which in his opinion could be grown profitably in the Old World. Besides his *Histoire de la Louisiane* (three volumes, 1758), the product of his trip was about three hundred medicinal plants which he sent to France carefully planted in cane baskets.

Trained naturalists who were able to find their way to the New World were in a unique position to make monumental contributions to European knowledge. Carrying with them a knowledge of what was known, they could exercise more selectivity in their collections than the unschooled explorers, and they could apply systematic thought to their work. Georg Marcgrave, a German, went with a team of scientists and artists to Brazil in 1638, where he spent the next several years making observations of all kinds and gathering a large collection of the flora and fauna. Although he died before he could put his material in shape for publication, others took up the task. De Laet, of the West India Company, edited some of Marcgrave's work, which appeared in Leyden under the title *A Natural History of Brazil*. More than 650 forms, nearly all new to science, were described in the book. Other specialists on this methodical exploration of the overseas territory studied the country, the natives, their illnesses and remedies; drew maps; made astronomical observations; and made colored sketches of natives, animals, and plants.[14] Similarly, the botanist Joseph de Jussieu traveled to Peru in 1735 attached to a commission sent out by the French Academy to measure a degree of the meridian. Jussieu was so fascinated by the new land and its flora that he remained in South America for thirty-five years, during which time he sent notes, lists of plants, and seeds to his brothers in France. Some of these materials were later incorporated by his brother, Bernard, into the systematic arrangement that became the basis for the "natural system" of classification.

Still later André and François André Michaux, father and son, were sent on an official mission to the United States, between 1785 and 1808. Charged in general with introducing into France promising species of American trees, they were instructed to be particularly on the alert for fast-growing species suitable for naval construction. The Michaux traveled all over the country and sent back to France more than sixty thousand living plants, as well as many boxes of seeds. Demonstrating the correspondence between the practical and the purely scientific in the new area, their own discoveries included over three hundred species of flowering plants. French interest continued well into the nineteenth century, when Pierre Paul Saunier, who had arrived with the Michaux, inherited and extended the garden they had established in Bergen County, New Jersey. Other French collectors included Jacques Gérard Milbert, who came to America in 1815, and Élias Durand, who arrived the following year, ran a pharmacy, and became a major collector of botanical specimens, supplying samples of about fifteen thousand North American species to the Jardin des Plantes in Paris.[15]

Outside of England and France, the cultivation of foreign plants which might contribute to the economy of their country was a major part of the program of many scientific societies, such as the Prussian, the Russian, and the Swedish academies. The German economist J. B. von Rohr, who in his *Compendienses Hanshaltungs-Bibliotheck,* published in 1716, recommended the establishment of "economic societies," considered it important that members of these societies should experiment in cultivating different foreign plants which could be of economic use. Later in the century, a Hessian officer in the American Revolution, Friedrich Adam Julius von Wangenheim, found time between campaigns to make an extensive study of American trees which he thought might be useful introductions into the forests of Germany. While the Revolution was still in progress, he published his heavily illustrated *Beschreibung einiger Nord-Amerikanischen Holzarten* (Göttingen, 1781), which contained descriptions of 168 American trees and shrubs.

It was this motivation on the part of Sweden that sent Peter Kalm, one of colonial America's best-known travelers, to North America in 1748. A realization of the dream of Linnaeus, Kalm's trip was planned by the Swedish Academy of Sciences and financed by the five universities of Sweden and Finland. Kalm, a

prize pupil of Linnaeus who had spent many years preparing himself for this task so dear to the heart of his master, was given an
official directive by the Academy of Sciences, in which his mission
was defined in detail. Kalm was to travel in those parts of North
America which were most like Sweden as regards temperature; he
was to travel especially in Canada, because it was as cold there
as in Finland and the few Canadian plants that had been brought
to Sweden stood the cold as well as native Swedish plants. When
he found plants which were known for their useful properties in
the preparation of food, in dyeing, or as timber, he was to collect
seeds from as far north as the plants were met with. Although he
was charged in general with collecting the seeds of plants which
would "improve the Swedish husbandry, gardening, manufactures, arts and sciences,"[16] Kalm was directed to seek in particular
the seeds of certain species, already known to be potentially useful
in the Swedish economy, such as the mulberry; different varieties
of oak; walnuts; the medicinal *Radix ninsi; Polygala;* and several
other plants.[17] Although Kalm returned with a wealth of information and a delightful diary of his stay in North America, the seeds
that he so carefully took home were uniformly disappointing.
They did thrive during the summer but the long Swedish winters
made it impossible for them to come to maturity.

The other impetus for practically oriented scientific exploitation
was, of course, the physicians. The prominence of physicians
in natural history circles on both sides of the Atlantic was,
in fact, one of the most striking characteristics of the pursuit
of science during colonial times. In England, surgeons and
physicians formed the largest single group among the scientific
Fellows of the Royal Society; most of the leading naturalists on
the Continent were physicians, including Boerhaave, Haller, and
Linnaeus; in America, the most active members of the natural
history group were European-trained physicians. In part, this
predominance of physicians followed naturally from the fact that
leading medical schools, such as the ones at Padua, Edinburgh,
and Leyden, offered the best scientific education of the day as a
regular part of the medical instruction. Students at the leading
medical schools took formal courses in botany, chemistry, and
anatomy, often including comparative anatomy. But even infor-
mally trained or self-educated physicians generally developed
strong scientific interests, for natural history—especially botany—

and medicine were closely connected during the seventeenth and eighteenth centuries. The most commonly used medicines were botanical, and the characteristic treatise on botany was an "herbal": a catalogue of common medicinal plants, describing their growing conditions, which parts of the plants to use, when to gather them, and what they were to be used for. An early example of this type of work was William Hughes's *The American Physitian; or a Treatise of the Roots, Plants, Trees, Shrubs, Fruit, Herbs &c. growing in the English Plantations in America,* published in London in 1672. Plants of Jamaica, the West Indies, and Virginia were described and classified by Hughes according to their Galenical qualities; i.e., they were hot and dry, cold and moist, and so forth.

Additional impetus to the search for indigenous plants was given by the prevailing medico-religious doctrines of the period. It was widely believed that God in His wisdom had provided both diseases and remedies in the same general locality. To have made a disease flourish in any part of the world and not to have provided a local remedy which could be discovered by the persistent and careful searcher, would have made Him an unfair God, and it would have furthermore been counter to the general teleological assumptions about the economy of nature. The belief was so pervasive throughout the eighteenth century that Benjamin Smith Barton's *Collections for an Essay Towards a Materia Medica* (1801 –4) described as "trite" the theory "that every country possesses remedies that are suited to the cure of its peculiar diseases . . . that the principal portion of indigenous remedies is to be found among the vegetables of the countries in which the diseases prevail." In the period of exploration, the belief accounted for the many expeditions in search of a cure for syphilis, a disease generally held to have been imported into Europe on Columbus's second voyage. In a later period, it accounted for much of the botanizing done by physicians, who did not look upon their excursions as pleasant afternoon diversions, but as a vital part of their business. Nothing could have been more natural than for European-trained physicians, finding themselves in a new land with many unfamiliar plants, to seize the opportunity for botanical discoveries. Given the wealth of material at hand, it was equally natural that even laymen should study American flora in the hope of adding to medical knowledge.

The same doctrines also led to an absorbing interest in Indians and to a belief that they knew a great deal more medicine

than they actually did. Sometimes, the reliance on Indian remedies worked well. Cartier, for example, reported on the "grosse
maladie" that he and his men caught on the way to Quebec
during the second voyage. His use of an Indian remedy, the
leaves of the "anedda" or Canadian cedar, was successful in
curing his men and it constituted the first recognition of the
therapeutic value of vitamin C, contained in the leaves.[18] Other
efforts were not so happy, as that of John Tennent, who came to
the colonies in 1725 and found that the Seneca Indians used
rattlesnake root for snake bite. Tennent, carried away by the
Indian claims, used it as a specific for many diseases including
pleurisy, pneumonia, gout, and intermittent fever.

Nevertheless, the physicians, however misguided they
generally proved to be in their desire to rely entirely on locally
available remedial agents, played a continuing and important part
in colonial science from the very beginning. It was a physician,
Dr. Lawrence Bohun, who appears to have been the first experimental scientist at Jamestown. Bohun, who came to the colonies
with Lord Delaware in 1610, finding himself in the midst of an
epidemic of "strange fluxes and agues" and faced with dwindling
medical supplies, looked for the possible medical uses of local
plants. He collected specimens, tested the medicinal value of
native plants and mineral substances, and recommended sassafras so successfully that returning ships carried cargoes of the
aromatic shrub to England. He also found in the gum of white
poplar a balm which would "heale any green wound." Physicians
who came with William Penn in 1682, stimulated by the same
practical objective, made important discoveries in regard to the
value of indigenous plants. The search for indigenous medicinal
specifics remained an important stimulus to botanical exploration
well down into the nineteenth century, reaching a new height with
the rebirth of "botanic medicine" as a sectarian school early in the
century.

For the most part, possibilities in the New World during the colonial period gave a powerful stimulus to the pursuit of cabinet specimens, already the chief activity of late-seventeenth-century
naturalists. Their program directly followed Francis Bacon's
injunction to study nature fact by fact. Part of the vision of
Bacon's *New Atlantis* (1624), taken as a model by the Royal
Society, was of an army of collectors ranging the earth, seeking

specimens for the cabinet of a central repository where they could be sorted and arranged into classes. Bacon's *New Atlantis,* indeed the whole Baconian program with its emphasis on fact gathering and systematic classification, was directly inspired by the New World and the unusual opportunities it offered. His program can best be understood as one of the first efforts to grapple systematically with a superabundance of novelties. According to Bacon, of course, correct generalizations would inevitably emerge from the miscellaneous data so assembled. British naturalists, wholly captivated by the Baconian vision, therefore set out on what the master termed "this labor, investigation, and personal survey of the world," the goal being a giant catalogue setting forth all nature in a rational system.

Many Americans willingly participated in the program, confident that their peculiar situation ideally fitted them to contribute to the Baconian quest. Dr. John Morgan expressed the advantageous situation of Americans in terms that must have occurred to many Americans, he said, live on a great continent of which only a small portion had yet been explored. Even the inhabited part was still imperfectly known. The American student of natural history was in an ideal situation where he had many advantages over the European:

> The woods, the mountains, the rivers and bowels of the earth afford ample scope for the researches of the ingenious. . . . The countries of Europe have been repeatedly traversed by numerous persons of the highest genius and learning, intent upon making the strictest search into everything which those countries afford; whence there is less hopes or chance for the students who come after them to make new discoveries. This part of the world may be looked upon as offering the richest mines of natural knowledge yet unriffled, sufficient to gratify the laudable thirst of glory in young inquirers into nature.[19]

As Morgan's statement indicates, the time was past when the principal scientific work would be done as an appendage to exploration by Europeans. It was a call to young Americans to take their place in the world of science by exploiting the opportunities open to them. It would be many years before the European explorer was completely replaced, and still more before exploration ceased to be an important scientific activity, but from Morgan's time on the bulk of the work was done by Americans.

III

Science in British America: From Colony to Province

ℭ Largely because of their situation, by the middle of the eighteenth century British Americans had been integrated into an international natural history circle that was one of the most dynamic intellectual forces in western Europe. Naturalists in England, France, Holland, Sweden, Germany, and Italy kept up a regular correspondence, visited each other, exchanged specimens, and frequently accepted posts in foreign countries. This cooperation between different individuals had become a prominent feature of science in the sixteenth century. A step leading toward the formation of scientific societies and academies, it was no doubt in part an offshoot of the letter writing of the humanists of that period. But it was also a deliberate organizational strategy for handling the opportunities offered by the New World. Most of the information about the New World by the eighteenth century was being channeled through England, which, despite its relatively late start, had become the most active member of the community.

Even by the end of the seventeenth century, French botanists, who had not been personally involved in the scientific reconnaissance, had succeeded in obtaining rich collections of specimens from British North America through their British colleagues. The most active liaison agent in this traffic was Sir Hans Sloane, who not only built up the nucleus of the British Museum as a private collection before his death at ninety-three, but also maintained a regular correspondence with Tournefort, Pierre Chirac, Magnol, François Geoffrey, Antoine de Jussieu,

Abbé Bignon, and virtually every important British colonial.
With them he exchanged seeds, plants, and information, and
shared his high hopes for the future of natural history.[1] The
strength of the Anglo-French scientific tie was made strikingly
clear during the frequent conflicts between the two countries,
when it was usual to provide that, in case of capture by a French
ship, the communications between the American colonies and
England be sent to Jussieu, Buffon, or the Académie des Sciences.
Scientific communications, in that more innocent age, were
deemed to be exempt from the hostility between nations.

Unquestionably, the naturalists hoped for too much from
their method, and in consequence a great deal of misguided
work was accomplished. The very richness of the New World and
the relative ease with which specimens could be collected resulted in a worldwide collection apparatus, without central
direction or a system of priorities, which began to pour forth
torrents of shells, fossils, insects, dried plants, and other "curiosities." The pains which John Winthrop, Jr., took to forward
an earless hog to England in 1673 is only one example among
a great many of the occasional misdirection and concentration
upon trivia that was the result.[2] The fact that Baconianism came
in with the age of exploration resulted in a much more widespread interest in collecting data, but it was often unsystematic
and uncritical. The assumption, quite often, was that one should
report everything—"just in case." The extreme lengths to which
this attitude could be taken were exemplified in an account sent
to the Royal Society by the Reverend John Clayton—and printed
in the *Transactions*:

> . . . There's a tradition amongst them [Virginians], that the
> Tongue of one of these wood-peckers dried will make the
> teeth drop out if picked therewith & cure the tooth ache
> (tho' I believe little of it but look upon it ridiculous) yet I
> thought fitt to hint as much that others may trie for sometimes
> such old-stories refer to some peculiar virtues tho' not to all
> that is said of them.[3]

This was, indeed, the safest attitude; nevertheless it suggests that
a great deal of the satire directed toward the eighteenth-century
virtuoso was deserved. In 1710, Addison's *Tatler* essays described
Sir Nicholas Gimcrack, the standard comic figure of the virtuoso.
In 1728, Swift's *Gulliver's Travels* satirized witless scientific
investigation; and in 1742, Pope's *New Dunciad* criticized the
latter-day virtuosi as a "tribe, with weeds and shells fantastic

crown'd," who saw "nature in some partial narrow shape."

What little direction there was for the activity was provided by important individuals within the societies who had
their correspondents in all parts of the New World, and of the
Old World as well. John Banister, who came to Virginia in 1674
after taking an M.A. at Oxford, at first collected plants for
Robert Morison, the first Professor of Botany at Oxford; later
he began sending his collections to John Ray, who published 147
of Banister's species in his *Historia Plantarum*. When the third
volume of Morison's *Historia Universalis Plantarum* (1698) was
finally completed by Jacob Robert, keeper of the Oxford Botanic
Garden, Robert incorporated what he had learned from Banister's
notes, specimens, and seeds. Leonard Plunkenet, who published
a work (*Phytographia*) which became an important source for
Linnaeus fifty years later, had eighty-nine drawings of Banister's
plants finely engraved and reproduced.[4]

Banister, to whom Ray gave full credit for his contributions, he termed a "most erudite man and consummate botanist"
(*"eruditissimus Vir et consummatissimus Botanicus"*), indicating that Banister was no mere random collector of plants. On
the contrary, Banister had the broad grounding and technical
mastery of natural history that made him eminently deserving of
Ray's praise. At one time he spoke of writing a natural history of
New England and mentioned his need for some kind of system
to handle the data. Apparently he had in the planning stage a
comprehensive natural history on the basis of the four elements,
which was perhaps the closest thing to a system available at that
time.[5] His accidental death in 1692, however, put an end to
whatever plans he had. David Krieg, a German, and William
Vernon, a Cambridge graduate, were actively collecting for Ray
in Maryland in 1698, although neither had the talent of Banister.
Vernon was sent by the Royal Society on the suggestion of William
Byrd; the Society paid his fare and a salary while he worked in
America. A man could not do a greater good to mankind, Byrd
said, "than to bestow a handsome stipend yearly upon a well
qualifyed Naturallist to come and make Discoverys in these Parts
of the World."[6] Specimens collected by Vernon and Krieg eventually came into the possession of Sir Hans Sloane. Sir Hans'
collection, which formed the nucleus of the British museum, was
also enriched by William Houstoun, who gathered plants in the
West Indies and in Mexico.

The Swedish naturalist Linnaeus likewise had his loyal

troops dispersed all over the world and carried on an extensive scientific correspondence with them. Cadwallader Colden in New York, Alexander Garden in South Carolina, and John Mitchell in Virginia are only the best known of the dozens of men who co-operated with Linnaeus and with the Swedish Academy, with which he was associated. ". . . Your agents bring you tribute from every quarter," wrote Collinson to Linnaeus.[7] The French Academy, for its part, in 1699 appointed a Canadian physician, Michael Sarrazin, a corresponding member in the hope that he could coordinate Canadian botanical work and serve as the Academy's chief liaison agent in America.[8] In the same manner the French Academy was also responsible for the work of Father Charles Plumier and of Jean-Baptiste Labat in the West Indies, both of whom wrote several books on the natural history of the islands.

The reward for faithful service was, generally, recognition in a work published by the master, quite often membership in a society, sometimes—the most coveted award of all—the naming of a species for the naturalist. The gardenia is an example of Linnaeus's effort to reward his most gifted American correspond-ent. How eagerly such a reward was sought after is indicated by the following comment of Peter Collinson in a letter to Linnaeus:

> Something I think was Due Mee from the Commonwealth of Botany: for the Great number of plants & seeds I have an-nually procured from Abroad and you have been so good to pay It by Giving Mee a Species of Eternity (Botanically Speak-ing). That is a name as long as Men and Books Endure— this layes mee under Great Obligations, which I shall never Forgett.[9]

Forty-five colonial Americans, including several from the West Indies, received the lesser reward of membership in the Royal Society of London and a great many others were appointed "cor-responding members" of other European scientific societies.

Without the active support and encouragement of influential individuals in England, the remarkable natural history activity of the eighteenth century would not have been possible. The generally high level of their support was largely due to a craze then sweeping through England. At about the turn of the century, it had become a fad among wealthy English gentlemen to collect extensive gardens of rare plants—exotic American plants were in particular demand. Even as early as 1587, in fact, when a new edition of a book on gardening was published, the editor added a spirited passage about the introduction of herbs, plants, and fruits which even at that early date were being brought in from India, America, the Canary Islands, and other parts of the world. With some exaggeration, he observed that there was scarcely one nobleman and hardly a merchant who did not possess a stock of the exotic flowers or medicinal herbs. He had, he claimed, seen three hundred, four hundred, or even more of these novelties in many gardens. The new gardens were so superior to the old, he said, that people could only think of their old gardens as dung heaps or morasses.[10]

Francis Bacon's essay "Of Gardens," calling for a "natural wildness" with no imposed order as a part of the garden, and strongly stressing the educational value of the exotic plants, was an important stimulus both to the gentlemanly interest and to scientific interest. The establishment of the botanical garden at Oxford by the Earl of Danby in 1632 further heightened the interest in botany and, in particular, the awareness of American plants. Such gardens not only provided opportunities for botanical study, but often provided jobs and publication opportunities for botanists both in England and America. Botanists were frequently hired to catalogue the gardens of the collectors, who were generally happy to finance an elaborate publication of the catalogue as a means of advertising their extensive collections. In like manner, the collectors were indispensable to American botanists. John Bartram, for example, was able to make a full-time living as a collector of seeds and plants for his European patrons; others, not quite so successful as Bartram, received partial support from their collecting activities.

The European correspondents of the men on the frontier primarily served as coordinating agents for their scientific work, but they also rendered other indispensable services for them. They bought books and instruments at the request of their colonial correspondents, wrote letters of encouragement, and provided directions. Most of all, by providing contacts with an older scientific culture, they contributed a sense of membership in an international community to men who keenly felt their provincial limitations, and made them feel that they were participating in an important intellectual enterprise. With such a reward in view, the eagerness of British American intellectuals was understandable. The very hint that something in particular was wanted was enough to send them on a long and often hazardous trip. They were so anxious to please, as one appreciative Englishman explained, that they took no thought of "exposing themselves to innumerable dangers and painful travels, in vicissitudes of climates, rigours of seasons, and abandoning themselves to the inhospitable regions and inhumanity of savages," all this simply to furnish new plants for their European correspondents.[11] "I am very proud it is in my power to gratify any curious gentlemen in this way," said John Custis to Peter Collinson, "being myself a great admirer of things of that nature."[12] Custis had merely heard second-hand that Collinson wanted some mountain cowslip. With the letter he enclosed some roots which he had traveled some distance to obtain, and he promised to try to get the seed at the proper time. Alexander Garden, a Charleston physician and the most gifted systematic botanist in the colonies, explained to his English correspondent, John Ellis, that he thought there was no greater stimulus to scientific inquiries than demands from correspondents:

> I know that every letter which I receive not only revives the little botanic spark in my breast, but even increases its quantity and flaming force. Some such thing is absolutely necessary to one, living under our broiling sun, else . . . we should rest satisfied before we had half discharged our duty to our fellow creatures, which obliges us, as members of the great society, to contribute our mite towards proper knowledge of our common father.

Even though Garden was a physician, his interest in botany was dictated more by his sense of the necessity to participate in a common scientific enterprise than by any hoped-for medical uses of the plants. In fact, unlike most physicians of his time, Garden's interest in botanic medicine was slight, to say the least.

It is equally clear that Garden's enormous concern with natural history amounted to far more in his mind than a mere gentlemanly hobby or pleasant diversion. As his letter to Ellis indicates, he and others like him in the eighteenth century regarded natural history inquiry as something approaching a sacred duty. It was a duty the discharge of which transcended any considerations of comfort, convenience, or even personal fame. Garden's letters reveal him at one point in the midst of a smallpox epidemic and faced with an outbreak of Indian up-

risings (1760), stealing two hours to botanize with John Bartram, taking time—after fourteen hours' work among the sick—
to pack a box of fish for Linnaeus and a box of seeds for John
Ellis. Later, writing to John Ellis, he interspersed natural history
details with accounts of a current Indian uprising. Discovering
that John Bartram had represented one of Garden's discoveries
as his own, Garden easily forgave him on the grounds that "it is
a matter of little moment who declares the glories of God, provided only that they are not passed over in silence."

Moreover, when his medical practice and his family re-
sponsibilities did interfere with his investigations in natural
history, Garden refused to accept payment for his scientific work
and, at the same time, felt intensely uncomfortable, even apologetic, for his failure to take advantage of the opportunities with
which Providence had blessed him by placing him "in a land of
wonders." Writing to Ellis after one such period of inactivity, he
pleaded that even after three years of neglect, it was not yet too
late to redeem himself and "return to the ways of well doing."
He only had to beg that Ellis and Linnaeus would stimulate him
to fresh exertion. "To you and Linnaeus I owe my all in that
way," he said, "and you must continue, by a continuance of your
correspondence, to impel me to do you any services in my power,
by making collections for you here."*

Although Ellis was of undoubted importance in stimulating
the work of Garden and some other American naturalists, the
greatest of these indispensable scientific dilettantes were James
Petiver, a London apothecary whose shop was for over thirty years
after 1685 the world's largest botanical clearinghouse, and
Peter Collinson, a Quaker merchant of London, who utilized his

---

* J. E. Smith: *A Selection of the Correspondence of Linnaeus and
Other Naturalists from the Original Manuscripts,* 2 vols. (London,
1821), I, 362-3, 531-2, 572-3. In Mexico, the world-renowned scholar
Don Carlos de Sigüenza y Góngora felt the same yearning for contact with kindred spirits that American provincials did. At one point
he confessed to an "insatiable desire" to know and communicate
with outstanding men of learning, particularly in Europe, with
whom he might share his observations and information. In his
circumstances, the desire could be only slightly appeased. "If some
mathematician," he wrote in his *Libra Astronomica* (1690), "wishes
to communicate with me observations on eclipses, either his own or
those of some one else, and especially of the Moon from 1670 on,
I shall send him mine from the same date with utmost liberality."
Irving A. Leonard: *Don Carlos de Sigüenza y Góngora: A Mexican
Savant of the Seventeenth Century* (Berkeley, Cal., 1929), p. 56.

worldwide business correspondence as a way of promoting natural science.

By entreaty, flattery, and the promise of seeing their names
in print, Petiver tried to persuade every traveler he came in
contact with to collect for him and to persuade others dwelling
in the lands they visited to do likewise. No scientific education
was required, for the enterprising apothecary provided detailed,
printed instructions as to what to collect and how to preserve and
pack the specimens for shipment to England. He would even
furnish brown paper for preserving plants, wide-mouthed bottles
for insects, small animals, and fishes, and "an easie method" for
pickling "Fleshy bodies capable of corruption" in brine or in
spirits. His instructions to his "butterfly boy" Isaac, whom he sent
to Maryland in 1699, are typical:

> Whenever you goe ashore take with you a Quire of Brown
> Paper or Collection Book, an Insect Box, Pins & a small Viall
> half fil'd with Spt. in which draw all your supernummery
> Flies, Beetles, Catterpillars, & other insects expecially such as
> you shall find in water. Also a Booke for Butterflies & Moths
> of each of which get all you can find; with a paper bag or
> two to put all ripe seed, Fruit & berries as also all the shells
> you meet with both land & water & as many of each sort as
> you can find; such as are thin & brittle you must put into a
> Pocket by themselves with moss or any soft leaves to keep
> them from breaking.

The young Scot, Cadwallader Colden, while contemplating
settlement in America, came to Petiver for "advice & assistance
in his studies" and to learn "the way of making Specimens of dry
plants & of collecting other natural curiosities."[13] Although
Petiver's returns from Colden were disappointingly small, Colden
became one of the more than eighty American correspondents of
the avid collector. Cotton Mather's short correspondence with
Petiver included an account of his now-famous observations on
hybridization in corn.

To such correspondents as these Petiver sent newssheets,
scientific journals, and the latest books on natural history; provided them with scientific instruments, gave free medical advice,
often accompanied by free drugs, and in several cases looked
after the welfare of his correspondents' relatives in England.
In the beginning he did not pay for specimens, although he frequently urged his correspondents to employ native collectors,
at his expense, and he reimbursed any expenses incurred in his

behalf. To the most talented, Petiver offered fame and recognition in the new Republic of Natural Philosophy by his "grateful & just" acknowledgment of their work.[14]

Peter Collinson, Petiver's successor as coordinator of colonial science, introduced—from all sources—over 175 different plants into England. His major difficulty—and one which characterized science throughout the colonial period—was in getting Americans to think consistently in terms of English needs. Time and again he objected that when correspondents sent him something from the colonies, they sent only what was uncommon to them; failing to realize that even their "common" might be "uncommon" to Englishmen. This misunderstanding naturally followed from the use of amateurs and the effort to direct their work at long distance.

Although Collinson was primarily interested in adding to the extensive collection of exotic plants in his private garden and in those of his friends, he by no means confined himself to promoting botany. It was he, for example, who was responsible for the publication of Benjamin Franklin's classic *Experiments on Electricity*, the most important theoretical contribution by a colonial American. This notable book, on which Franklin's scientific fame largely rests, originated as a series of letters from Franklin to Collinson.

Collinson, James Petiver, and a few others were only the best known of the private collectors of exotic plants. As Collinson wrote to John Bartram, "There is a [great] spirit and love of [gardening and planting] amongst the nobility and gentry, and the pleasure and profit that attends it will render it a lasting delight." Some of the nobles who could afford it literally transplanted a bit of America to England. Describing the collection of Lord Petrie, one of Bartram's patrons, Collinson observed that during the previous year he had planted about ten thousand American specimens. Walking among Lord Petrie's nurseries, Collinson said, "one cannot help thinking he is in North American thickets, there are such quantities."[15]

Publication of Mark Catesby's magnificent two-volume work (1731) on the *Natural History of Carolina, Florida, and the Bahama Islands*, termed by many the most beautiful work printed in England up to that date, gave a tremendous impetus to interest in American nature. Catesby had spent seven years in Virginia, returning to England in 1719 with a quantity of paintings and a collection of natural history objects that quickly brought him to the attention of Peter Collinson and the Royal Society. His second trip to America (1722–6), when he did the paintings that later appeared in his two-volume work, was financed by Sir Hans Sloane, Collinson, and a group of other British naturalists. While in America, Catesby wrote frequently to his British patrons, keeping them informed about "one of the Sweetest Countrys I ever saw" and periodically sending them large quantities of plants and seeds.[16]

Sometimes, when the British enthusiast for "exotics" was in an important position in the colonial service, he was able to render an even more direct aid to natural history. Henry Compton, who in his capacity as Bishop of London appointed ministers to the colonies, was one such interested amateur who developed the famous botanical garden at Fulham Palace. During his tenure, ministers sent to the colonies were often questioned more on their botanical knowledge than on their theological understanding or their clerical aptitudes. It was he who sent John Banister and John Clayton to Virginia and Hugh Jones to Maryland, all of whom made notable contributions to the botany of North America.

When John Clayton left for Virginia to become rector of James City Parish in 1684, he carried with him the best scientific education obtainable at the time, along with microscopes, barometers, thermometers, chemical instruments, reference books, and notes on experimental work he had been doing in England. His intention was clearly to spend a large part of his time in Virginia making serious scientific investigations. Unfortunately, all but his education was lost when the ship following him, on which the instruments had been shipped as a way of dividing the risk, was lost at sea. The instruments could not be duplicated in America, and their lack remained a serious problem long after Clayton's time. For example, as late as 1720 Dr. William Douglass was unable to find a single thermometer or barometer in Boston.

Despite Clayton's loss of equipment, however, after a two-year stay he returned to London with drawings of plants, "probably some collections, and certainly a vast fund of information."[17] His accounts to the Royal Society on the geography, soil, meteorology, and birds of Virginia, which have recently been republished by Edmund Berkeley and Dorothy Smith Berkeley, make fascinating reading today. His summation of Virginia and its inhabitants may stand as an example:

> Its a place where plenty makes poverty, Ignorance ingenuity,
> & coveteousnesse causes hospitality that is thus evry one covets
> so mch & there is such vast extent of land that they spread
> so far they cannot manage well a hundred pt of wt they
> have evry one can live at ease & therefore they scorne & hate
> to work to advantage themselves so are poor wth abundance
> They have few Schollars so that evry one studys to be halfe
> Physitian halfe Lawyer & with a naturall accutenesse would
> amuse thee for want of bookes they read men the more Then
> for the third thing Ordinarys ie our Inns are extreame ex-
> pensive wherefore with a comon impudence they'le goe to a
> mans house for diet & lodgeings tho they have no acquaintance
> at all rather than be at the expence to lie at an Inn & being
> grown into rank custom it makes them seem liberall.[18]

Students of Virginia history have much cause to regret that
Clayton's promised report to the Royal Society on the "state of
the Inhabitants" was never submitted.

A second John Clayton, of no known relationship to the
first, also contributed notably, with the help of the Dutch
botanist Gronovius, to the understanding of Virginia botany. Of
the work by which the American Clayton is remembered chiefly,
the *Dictionary of American Biography* says that the first edition
of the *Flora Virginica* is of "merely historical interest," but that
the second edition takes rank as "true, modern, systematic work."
The explanation of this statement lies in the fact that Linnaeus's
Species Plantarum, 1753 edition, was published in the interval
between the first and second issues of Clayton's work, and that
in the 1762 edition of the *Flora Virginica* the plants described by
Clayton were classified according to the Linnaean system. Earlier
he had used John Ray's scheme of classification. The Linnaean
classification, however, is believed to have been the work of the
younger Gronovius, who edited the second edition of Clayton's
work.

Clayton himself, however, was no mere dabbler. On the
contrary, his work in the *Flora Virginica* shows considerable
evidence of interest in the ecology of the plants. The comments,
undoubtedly furnished by Clayton, even though he may not have
done the work of classification, often include the character of
the habitat in which the plant thrives, the time of flowering,
other plants associated with it, methods of propagation, effects
of weather on the plant, and its relations with animals.

Spurred on by the demands of their European correspondents, American colonials were able to make large contributions

to European knowledge of natural history—although, from the frequent complaints that one finds, the contributions did not always flow in fast enough to satisfy the demand for knowledge about America. Henry Oldenburg, for example, in his capacity as secretary of the Royal Society, frequently found it necessary to administer a friendly slap on the hand to American correspondents:

> You will please to remember, that we have taken to taske the whole universe [he reminded John Winthrop, Jr., in 1667] and that we were obliged to do so by the nature of our Dessein. It will therefore be requisite that we purchase and entertain a commerce in all parts of the world with the most philosophical and curious persons to be found everywhere.

Winthrop, despite his early promise, had not done all that was expected of him by the Royal Society, having been too much taken up with politics, wars against the Dutch, and efforts to increase the family holdings. Four years earlier, Oldenburg had confidently informed him that the Society expected from him a better description of:

> The remarkables than is any yet extant, concerning the mappe of the country, the history of all its productions, and particularly the subterraneous ones . . . likewise a relation of the Tides upon your coast, together with the course of your rivers, but especially and above all, a full account of your successe in your new way of saltmaking, whereof we could not compasse the experiment here, as was much desired.

Later he urged that Winthrop strive to foster in New England "this reall Experimental way of acquiring knowledge, by conversing with and searching into the works of God themselves," and he suggested that Winthrop might write a complete natural history of New England.[19] That the insatiable desire for American seeds, cuttings, and other treasures persisted into the following century is illustrated by a series of letters from John Fothergill to Humphrey Marshall, written between 1767 and 1771 and preserved at the Historical Society of Pennsylvania, each pleading for more. Fothergill ordered tortoises, insects, snakes, birds, plants for medicinal purposes—carefully including instructions for handling. He also prodded John Bartram to become more systematic in his work—asking him to send drawings of North American land tortoises—and at the same time to ask the artist

(William Bartram) "to make some short notes with respect to their natural history, way of life, places of abode, generation and whatever occurs to him."[20]

Both in quantity and in quality, scientific research in the British colonies continued to show marked improvement throughout the eighteenth century. After the publication in the *Philosophical Transactions* of two articles by Thomas Brattle on astronomy in the first decade of the century, scientific interests were continually broadened out from the exclusively natural-history orientation of the previous century. A spurt of vigorous activity in the British colonies beginning about 1720 did not die down until the Revolution. But despite all the advances that were made, science in the colonies remained definitely *colonial;* that is to say, the primary determinant of the direction of scientific work was the relation of the colonies to Europe. There was no real indigenous scientific community during the colonial period; the colonial scientist was a member of the scientific community of Western Europe and his activities were mostly determined by his ability to contribute to that community. The problems were suggested by that community and the colonial scientist depended upon it for recognition.

As the Marquis de Chastellux observed in 1782, the Americans had unique advantages in certain areas:

> The extent of her empire submits to her observation a large portion of heaven and earth. What observations may not be made between Penobscot and Savannah? between the lakes and the ocean? Natural history and astronomy are her peculiar appendages, and the first of these sciences at least, is susceptible of great improvement.[21]

Certainly Chastellux had a point, and it was not one that escaped many observers. In those areas of the American's special competence, his unique geographical position gave him access to data which could not be duplicated elsewhere and which therefore won for him the prompt support and active encouragement of Europeans. American specimens and seeds could be collected only in America, either by Europeans who made the dangerous and expensive trip across the Atlantic or by Americans already on the scene. Since it was obviously advantageous to have Americans collect the desired objects, colonials were used whenever possible, and they were given every possible encouragement.

Nevertheless, the colonies had to reach a certain level of maturity before American collectors predominated over the European visitors, and the former did not entirely displace the latter until long after the end of the colonial period.

Even though some Americans, such as Alexander Garden or John Clayton, developed considerable skill in systematic botany, it is significant that neither of these men achieved the reputation of John Bartram, who never quite learned to read Latin and who never fully mastered the Linnaean system. Garden, hearing of Bartram's planned trip to the Mississippi in search of plants, wrote that he would undoubtedly return "loaded with great spoils," but thought it was a pity that Bartram did not know the method of characterizing the specimens "which he must meet with that are entirely new."[22] Although Garden was correct in his assessment of Bartram, his criticism entirely missed the point. Bartram's perception of the *new* was all that mattered; it was not necessary—and really not even desirable—that he be able to classify the new plants he found, for this could be done more satisfactorily by the better-trained Europeans who had access to herbaria and specimens from all over the world and who were therefore able to integrate Bartram's specimens into the general body of Western science. The superiority of Europeans remained a fact even though dried specimens were notoriously an uncertain dependence and they, consequently, at times made errors that Bartram would not have made. John Bartram, simply because he fit so well the expectations of European naturalists, was well known to all of the leading European botanists, and many of them acknowledged their indebtedness to him. "The best natural botanist in the world," Linnaeus called him. Bartram's work was in such high demand that he was able to make a substantial income from his collecting activities, thanks to arrangements made by Peter Collinson. In the 1750's he was employed on an annual basis to furnish seeds and plants to at least fifteen patrons, including the Prince of Wales, at a charge of five guineas each. The high regard in which he was held by scientists all over western Europe was a simple recognition of the pre-eminence of his type of work.

It was not merely the facts of colonial *birth* and a limited scientific education which placed certain disadvantages on Bartram, however; the very colonial *location* made even highly trained European investigators cautious in their scientific judg-

ments. Operating alone, for the most part, and having no opportunity for those tentative discussions with like-minded men that
are so important in reaching an opinion, European-trained immigrants, too, generally contented themselves with relating the
facts, submitting specimens, and leaving the analysis to others
more favorably situated. Thus John Clayton, coming to America
with the best education as an experimental scientist which
England could offer, nevertheless turned descriptive natural
historian when he reached this continent. And this transformation was not due simply to the loss of his instruments. Submitting
some petrified oyster shells from the Virginia tidewater to the
Royal Society, Clayton suggested the possibility that these might
be crystallized forms rather than petrified oysters. There was
nothing, he said, "too difficult or wonderful for nature." Yet he
would prefer to simply raise the question and "leave to the
Honourable Society to determine" the answer.[23]

In fields where the American's geographical position did
not confer an advantage, his inadequate training and his remoteness from contact with other scientists made it impossible
for him to compete with Europeans. As raw materials for physical
science, his observations were welcome, just as they were in
natural history. Even in the seventeenth century, for example,
Thomas Brattle's observations on the comet of 1680 (Halley's)
were made use of by Newton, who spoke with praise of the
observations from New England. Brattle was using a 3½-foot
telescope brought to the colonies in 1663 by John Winthrop, Jr.,
who had donated it to Harvard College. Winthrop himself had
sent observations of three earlier comets to the Royal Society—
observations which had proved valuable because of the observer's
vantage point. He had barely missed another chance to contribute
in 1664, when word from the Royal Society of a predicted transit
of Mercury arrived too late for him to observe the phenomenon.[24]
The dependence of colonials upon instructions from Europe is
nowhere better illustrated than in this delayed communication.

During the next century, other colonials joined the observation team. Cadwallader Colden and James Alexander sent several
observations on the eclipse of the first satellite of Jupiter in an
effort to determine the longitude of New York; Paul Dudley sent
a history of New England earthquakes; many—such as Thomas
Robie in Boston—sent daily weather reports for analysis by their
European correspondents. But with a few notable exceptions,

work was not conducted independently in the colonies, and the typical colonial scientist obtained his problems either by direct instruction from Europe or from practice. Paul Dudley's papers in the *Philosophical Transactions* are exemplary of this second mode. They deal with the making of sugar from maple trees and molasses from apple trees; with vegetables, plants, and animals of commercial importance; and there is a study of the spermaceti whale and the nature of ambergris. In 1724, he contributed to hybridization studies by describing the spontaneous crossing of different varieties of New England corn. Cotton Mather, as early as 1716, had pioneered in this last subject, recording and transmitting to the Royal Society the facts of wind pollination, of variety crosses, and the resemblance of some of the progeny to the male parent.[25]

Despite such occasional contributions as these, the scientific side of the agricultural revolution, for reasons which will be dealt with in a following chapter, never really took hold in America during the colonial period, and very little experimental work was done in the colonies.

Even as talented a scientist as John Winthrop IV, Professor of Natural and Experimental Philosophy at Harvard from 1738 to 1779, keenly felt his limitations as a colonial. For example, beginning in 1742 and continuing until four days before his death, Winthrop kept a comprehensive journal of the weather at Cambridge, including twice-daily observations of such things as barometric pressure, temperature (from two thermometers), wind direction, force, cloud cover, and precipitation. Monthly and annual calculations of means, highs, lows, and totals, as appropriate, completed his work, which he dutifully shipped off to the Royal Society, explaining that his work was "agreeable to Dr. Jurin's admonition in Philosop. Transact. No. 379." Despite the mass of material, Winthrop never attempted to generalize, to draw trends, or to draw any kind of conclusions; that was for others to do—others who had more complete information. Again, even though Winthrop was apparently the first to suggest the wave motion of earthquakes, he did not attempt to argue his point, but left all judgment up to the Royal Society, which was in possession of more complete information. Winthrop, realizing that he had neither the instruments nor the opportunity to explore his surmise, never took the matter any further.[26]

When Winthrop compiled his observations on the transit of Venus in 1769, out of long-standing habit he sent his work directly to the Royal Society, even though a reinvigorated American Philosophical Society was collecting such observations for publication. Such deference patterns, which are established in the first instance by real limitations, generally persist because of leading colonial scientists' natural desire to publish in more prestigious, long-established journals where they have a better chance of being read by their peers.* Overcoming this pattern of behavior is one of the most difficult tasks to be faced by those attempting to establish an indigenous scientific tradition.

Winthrop, therefore, despite his obvious talents, to the end of his life displayed the characteristics of a colonial. Almost without exception, his scientific writings arose either from a direct request or from some immediate natural phenomenon or catastrophe—a lightning stroke, an earthquake, the appearance of a comet, a lunar eclipse—which could be observed in America. In consequence, his scientific writings remained descriptive, fragmentary, and disconnected.

It could hardly have been otherwise in Winthrop's time. Scientists, above all other men, need the stimulation provided by interactions within a group. The Royal Society of London provided the program and the coffee houses of London provided an indispensable meeting place where savants in pursuit of a common program could meet and discuss the problems they encountered. It is now well understood—and the evidence suggests that it was then—that the scientific judgment can usually be sharpened through such group interactions. The lone investigator may think deeply about a problem but fail to solve it simply because of some small blind spot: perhaps the lack of a single piece of knowledge that another member may supply, perhaps a slight failure in analysis which can be corrected by another. This was the need that Americans tried to meet with their voluminous correspondence, but even under the best of conditions correspondence fails to supply the immediacy that comes from face to face contact. And under the poor communications conditions that existed in the eighteenth century, it was especially inadequate. With such a limitation, it is hardly surprising that

---

*Even in the early twentieth century, leading American physicists preferred to publish in English journals.

[only a few colonials ever rose above the level of natural history *reporting*] When they did attempt to go beyond the reporting of facts, the results were almost uniformly disappointing at best and at worst, ludicrous.

Even Benjamin Franklin, despite his world-wide reputation and his truly impressive accomplishments, did not constitute a real exception to the general features of colonial science. The only scientific field in which Franklin made discoveries of lasting significance was in electricity. His reputation as a physicist was earned because of a series of letters to Peter Collinson, later published as a book entitled *Experiments and Observations on Electricity, made at Philadelphia in America.* Franklin's book was a work of genius and it was certainly the most important contribution made by an American in the colonial period, but it, too, reveals the limitations of colonial science.

Franklin probably received his first knowledge of electricity from the itinerant lecturer Adam Spencer, whom he heard speak in Boston in 1743. The following spring, Spencer provided more information while he was lecturing in Philadelphia. He also discussed the subject with William Claggett, who was lecturing in Newport when Franklin visited the city, and he probably read accounts of French and German experiments in the *American Magazine* during 1745 and 1746 and in the *Philosophical Transactions* dating from 1739. Electricity, in a word, was in the air, and any alert colonial could have had access to the basic facts.

It was Peter Collinson, however, who proved crucial in Franklin's experiments. Sometime before 1747, when Franklin was on the verge of retirement from business, Collinson sent to the Library Company of Philadelphia a glass tube for conducting electrical experiments and an account of German experiments in electricity, with directions for repeating them. To the Collinson gift, Franklin added all of Adam Spencer's apparatus, which he purchased when the itinerant lecturer left Philadelphia; and the proprietor of the colony graciously presented the Library Company with "a complete electrical apparatus." There was now no lack of equipment; tubes were turned out in quantity by a Philadelphia glass house and there were enough skilled mechanics in Franklin's circle to contrive more elaborate apparatus. His curiosity already kindled, Franklin then threw himself into the study of electricity to a point where it "totally engrossed" his attention as it had never been engaged before.

At the heart of the "Philadelphia experiments" was an important theoretical conception which Franklin was able to demonstrate with an impressive set of experiments. He was, first, able to show how a single-fluid concept fitted the observed conditions more satisfactorily than the two-fluid theory then widely accepted because of Charles Dufay's writings. Franklin also suggested the terminology that became so widely useful, calling one charge "negative" and the other "positive"—the presumption being that a deficiency of electric fluid caused the negative state. Even though he was wrong about the direction of flow, the Franklin terminology and the concept of a flow from positive to negative are still used by electricians. Time after time in his experiments, Franklin showed the ability to construct brilliant hypotheses which grasped the essential nature of the phenomena, and to construct adequate experiments for checking them.

Obscuring his more basic conclusions, however, were the dramatic suggestions Franklin made concerning lightning. First, on the basis of a careful comparison of properties, Franklin suggested that lightning was an electrical phenomenon. This was not a new idea, but Franklin devised two impressive experiments to test the hypothesis. The guard-house experiment, in which a man stood inside an insulated house on top of a steeple surmounted by a lightning rod, from which he could draw off electricity, was performed first in France; but Franklin was the first to perform his more famous kite experiment. The hypothesis being proven, Franklin then moved on to the utilitarian accomplishment of his study—the lightning rod—a suggestion that he was led to by his study of the electrical effect of points.

Despite the currency of the electricity fad, Franklin and his associates at Philadelphia began with a minimal knowledge of current European progress—a common failing of colonials, which usually kept their science from advancing beyond the elementary stage. As usual, they wasted some time in rediscovering what was already known. Philip Syng, for example, mounted a glass tube on an axle, which he turned with a crank to save the fatigue of rubbing. No one in Philadelphia then knew that similar machines had long since been used in Europe. Even when Franklin, in May 1747, reported to Collinson on "the wonderful effect of pointed bodies, both in *drawing off* and *throwing off* the electrical fire," he was unaware that Von Guericke of Magdeburg had begun the study of pointed conductors in the seventeenth century. Nevertheless, it was Franklin's independent study of points that led both to the colonial scientist's greatest experiment and to his most famous invention.

Although some, even among his own contemporaries, have credited Franklin's very naïveté with his success in electricity, more pertinent is the fact that there was very little to be naïve about—and most of that was wrong.[27] Franklin's apparent ignorance of Dufay's distinction between the two types of electricity—accepted by the most advanced European thinking—was certainly not a disadvantage. In his ignorance that a "satisfactory" theoretical framework already existed, Franklin was able to proceed directly from his own observations to what seemed to him the obvious assumption that electricity was a single fluid. The common disadvantages of a colonial simply did not apply here. Furthermore, it was still possible to carry on important electrical experiments using kitchen equipment; electricity had not become mathematical; and there was no body of theory with which one had to be familiar. The conditions, in other words, were roughly the same in electricity as in natural history, another area in which Americans could excel. With the most primitive equipment, the observations could be made anywhere; any observation had a strong likelihood of being a contribution to knowledge; and no special learning was required. All this is not intended to downgrade Franklin's achievement, for he did apply one of the most powerful minds of his time in interpreting the experiments. Nevertheless, it is not likely that he could have made such a fundamental contribution to any other physical science; for all of his advantages would have turned into the same kind of limitations that caused Cadwallader Colden's dismal failure in his effort—far from the latest research and the conversation of the learned—to "improve" Newton's hypothesis.[28]

Since the colonial American scientist identified with the European scientific community, early efforts to found scientific societies were generally failures or, at most, only temporary successes. There were a great many such attempts, beginning with Increase Mather, who gathered together a small group to discuss philosophical problems and to add "to the Stores of Natural History."[29] Abraham Redwood made a similar effort in Newport in 1729, and William Douglass in Boston tried several times to found a medical society which would also have broad natural science functions.

Cadwallader Colden was apparently the first American to suggest, in a 1728 letter to Dr. William Douglass, the formation of
an intercolonial scientific community in the form of a society
or academy:[30] since the academy was the primary focus of scientific activity in the eighteenth century, it was thought to be the
logical starting point for developing an indigenous scientific com
munity. Also logical at that time was Colden's suggestion that
leadership in the society come from Boston, then still the leading
cultural center of the colonies. As the outcome was to demonstrate, however, a functioning scientific community was a pre-
requisite for the formation of a lasting academy, for in no case
did such early efforts result in a continuing society. The great
complaint of Thomas Brattle in 1705, and of Paul Dudley a few
years later, that even in Boston they could find none of their
neighbors interested in experimental science was characteristic
of the colonial situation. As late as 1755, Alexander Garden of
Charleston was exclaiming that South Carolina was "a horrid
country, where there is not a living soul who knows the least
iota of Natural History."[31] Also characteristic of the European
orientation of colonials was the fact that, in 1760, John Clayton
reported that he often had to rely upon Peter Collinson's letters
from London for news of other Americans.[32] It was only natural
that communication between a colonial scientist and English correspondents should have been more frequent than among colonials, for within the colonies the forest was a formidable barrier
to communications and travel. Water routes were always preferred, even at the price of indirection. In 1723, for example,
Benjamin Franklin, leaving Boston to try his fortune in Philadelphia, traveled by water to New York City, then overland to the
Delaware, where he took another vessel to Philadelphia. The
communications patterns before the development of more adequate routes during the latter half of the eighteenth century both
reflected the principal relationship between the individual colony
and the mother country and reinforced that relationship.

Only in Philadelphia was there a large enough group of
peers having similar interests to make organizational efforts
worthwhile, and definitive success even there came only as a
consequence of the pre-Revolutionary agitation. In 1739, when
John Bartram suggested founding a learned society at Philadelphia, Peter Collinson felt that this type of endeavor was beyond
the capability of colonials. But apparently by 1743, when he heard

about the formation of the American Philosophical Society, Collinson felt the colonial culture had finally generated a scientific community approaching the critical size, for he became excited.[33] Even this society, however, proved premature, despite the worldwide encouragement it received; for the best efforts of Franklin and Bartram proved inadequate to generate and sustain the kind of broad local support necessary for its success. It collapsed after a few meetings and was revived only on the eve of the Revolution.

IV

The Transit of Ideas

The most famous ship in American history, the *Mayflower*, after a voyage of sixty-four days from Southampton, England, anchored off what is now Plymouth, Massachusetts, in November 1620. The majority of the 102 English passengers who came ashore to found the first permanent colony in New England were Puritans who, for a complex of reasons including religious zeal, economic aspirations, and social discontent, had decided to take up residence in the New World. A less famous ship, the *Arbella*, ten years later led a flotilla of Puritans to found the colony at Massachusetts Bay.

These events mark the establishment of what was to be an extremely homogeneous society for the next 150 years. The culture of the Puritans, along with their essentially religious view of life, dominated New England all during the period known as the scientific revolution. The generation of Europeans who first settled the eastern seaboard of North America was the same generation that first discovered the universe revealed by modern science. This was the generation of Kepler, Galileo, Bacon, Harvey, Boyle, and Descartes—the men whose work prepared the way for such a later giant as Isaac Newton, whose work, in turn, formed the intellectual basis for a wholly different world view which has given its name to a period in history: the Enlightenment. Although the American enlightenment reached its height outside New England, it was the New Englanders who first encountered the new currents of thought and had to come to terms with them. When Isaac Newton wrote, the city where the Enlightenment reached its highest colonial development—Philadelphia—was not yet in existence. The Southern colonies, lacking urban centers of population, could not develop a flourishing intellectual life. Boston, an important seaport dominated by descendants of the early settlers, was still the intellectual, as well as the mercantile, capital of British America. Any account of the transit of ideas from the Old World to the New, and the consequent reshaping of thought that occurred as a result of the new ideas, will therefore have to concentrate heavily on these New England Puritans.

Ideas arriving from England did not strike a vacuum in America, for the colonists had been remarkably well equipped from the beginning. A great deal of intellectual baggage as well as the more conventional possessions came in with each shipload of passengers. By the original settlers, who were an unusually well-educated group, the learning especially of Cambridge in the 1610's was brought to New England. Before 1646, at least 130 university graduates arrived in New England along with a large, but indeterminate number of others who had been educated in the English grammar schools and therefore had the same general orientation as the university graduates. These men, mostly ministers, constituted an intellectual ruling class whose standards were accepted by the community and maintained by the college— Harvard—that they founded in 1636. A theological orientation among the early settlers should therefore be assumed as a matter of course, for the preoccupation of the English universities was theology in its various forms: ecclesiastical polity, the philosophical aspects of Christianity, the relation of man and nature to God, and the nature of God himself. These were the vital issues to men of the seventeenth century. There were no mathematical or scientific interests in the English universities until after the Restoration; medicine was not yet a university subject, but a profession learned by apprenticeship or independent reading; professors of astronomy still regularly cast horoscopes; earth, air, fire, and water were still the basic elements of the chemist, whose main work was trying to change base metal into gold. Moreover, practically all the scientific advances of the seventeenth century were made outside the universities, which resisted each innovation by an appeal to Aristotelian physics, Ptolemaic astronomy, or the Bible. The scientific training of the early Puritan leaders, even under the best of conditions, could have amounted to no more than a smattering of mathematics and one term of reading "natural philosophy," meaning Pliny, Aristotle, or some other classical author.[1]

Even had they looked outside the universities where they had had their training, the first generation of intellectual leaders in the colonies would not have found very much. The scientific revolution was barely beginning in England. Few had heard yet of Kepler and Galileo; Harvey's work was barely completed; Boyle's great work which eventually led to a new concept of the chemical composition of matter was not yet begun; and the ideas which later entered into Bacon's *Novum Organum* were only being formulated. The group which eventually formed the Royal Society did not meet until 1645; and it was not until after the Restoration that laboratories, microscopes, and other scientific "curiosities" became fashionable in the world of Charles II, Samuel Pepys, and the Earl of Rochester. The Puritans, having removed themselves to the edge of a "howling wilderness," of necessity missed all this. In a word, with the New England Puritans, we have a unique, laboratory-type situation. It was a small, homogeneous society—numbering no more than one hundred thousand by 1700—transplanted to a New World, accessible only with difficulty, and carrying with it the best of current learning. We are able to observe in New England, as nowhere else, the actual processes of intellectual change.

Even the best education eventually becomes obsolete—and the process of obsolescence is accelerated during a period of such intellectual ferment as the seventeenth century. The books brought with settlers likewise grew out of date, and even the new ones imported did not give the immediacy and the stimulus of personal contact with persons of similar interest. Nowhere in America was there the equivalent of the London coffee houses, favorite meeting place of the "curious"; the few devotees of science were isolated, travel was difficult, and the meeting of like-minded men infrequent. One should therefore expect a certain amount of retardation of intellectual development—perhaps not the "stagnation of social evolution" that it has often been termed, but at least a minor case of intellectual lag.

Correspondence partially filled the gap in communication, but this was slow and uncertain, and under the best of conditions not nearly as satisfactory as daily intercourse. One therefore is not surprised to find a lack of current information on the most common subjects. The letters to England of colonial intellectuals are filled with requests for information about current scientific happenings. John Winthrop, Jr., for example, writing in 1659, appends a list of "quaeries" to a letter to Samuel Hartlib, a member of the group which formed the Royal Society. He wants to know whether Glauber really has a liquor called "Alkahest" with the properties he ascribes to it?—whether those medicinal and chirurgical balsams mentioned in Digby's treatises may be obtained for use in America?—whether any real invention has been made known for dissolving or breaking the gallstone?—what perfections have been added to the telescope since Drebles and Galileo?—what new discoveries have been made in the celestial bodies?—what is there new about perpetual motion? In short, Winthrop wanted to be filled in on the everyday gossip of the coffee houses. Hartlib, a distinguished virtuoso who was acquainted with practically every scientific figure in Europe and made it his business to keep abreast of the latest "philosophical" developments, was a favorite source of information to Winthrop. The Englishman not only apprised Winthrop of the activities of the renowned, but he also supplied large quantities of literature for the colonial savant.[2]

The isolation of the colonials was real enough, and so was the consequent lack of current knowledge; however, the most surprising thing is the extent to which they were able to overcome their condition. The intellectual giants of the day were avidly studied by colonials, and their works eventually incorporated into the structure of thought. Later in the seventeenth century, one frequently encounters such names as Galileo, Kepler, Gassendi, Boyle, Hooke, and Descartes. Gassendi, the leading reviver of atomism and an important precursor of Newton, was a particular seventeenth-century favorite in America as he was with the learned in England. Harvard students were using his astronomy textbook, which incorporated Galileo's discoveries, by 1675, and they continued to use it for over half a century. The Harvard published almanacs of 1659 and of 1661 both mentioned Gassendi and explained some of his doctrines.[3] John Foster's *Almanack* of 1681 spoke of what he termed the "Common Saying," which could be traced to Gassendi: ". . . nothing is in the mind which was not first in the senses."[4]

The almanacs, as ubiquitous in New England as the Bible, and nearly as popular, were the chief written agencies in propagating the new learning of the seventeenth century. Because a standard sheet of paper was generally folded to make sixteen pages of which only twelve were needed for the months of the year, the almanac-maker had four extra pages to fill with a variety of useful material. Representative almanacs contained such information as the condition of the roads; distances between major towns; proper forms of legal papers such as bonds, indentures, or wills; bits of New England history; tables of "remarkable events"; times of meeting of the courts, tables of weights, measures, and coinage; and medical information. The medical material was frequent and often, of necessity, quite detailed. Colonial physicians were unable to serve the widespread population and most people had to shift for themselves with the best information they could get. Even in the middle of the following century, Benjamin Franklin was still publishing detailed prescriptions that he copied out of European medical journals.[5]

A surprisingly large number of the almanac-makers, having mathematical or astronomical interests, and often some talent, indulged in a page or two of technical instruction in their favorite subjects. It was in such an almanac that the first written American argument for the Copernican system appeared. This was Zechariah Brigden's "A Briefe Explication and Proof of the Philolaick Systeme," appended to his almanac of 1659. Brigden's essay was based on Vincent Wing's *Astronomia Instaurata*, the first satisfactory popular explanation of the Copernican system by an Englishman, published only three years before. Others contained such matters as "The Suns Prerogative Vindicated" (an argument that the planets had no light of their own); semi-popular accounts of the history of astronomy—after 1665 always ending triumphantly with Copernicus, Kepler, and Galileo; arguments against various parts of the Ptolemaic system; discourses on comets, the tides, and phenomena of climate.

Throughout the seventeenth century, the information contained in the almanacs generally remained at least a decade out of date, and what information they did contain was not perfect, even for their time. Nevertheless, as one historian has said, "they did good work in persuading the New England farmer that he lived on a revolving planet."[6] In many cases they did a great deal more. In 1663, for example, Nathaniel Chauncey's almanac marshaled all the evidence to disprove the theory of solid planetary orbs (which even Copernicus had retained). In 1674, Jeremiah Shepard outlined Kepler's discovery that the planets moved in

ellipses rather than in perfect circles. John Foster's almanac for
1681 pictured what was perhaps the first American diagram of
the infinite universe, with the "fixed stars" scattered beyond the
planets rather than on the same orb. In the following year, William Brattle explained the mechanism of an eclipse of the sun
and the moon, events which still struck terror into some at that
time. In 1686, Nathanael Mather gave an account of discoveries
made with the telescope, including the rings of Saturn, solar spots,
and hollows on the moon. A particularly ambitious attempt at
instruction was Henry Newman's *News from the Stars, an Almanack for 1691* (Boston), which amounted to an elementary
astronomy textbook. Running to twice the length of an ordinary
almanac, it contained full-page treatments of the latest discoveries about the various heavenly bodies.

Sometimes it is possible to trace the change of mind of an
almanac-maker as knowledge of the New Science became available to him. Under conditions of arrested development like those
which pertained in New England, change when it does come,
comes quickly and often dramatically. Colonials missed the preliminary discussion that generally attends the acceptance of a
new idea; the books that they imported did not contain the gradual
introduction of parts of it into older frameworks that helped men
become accustomed to it. Instead, following their custom of importing only the "best" book or the "most up-to-date" works, they
would not generally be confronted with the idea until it had been
fully developed and accepted. For example, "T. S. Philomathemat,"
who published almanacs at Cambridge between 1656 and 1661,
was evidently influenced by his reading during that brief interval;
the assumption is that it was the same book that Brigden had
read in 1659. In his almanac for 1656, he appended a two-page
exposition of Ptolemaic Astronomy, with no hint that the system
even had any rivals. In 1660, he mentioned in passing that the
Ptolemaic equations were "not at all fortified by observation,"
and acknowledged the dissenting opinion of Copernicus. The
following year, in an article entitled "A Brief discourse of the
Rise and Progress of Astronomy," the writer ridiculed the Ptolemaic system with its necessity to be forever adding circles,
lauded Copernicus for his discovery of a "more easy and rational
way," and celebrated the genius of Galileo, Kepler, and Gassendi,
among others, who had confuted the Ptolemaic and Tychonic
systems and demonstrated the truth of the Copernican system.

In reading the article, one can almost feel the excitement of the writer at his discovery, and his anxiety to correct the erroneous opinions of his countrymen. Handed to him as a finished product, with the earlier difficulties removed and the inconsistencies smoothed out, the new system carried such conviction that acceptance of it virtually amounted to a conversion experience. There was no sense of an inner struggle, no long, agonizing period of attempted reconciliation with elements of the old views that marks the initial establishment of a new concept. It came to the writer as a neatly packaged truth from the center of learning, and it was simply accepted.

Imperfect as they were, the colonial almanacs carried the names and some of the doctrines of Copernicus, Gassendi, Galileo, Brahe, and Kepler into remote villages all over the colonies. Surely there was no hamlet in New England where the local minister, at least, did not possess an almanac; and we know that the material from the minister's almanac quickly found its way into his sermons. Through this medium, information filtered down to the "second level" of intellectuals and below, and played an important part in creating a climate of receptivity to new ideas.

During the following century, the almanacs of Ames and Franklin carried Newtonianism and scientific deism in the same manner, and made it convincing to those who had already become familiar with the giants of the early seventeenth century.[7] *Poor Richard Improved* (1748), for example, contained verses adapted from John Hughes's Newtonian poem, *The Ecstasy* (1720), and the 1756 issue quoted a tribute to astronomy from Moses Browne's *Essay on the Universe* (1739). This work, however, was subject to the same limitations as earlier popularization in the almanacs. For example, as late as 1737, Nathaniel Bowen in the *New England Diary* could write an article on the "Discoveries in the System of the World, made by Newton," without even hinting at the law of gravitation. In an earlier issue (1731), however, he had stated clearly, but without comment, Newton's three laws of motion "by which we think that every thing that relates to motion may be explained."

Continuing traces of astrology, of course, remained in the almanacs—as they did in Europe. There were directions for prognosticating the weather from natural signs, the astrological times for gathering various herbs, charts showing the reigning diseases at various periods of the year, and advice on the best times for bleeding. The New England almanac-makers, however, in marked contrast to their English counterparts, were generally wary of specific predictions, for Puritanism encouraged a skeptical attitude toward too close a prediction of God's providence. Whatever astrological advice they did offer was vague and general; many actually ridiculed the predictive efforts of astrologers.

The information used by the almanac-makers generally came from books ordered from England or brought in by late arrivals. Although immigration to New England tapered off dramatically after 1640, it never stopped altogether. Many of the late arrivals, being products of the dissenting English academies which were beginning to emphasize science, brought in newer ideas with them and often personal knowledge derived from their association with leading English scientists. The Reverend Samuel Lee, for example, arriving in Boston on August 22, 1686, brought with him a library unequaled in America for science and philosophy, even by the Mather collection.[8] When sold in Boston in 1693, it contained more than 250 titles in physics, philosophy, mathematics, and astronomy; the authors included Paracelsus, Riolau, Willis, Helmont, Descartes, Tycho Brahe, Gassendi, and John Evelyn. He also brought with him a deep personal knowledge and understanding of Robert Boyle's works and his philosophy, and a commitment to the scientific religion of progress to an extent that was previously unknown in New England. In his first book published in America, *The Joy of Faith* (1687), Lee expressed his great hopes for the future of scientific understanding in much the same terms that Thomas Jefferson would use more than a century later. Just as the learned of the seventeenth century looked back with pity upon the benighted condition of men in past ages, Lee said:

> Indeed so may posterity deride at these our ages, and the more ingenious of future times, may stand amazed at our dulness and stupidity about minerals, meteors, and the cure of diseases, and many thousand things beside. . . . The superfine Wisdom and Learned Wits of those acute times will discover vast regions of darkness and ignorance. There will be a plus ultra to the end of the world.

Since Lee was already a famous scholar when he arrived in New England, he was received with great honor, and he quickly took an important place in the intellectual life of the colony. He preached at the Old South Church, at both the Cambridge and the Boston lectures, and in April 1687 he received a call to the church at Bristol, Rhode Island. Several of his early books were reprinted in New England, both before and after his death. Consequently, Lee lent respectability to the growing interest in Cartesian natural science in New England at that time, and he played an important role in introducing the latest English modification to Cartesian philosophy. After Lee's arrival, strong traces of the thought of Robert Boyle, Lee's friend, began to appear especially in the work of Cotton Mather, the Puritan divine who has been unjustly remembered primarily for his role in the witchcraft episode of the 1690's.

Mather's complex character is indicated by the fact that he has been variously described by historians as a "witch-burner," a "Puritan priest," and a "deist." His election to the Royal Society in 1713 was a testimony to his many contributions to natural history, including observations on "giant bones," which he presumed to be antediluvian remains of the giants mentioned in the Bible (an interpretation that had been current since the time of Acosta), weather observations, and a pioneering article on hybridization in corn. His championship, against vigorous opposition, of smallpox inoculation during a Boston epidemic and his manuscript *Angel of Bethesda*—a medical work that included acceptance of a rudimentary germ theory of disease—has earned him the title, from two historians, of "the first significant figure in American medicine."[9]

In addition, Mather was the first thoroughgoing Newtonian in the colonies. He had become a convert by 1689, when one of the earliest indications of his acceptance of the new philosophy occurs in a biography of his younger brother, Nathanael, who, he says, at the Harvard commencement, "maintained for his position *datur vacuum;* and by his courses upon it (as well as by other memorials and experiments left behind him in manuscript) he gave a specimen of his intimate acquaintance with the Corpuscularian (and only right) philosophy."[10] That Mather in 1689 thought the corpuscularian the only right philosophy is a matter of great significance, especially when one recalls that, as recently as 1678, Ralph Cudworth, one of the Cambridge Platonists, had said: "the atheistical system of the world . . . is built upon a peculiar physiological hypothesis . . . called by some 'atomical' or 'corpuscular.' "[11]

It was the corpuscularian philosophy, as named and described by Robert Boyle, which sought to unite the atomism of Gassendi and the vortices of Descartes into a description of the universe as composed of minute material particles, through which motion might be transmitted from one body to another, and by which such phenomena as magnetism might be at least partially explained. Dating only from the 1660s, this philosophy was widespread during the remainder of the seventeenth century, for, along with the effluvium theories of William Gilbert and Boyle, it synthesized the non-mathematical scientific thought of the period.[12] That it was quickly received in New England is indicated by a comment written in 1686 by Nathanael Mather to his father. In examining the Harvard theses for that year, he said, "I perceive that the Cartesian philosophy begins to obteyn in New England & if I conjecture aright the Copernican system too."[13] In the following year, Harvard student John Hancock was entering lengthy extracts from Boyle's tracts into his commonplace book and drafting an argument to explain why Copernicus's system was more reasonable than that of Ptolemy.[14]

Certainly there were other sources besides Lee by which Mather could have become acquainted with the corpuscularian philosophy. Five years earlier, for example, his father, Increase Mather, had extracted material from Boyle's *Usefulness of Natural Philosophy* (London, 1663) for his *Essay for the Recording of Illustrious Providences* (Boston, 1684). Although this work was, as the title suggests, concerned with the manifestations of the supernatural in New England, and although it has been connected with the witchcraft episode of the following decade, Increase Mather's *Essay* nevertheless in its own day would rate rather high as a work of popular science. The accounts of imps, changelings, diabolical possession, interventions of God in nature, and other miraculous events were based as much as possible upon the testimony of eyewitnesses, and Mather demonstrated no more credulity than any other learned man of the period would have regarding these matters. It was, in fact, in the preface of this work that Increase Mather spoke of his "philosophical club" and expressed his hope that it might provide leadership in compiling a natural history of New England according to the "rules and methods described by that learned and excellent person Robert Boyle, ESQ." The providential interventions, as well as the instances of diabolical actions in the *Essay*, were regarded as essential parts of the planned natural history, for God and the Devil had not yet been as effectively separated from nature as would be the case with later generations.

But despite this prior interest in the work of Boyle, the timing of Cotton Mather's first publications on the subject of the corpuscularian philosophy, and the close connection of other works of this period with the interests of Lee and Boyle, argue that Lee's arrival was more than a coincidence. In the same year of the memoir of his brother, Mather published *Work Upon the Ark*, in which he discussed the problem of how Noah stored away all the animals and their food in the ark. This was a problem that had been recognized since the time of Acosta, but which English intellectuals had been particularly preoccupied with in the past two decades. Samuel Lee's teacher at Wadham College, Warden John Wilkins, had contributed a well-known dissertation on the subject in 1669. In 1690, Mather published his sermon on *The Wonderful Works of God Commemorated,* the first American work to contain a full-blown Boyle-type argument from design for the existence of the Creator*—later a hallmark of Enlightenment thought. Two years later, Mather prefaced Lee's *The Great Day of Judgment* with a sermon of his own called "Preparatory Meditations Upon the Day of Judgment," in which he went into a long digression over the physical possibilities of having enough room on the earth after the Resurrection.

Mather was permanently affected by his exposure to modern science in the 1680's, for until the end of his life he studied intensely, read widely, and began to inject bits of the latest science into the sermons, lectures, and tracts that flowed from his pen in a seemingly endless stream. In *Thoughts for the Day of Rain* (Boston, 1712), for example, he discussed rainbows, the laws of optics, and called Newton the "perpetual dictator of the learned world"—a title again used in his *Manductio ad Ministerium,* a handbook prepared for the guidance of young ministers. In *Christianus Per Ignem,* a sermon of 1702, he explained

---

*The argument from design, developed by Robert Boyle and other seventeenth-century English scientists, holds that, if one can find evidence of a design, or pattern, in the universe, then the existence of a designer may be inferred. Since nothing known to man could be responsible for such observed phenomena, say, as the regularity of the planets in their courses, it follows that a Divine Designer—God—must have been at work. This argument was applied to every area of science until late in the nineteenth century.

fire in corpuscularian terms, explained the dependence of fire upon "air," and described experiments by Robert Boyle showing the dependence of life upon "air."

As early as 1693, Mather had given clear expression to the typical eighteenth-century concept of the universe as a marvelously regulated machine: "The *Wandring* Stars, the *Fixed* Stars, and the *Satellites* of each, how inexplicably circumstanced are they? How regular to the Hundredth part of a Minute, are they in their motions?"[15]

Mather was able to accommodate naturalistic explanation with his fervent religious beliefs in the same convenient manner worked out by earlier Puritan theologians, including his own father and grandfather. In a 1695 sermon, for example, he explains:

> It is true, that the *Thunder* is a great *Natural Production*, and by the common laws of *Matter* and *Motion* it is produced: There is in it a Concourse of divers weighty Clouds, Clashing and Breaking one against another, from whence arrises a mighty *Sound*, which grows yet more mighty by its Resonancies. The Subtil and sulpherous *Vapours* among these *Clouds* take Fire in this Combustion, and *Lightnings* are *Fulminated* with an Irresistible Violence upon our Territories.

Although the modern meteorologist might question the adequacy of Mather's explanation, it is, nevertheless, an indisputably naturalistic one. But in the next sentence, Mather asks: "But still who is the Author of those Lawes, according whereunto things are thus *Moved* into Thunder?" This, obviously, was the important question, and the answer to it maintained Mather's theism intact in spite of his acceptance of modern science.[16]

Furthermore, Mather remained convinced that the revelations of science were positively helpful to religion: "There is not a Fly, but what would confute an *Atheist*," he said in a 1690 work. "And the Little Things which our Naked Eyes cannot penetrate into, have in them a *Greatness* not to be seen without Astonishment." he continued. "By the Assistance of Microscopes, have I seen *Animals* of which many Hundreds would not aequal a Grain of Sand. How Exquisite, How Stupendous must the Structure of them be!"[17]

The large things were just as convincing, he explained three years later. The Sun, he observed, had been calculated by one astronomer to be 3,462 times larger than the earth: "But how much does it then, *Declare the glory of God, and shew forth his Handy work?"* Such was the strength of Mather's exuberant belief in the awe-inspiring majesty of God, that he was even able to contemplate the concept of a plurality of worlds with equanimity. How wasteful it otherwise would have been for God to have created planets with no inhabitants! Again, since it would have been grossly inefficient for God to have created the stars merely to contribute toward heating the earth (other means would have served better), they must each be shining on a populated world, he thought.[18]

The new-type natural theology was "in the air" by the time that Mather wrote, and certainly no originality can be claimed for him. Robert Boyle in 1663 and John Ray a few years later had been the first writers to give a new impetus to natural theology in England through their efforts to turn to the use of religion the new discoveries of seventeenth-century science. William Derham, George Cheyne, Theophilus Grew, and others were making a great deal of use of the new knowledge in restatements of the argument for theism from the design of the universe. Their books, which Mather possessed, and which quickly became known in America, were filled with enthusiasm for the new facts being established by their contemporaries, and by a pious wonder at the complexity and immensity of the world thus revealed. In his previously cited 1690 sermon, Mather drew primarily on Boyle; in *Winter Meditations* (1693), he used Ray's count of species published only two years earlier. But even before Mather wrote, John Hancock, then a student at Harvard, had entered the following thought into his commonplace book (1687): "If I had 1000 tongues, & should live a 1000 years, I should not be able to describe the admirable order of Creation . . . and if so much wisdom be requisite to observe ye wonderful & design in ye structure of ye world how much more was necessary to form it. Minutim felix!"[19]

Mather's particularly strong desire to find scientific support for his religious beliefs provided the dominant motif for his most ambitious work, his six-volume "Biblia Americana," compiled between 1700 and 1720 but never published. Mather seemed driven to explain naturally all of the marvelous occurrences in the Bible. His method in this verse-by-verse examination of the Bible is admirably illustrated by his comments on Joshua's command to the sun and the moon to stand still. As is well known, he said, the sun is the center of the system:

Ye motion of the earth, may be argued, even from this very
speech of Joshua. For Joshua had no occasion for any service
from the *moon*. Why did he command ye *moon to stand still,*
as well as ye *Sun*? Why a Stop given to ye Diurnal motion of
ye Earth, unavoidably produces ye Phenomenon of ye *moon
standing still* as well as Ye *Sun*.

Mather added to this thought the information that, according to
one theory, the earth's motion depends upon the motion of the
sun about its own axis; therefore Joshua gave the correct
command.[20]

Although Mather's *magnum opus* was never published,
many of the ideas in it did appear in his more modest *Christian
Philosopher* (1721), originally designed as an addendum to
the larger work. This was the first general Newtonian approach
to physical science published by an American.

The *Christian Philosopher* shows more clearly than any
other work the entire compatibility between Puritanism and
rationality, for the point of view would have been congenial
to Thomas Jefferson or any other Enlightenment philosopher.
It expresses the typical eighteenth-century belief that the world
is well planned and well ordered, that it is beautiful; that to
study nature is to realize God's goodness; therefore, that man
can appreciate God by the exercise of observation and reason.
One of the most striking characteristics of the work was an
attitude that could only be termed extreme—perhaps even
"excessive"—functionalism. Mather seemed positively driven
to find a reasonable use for everything. Anything less would,
presumably, have been an insult to God, a suggestion that He
had created something meaninglessly. For example, the nip-
ples on the breast of a male gave Mather some difficulty, but
he overcame this with a remarkable exercise of ingenuity,
discovering that in a few cases of extreme emergency fathers
were reported to have suckled their infants. What had seemed
an insuperable difficulty before investigation turned out to
be simply another evidence of the Creator's wisdom in foreseeing
every contingency. At another point, he suggested that the
earth's atmosphere was replenished by vapors emitted from
the tail of comets; the irregularity of comets distressed him,
and he seemed under a strong compulsion to find both a pattern conformable to reason and a function for them. As Thomas
Rodney, a later product of the Enlightenment, wrote, the idea

that comets were wandering at random was both an absurdity and a blasphemy.[21]

It is difficult to overemphasize the importance of Mather's *Christian Philosopher,* which quickly passed into general use in New England. After its publication, the almanac-makers had a ready source for data. For example, Daniel Travis, in his *Almanack for 1723* (Boston) copied extracts from explanations of Thunder, Lightning, Hail, and Revolutions of the Planets, carrying some of the concepts to those who would never have occasion to use the *Christian Philosopher* itself. Mather's work quickly came into use as a scientific text. John Winthrop IV, who became a distinguished professor of natural philosophy at Harvard, as a student in 1729 copied eleven pages of extracts into his commonplace book.[22] Benjamin Franklin, a young printer in Philadelphia, read the work and thoroughly endorsed the rationalistic point of view it contained.

One must be careful not to make Mather appear too modern, too much a man ahead of his time. To understand how thoroughly he was a man of his age, one must understand both the scientific and the religious side of Cotton Mather. At the same time, one will understand the age itself a great deal better. For example, the same early sermon which celebrates the admirable order of Creation also contains page after page of incidents purporting to show the direct interference of God in the world, either to preserve New England from her Indian enemies, or to punish Virginia for her persecution of non-conformist ministers. In many works he placed sufficient emphasis upon the features of harmony and order to suggest that he did believe in a truly scientific universe, in which events invariably proceed from natural causes. At other times, he was positively insistent upon the "reserved cases" where God worked in mysterious ways to provide special interventions. His mind, in a word, continued to vacillate between two points of view, and in this very vacillation he was most typical of the times in which he lived.

"Vacillation," however, has a modern connotation that does not quite do justice to the complex relationship between the natural and the supernatural during that former transitional period. Even such a rationalist as Dr. Thomas Robie, who constantly sought naturalistic explanations for phenomena, nevertheless thought it worthwhile to enter in his commonplace book (1711) the conventional implications of earthquakes ("wars, slaughters of men, subversion of kingdoms, Fortunes of Estates"), eclipses ("death and destruction of Kings and Princes or Oppression by War"), and Hail ("war and Incursion of the Enemy, Especially if they come from the North").[23] Puritans believed, and had believed for a long time, in what we might term the "ordinary sufficiency" of second causes. That is, they believed that the natural laws created by God—the first cause— were ordinarily sufficient to determine events. However, it was important for them to maintain the belief that God *sometimes* demonstrates his sovereignty through ordering time and chance in such a manner as to frustrate second causes. This reservation leaves the way clear for an occasional miracle, an occasional visitation upon some sinner, an occasional special dispensation as a sign of God's favor. Praying was therefore worthwhile, even though it was assumed that God rarely intervened in the order of nature. Miracles, as John Cotton argued, were not to be multiplied beyond necessity. And as Samuel Nowell put it, "God can work miracles, but when ordinary means may be had, He will not work miracles." He is more likely to use skillful management of causes to produce proper effects by the laws of nature. God will still deliver His children out of Egypt, John Richardson explained in his Election Sermon of 1675, but not often by such a defiance of His own laws as the division of the Red Sea. "We in these days have no promise of such a miraculous and immediate assistance," he said pointedly to the assembled militia; "God works now by men and meanes, not by miracles."[24]

The growth of the concept that Providence worked by the laws of nature still reserved a special place for God, and it grew naturally out of Puritan thought; but it was a far cry from the arbitrary, even whimsical, behavior of John Calvin's God. It was but a simple, and obvious, step from such statements as those above to Jeremiah Dummer's declaration in his trial sermon during the first decade of the following century that, although God was omnipotent, had he commanded anything contrary to reason "it would have been an infinite stain and blemish on his unspotted purity." This statement, according to the historian Perry Miller, was one of the most important events in the transition from the Age of Faith to the Age of Reason.[25] It led immediately into the Enlightenment concept of deity— the God who settled the course of external nature for all time

at the Creation, and established His laws so that they might
operate throughout all time in such a manner as to best answer
His designs in the government of the world. "What I mean is,"
said Charles Chauncey in explaining this difficult concept,
"He might settle such *laws* with respect to natural causes, so
proportion their force, sphere of action, manner and degree
of operation, as that, under his all-powerful and alwise con-
curring influence, they should conspire together to produce
those effects, at such times, and in such places, as were best
fitted to their natural state, and might serve for warning and
correction, or total destruction and ruin, as he should judge
expedient."[26] The effort had been to achieve a theological justification for special providences which allowed for an uninter-
rupted, because predestined, order of nature. There is no better
example of a desire to maintain theological uses in extraordinary events and at the same time to avoid the supernatural.
The great difficulty, however, is that the Deity in this chain of
reasoning inevitably tends to become a remote and somewhat
unnecessary appendage to a mechanical universe—the precise
fate that He suffered in Enlightened thought.

Mather's earliest interest in the new philosophy is associated with his acquaintance with Samuel Lee, but still it would
be gratuitous to suggest that his work would have been substantially different had Lee never appeared in New England.
Although in Lee's case the evidence is indirect and circum-
stantial, there seems little doubt that he had something to do
with the complex process whereby the scientific revolution was
introduced to New England. There were other individuals of
equal, or greater importance, and in at least two cases there
is no doubt of their influence.

A second important immigrant of 1686 was Charles Morton, another dissenting minister who had conducted a famous
school in England, and who arrived in Boston with the expectation of becoming president of Harvard. Although disappointed
in this expectation, Morton did become a teacher there and
thereby introduced Harvard students to the most advanced
natural philosophy textbook of its time, his own manuscript,
Compendium Physicae. Although it is a transitional work, clearly
belonging to the 1680's, it was compiled largely from the *Philosophical Transactions* of the Royal Society and contained the
most up-to-date knowledge available.

The Aristotelian flavor retained by the thought of the

time is particularly evident in the opening part of his book. Morton begins, in the ancient manner, with definitions: substance, place, form, time, motion, divisibility, etc. Furthermore, he adopts Aristotle's four elements—earth, air, fire, and water —but adds Descartes' three elements for good measure: crass matter, celestial matter, and subtle matter (the "aether" which interpenetrates everything else). He also adds the Gassendist idea that atomic particles are of different shapes and sizes, thereby accounting for such things as different densities.

But despite the remnants of Aristotelianism in his work, Morton's students could not have remained unaware that important intellectual changes were taking place. For example, Morton reviewed all systems of astronomy, stating the advantages and disadvantages of each, and, without judging definitively, indicated that he leaned toward the Copernican. He rejected the doctrine of stars in fixed spheres, stating that they were simply floating in space. Again, through his discussion of the eye, Americans became, perhaps for the first time, aware of the corpuscular theory of light; and, through his discussion of hearing, they learned of the undulatory theory of sound.

A great many new, definitely anti-Aristotelian propositions appeared in other sections of the book. There was, for example, a clear explanation of the mechanical theory of heat —that heat consists in motion, cold in rest. Morton defended the proposition that matter is neither generated nor corrupted, a statement of the conservation of matter which was not generally accepted until the time of Lavoisier, almost one hundred years later. Newton's laws of motion were clearly set forth and explained.

On the other hand, the proposition that witchcraft is effected through an impious compact of men with demons was considered as valid by Morton as Newton's laws, and he included the doctrine, apparently from Boyle, that physical objects exert an influence on each other by drawing off a stream of volatile particles. This last is the theory of "effluvium" which a few years later was called upon by the Salem witchcraft tribunal to explain the operation of the "evil eye" in mechanical terms.

Even with all the limitations of Morton's text, the least that can be said is that those who studied it would be well prepared to receive the work of Newton and probably well disposed

toward it. Its use as the sole scientific text at Harvard and later
at Yale, at a time when instruction still consisted of the explication of a single text, guaranteed it an influence not even con-
ceivable for a present-day science textbook.[27]

The influences of Lee and Morton were felt most strongly
at Harvard, then the colonies' only institution of higher education. A few years later, Yale College afforded a similar example,
where the tremendous impact that a single recent arrival could
have was illustrated by the case of Jeremiah Dummer, who
came to Yale in 1715. Dummer brought with him a magnificent library containing the latest scientific works, including a
copy of the *Principia,* which had been given to him by Newton
himself. Samuel Johnson, a tutor at Yale, had completed his
manuscript synopsis of technologia on November 11, 1714. This
synopsis did not even demonstrate an acquaintance with Copernicus, to say nothing of Descartes, Boyle, or Newton. The
Dummer library brought the accumulated learning of the seventeenth century suddenly, and its impact was great. The next
year, a few months after Dummer arrived, Johnson recorded on
the margin of his manuscript that he had abandoned all this and
become "wholly changed to the 'New Learning.' "[28]

Johnson and a number of other faculty members and
students at Yale continued their studies in the Dummer collection, reading the works of Francis Bacon, Robert Boyle, Isaac
Newton, John Locke, and John Tillotson—works that the Harvard group had assimilated somewhat more slowly. On the basis
of their reading, they began to reconsider their theological position. One early product of the discussion was the conversion
to Anglicanism of Johnson and others, all of whom were dismissed from Yale for their conversion. Apparently the new
rational tendencies of the age made Anglicanism appealing—
the concept of man's perfectibility without the necessity of a
conversion experience harmonized readily with Enlightened
thought.[29]

Such seeds of thought as those brought to America by Lee,
Charles Morton, and the Dummer library would never have
germinated had they not fallen on fertile ground. The late arrivals and the books were merely vehicles for transporting the
ideas, and they served as focal points for discussion. To understand this process of change completely, one must pay great

attention to the complex religious phenomenon known as Puritanism. One of the most striking things about the new ideas is the almost total lack of opposition they encountered in the highly theological society of New England. The allegedly narrow-minded and straight-laced Puritan clergy played a commanding role there, for even those who personally rejected some portions of the new science were remarkably tolerant of it. For example, when Zechariah Brigden published his almanac for 1659—the work containing the first American exposition of the Copernican system—the ultra-orthodox Reverend John Davenport, vigorous persecutor of Quakers and other benighted souls, commented on the article in a letter to John Winthrop, Jr. Davenport did not agree with Brigden; he thought that the notion of the movement of the earth might be contrary to scripture, and he would continue to believe what he had been taught at Oxford in the 1610's. In terms of personal conviction, his ideas were all formulated and his mind closed. But surprisingly, Davenport, a trustee of Harvard College, did not suggest that Bridgen be fired from his fellowship, or even that his almanac be proscribed. "However it be," said Davenport, "let him injoy his opinion; and I shall rest in what I have learned, til more cogent arguments be produced then I have hitherto met with."[30]

This reaction by Davenport was characteristic of the Puritan clergy. Far from opposing science, they were the chief patrons and promoters of the new astronomy and other aspects of the scientific revolution in New England. As a matter of fact, by the following century they were frequently being accused by laymen of being far *too* interested in science. Speaking of a sermon he heard at King's Chapel in July 1744, one lay critic commented that the minister

> . . . gave us rather a philosophical lecture than a sermon. . . . We had a load of impertinence from him about the specific gravity of air and water, the exhalation of vapours, the expansion and condensation of clouds, the operation of distillation, and the chemistry of nature. In fine, it was but a very puerile physical lecture and no sermon at all.[31]

Essential to an understanding of Puritanism is the knowledge that it was never hostile to the intellect or to learning; reliance upon the supernatural did not cause Puritans to avoid naturalistic explanations—up to a point. Their religion was in most cases a positive stimulus to science and learning. The con-

centration of learned clergymen in New England helps to explain the early and widespread interest in science in that region,
for the clergy adapted science to their own pious purposes, not
usually realizing where such investigations might ultimately
lead them. There were, accordingly, more natural philosophers
in the Boston–Cambridge area than anywhere else in the colonies until about 1750, when Philadelphia took the lead. Ironically, it was an increasing secularization and consequent decline of interest in religion that helps account for the decline
of scientific interest in New England.

Through their peculiar view of God and the workings
of his Providence, Puritans were able to accept with equanimity
virtually any scientific explanatory scheme. To Puritans of the
seventeenth century, it was a matter of little consequence how
God performed His works; the important thing was that He performed them according to His own plan. In the foreground of
the Puritan consciousness was the task of explaining how God
always worked in nature through a settled order and yet se-
cured intelligible goals, how He constrained himself to observe
the laws of nature according to the covenant He had freely
entered into with man and yet contrived that whatever He
decreed should inevitably come to pass. It was essential to the
Puritan concept of God that explanation be in these terms,
but which system explained all of this was unimportant. One
system could gradually replace another without distracting attention from the main issue. The thought that there could be a
conflict between their understanding of nature, God's noblest
work and a living testimony to His constant operations, and
their concept of the majesty of God would have been inconceivable to seventeenth-century American Puritans, and it remained
so for almost as long as Puritanism endured. It was not until
well after 1700 that New England divines even began to wonder
whether in transferring their allegiance to the New Philosophy
they had not raised up dangers unknown in the old.[32]

Certainly there were dangers in the New Philosophy.
The new importance placed upon man and his potential through
the use of "right reason" had the natural effect of sharply reducing God's role in man's affairs. It increasingly came to be
believed that man, by the use of the same principles as Newton
in his exploration of the elements of the new science, could
discover the determinative and immutable laws of nature and

that by simple adjustments of human institutions to conform
with these laws man could reduce the friction between himself and the system of cosmic relationships governing the universe. By improving the environment, progress would be inevitable and man would ultimately achieve perfection; and all
of this was to come through the use of reason, not through a
spiritual rebirth. God, in this system, was increasingly considered as simply a Prime Mover who had created a universe
held together by harmonious laws: a perfectly operating universe which, once having been created, was destined to exist
indefinitely without further adjustment. Man, therefore, did
not appeal to God, but to the immutable laws of nature that
governed God's creation. Before the transformation in thought
ended, in other words, God and man had become remote from
each other. But by the time this was recognized—so naturally
had it developed out of Puritan concepts—the transformation
had so thoroughly occurred that there was no real retreat possible.

The peculiar Puritan viewpoint explains how a latter-day
Puritan like Thomas Clap, President of Yale College, although
staunch and doctrinaire in theology, could be so tolerant and
open-minded in matters of science. Clap was always ready to
confute a heretic in religion—be his heresy ever so slight—yet
he could read even the speculations of Cadwallader Colden
without offense, and he was responsible, after 1740, for creating an advanced curriculum at Yale, including astronomy,
mathematics, and Newtonian science in general.[33]

Even in such a man as John Cotton, a leader of the first-
generation clergy who was totally dedicated to the great experiment of planting a godly society in the New World, one can
find evidence of strong admiration for the science of his time.
Cotton's scientific information was old-fashioned: like Davenport,
he was opposed to Copernicus, he knew little of contemporary
ideas in physics, meteorology, or medicine, and he considered
the Bible an authoritative handbook of science. In his *A Briefe
Exposition of Canticles* (London, 1642), Cotton disseminated
knowledge of ancient physiology and botanical medicine; his
The Keyes to the Kingdom of Heaven (London, 1644) contains
an exposition of Aristotelian physics. But despite his old-fashioned information, Cotton at no time in his life was opposed to
science—in fact, he appeared at times to be willing to argue

against what appeared to be the "plain meaning of scripture" on behalf of science. For example, in his commentary on *Ecclesiastes,* apparently the most anti-intellectual book in the Bible, and in which all of worldly knowledge appears to be among the famous deprecated "vanities," Cotton managed to pose strong arguments in favor of science and learning. In commenting on Verse 29, Chapter 7 ("Lo, this only have I found, that God hath made man upright, but they have sought out many inventions"), Cotton explained that the Scripture

> . . . Meaneth not profitable inventions for the good of themselves or others, as Arts, Manufactures or Occupations, but such inventions whereby we seek to start away from God, and to corrupt ourselves.

And he continued:

> To study the nature and course, and use of all God's works is a duty imposed by God upon all sorts of men; from the King that sitteth upon the Throne to the Artificer. , , . It is a disgrace to a good workman not to look at his work, but to sleight it.
> But studying the nature of all things, which by observation and conference men might learn one of another, would enlarge our hearts to God, and our skil to usefulnesse to our selves and others.*

The point here is the same that later became well known in natural theology: Since God created everything, the more intricate the orderly relations that could be discovered, the more they testified to His wisdom and power. The study of nature becomes a religious duty; it becomes, in fact, an important form of worship.

The second-generation clerical leaders displayed, in general, the same receptivity to science as had Cotton and Davenport. Increase Mather's interest in science dates at least as far back as 1664, when he is known to have owned a copy of Francis Bacon's *Natural History.* In 1669, commenting on the passage in Matthew 25:29, "and the moon shall not give her

---

* Cotton's grandson was equally adept at explaining away scriptural passages that seemed to conflict with his belief in reason. For example, in the manuscript *Biblia Americana,* VI, at the Massachusetts Historical Society, he comments on Colossians 2:8, which reads: "Beware lest any man spoil you through philosophy." Mather simply explained that it was the Platonic philosophy, which had introduced corruptive elements into Christianity, that was meant.

light and the stars shall fall from heaven," Mather comments: "Yea, some imagine that the Stars shall really fall from the Heaven to the earth; but this cannot be, for how should the stars fall upon the earth, when one Star is greater than the earth."[34] Clearly, since there was an apparent conflict, it was the understanding of the biblical statement that must be adjusted to conform to the physical evidence, not the reverse.

But it was the comet of 1680 that started Increase on the serious study of science. After delivering two sermons, in 1680 and in 1681, on the significance of comets as signs of God, Increase Mather sat down to study astronomy. In 1683, he read Gassendi and cited Bacon's *De Augmentis Scientiarum;* he then read Robert Hooke's *An Attempt to Prove the Motion of the Earth from Observations.* From this study he emerged a champion of observation and mathematical reasoning. Although he announced that he now believed comets proceeded from natural causes, he did remain convinced that scientists would never get to the point where they could predict the arrival of comets.[35] Since this was several years before the publication of Halley's prediction (1705), and many more years before its dramatic confirmation (1759), Mather's caution in this respect is understandable.

Basic to the Enlightenment point of view—indeed, the very essence of it—was a new conception of science. The older conception of science as a dogmatic subject matter—as an extension of technologia—was replaced in the minds of educated New Englanders by a conception of science with a mathematical and experimental foundation, relying upon sense experience rather than intuition, innate ideas, or authority, and concentrating upon explanation in terms of mechanical processes rather than ultimate purposes.

The towering giant of that new age was Isaac Newton— "the perpetual dictator of the learned world," Cotton Mather termed him. Newton's *Mathematical Principles of Natural Philosophy* (generally referred to as the *Principia*), a five-hundred-page book published in 1687, was a concise account of the principles of dynamics underlying the three great empirical laws of planetary motion set down by Kepler, a further extension of the law of falling bodies discovered by Galileo, and a development of the theory of gravitation as applied to the moon, comets, and the planets. It also contained an important explanation of the irregularities known as perturbations in the movement of celestial bodies.

Newton's work was pregnant with implications for what appeared a total explanation of the world. One after another, formerly intractable problems fell before the penetrating logic of Newton and the power of his premises. One could, for example, finally explain the action of the tides—it was simply a matter of the gravitational pull of the sun and the moon, depending upon their positions relative to each other. One could deduce the exact shape of the earth—it was an oblate spheroid, with a bulge at the equator and flattened out at the poles. One could explain the precession of the equinoxes, and this, in turn, made possible an exact forecast of eclipses, removing former sources of awe and fear and reducing eclipses to the regularity of cosmic law. Using Newton's formula, one could even deduce the density of the earth; the knowledge that it was five and one-half times the density of water led to the important geological assumption that the core of the earth is composed of heavy minerals.

Most impressive to Americans who commented was the new method illustrated by Newton: the positivistic approach of reliance upon "facts" and eschewing speculation. Essentially, Newton's method is characterized by four "rules," which describe the method of reasoning from particular phenomena of experience to general conclusions about the causes of these phenomena:

1. Assign to any natural event only its necessary natural cause.

2. Always ascribe the same natural cause to the same natural event.

3. Generalize that properties which are common to all bodies within our experience and in respect to which our experiments reveal no difference of degree are common to all bodies.

4. Generalizations are to be considered highly probable until they are shown by later phenomena within our experience to be in need of correction.

The very obviousness of Newton's "Rules of Reasoning" to the modern reader is a testimony to his gigantic achievement, for the emphasis which he placed on the power of man's mind—

the unaided intellect operating according to carefully specified
rules—to achieve knowledge was overwhelming in its grandeur to men of the eighteenth century. Bacon had spoken of the
power of man to control nature, and he had been highly influential; but Newton, through a "sublime geometry," as David
Rittenhouse expressed it, deduced certain knowledge about the
fundamental operations of the universe "with the strictest
conformity to nature and observation." The pride which Enlightenment philosophers felt in Newton and, therefore, in the
science of their own times, was suggested by Rittenhouse in
his 1775 oration before the American Philosophical Society:

> Other systems of Philosophy have been spun out of the
> fertile brain of some great genius or other; and for want of
> a foundation in nature, have had their rise and fall, succeed-
> ing each other by turns. But this will be durable as science,
> and can never sink into neglect, until "Universal darkness
> buries all."
>
> Other systems of Philosophy have ever found it neces-
> sary to conceal their weaknesses and inconsistencies, under
> the veil of unintelligible terms and phrases, to which no two
> mortals perhaps ever affixed the same meaning. But the
> Philosophy of Newton disdains to make use of such subter-
> fuges; it is not reduced to the necessity of using them, be-
> cause it pretends not to be of nature's privy council, or to
> have free access to her most inscrutable mysteries, but to
> attend carefully to her works, to discover the immediate
> causes of visible effects, to trace those causes to others more
> general and simple, advancing by slow and sure steps towards
> the great First Cause of all things.[36]

Indeed, as Rittenhouse suggested, Newton's philosophy
was as striking for what it did not pretend to deal with as it
was for what it did deal with. The law of gravitation, for example, does not tell us *why* an apple falls to earth rather than,
say, to the moon. To say that it falls because the earth attracts
the apple is no better than the Aristotelian "the earthly nature
of the apple is seeking its natural place." The law, rather,
tells one *how* the movements are determined. And, again as
Rittenhouse suggested, success in dealing with the *how* inclined
men to dismiss the previous *why*-questions as so much mysti-
cism buried under the veil of "unintelligible terms and phrases."
This tendency was one of the more subtle, but still among the
most significant, aspects of the Newtonian revolution.

Of course, there was a great deal more to the philosophy

of Newton than his scientific method. There was, for example, the corpuscular philosophy—Newton's final view of the nature of matter adapted from Boyle, Gassendi and other predecessors, and published in his *Opticks:*

> It seems probable to me that God in the beginning form'd matter in solid, massy, hard impenetrable, moveable Particles . . . and that these primitive Particles, being Solids, are incomparably harder than any porous Bodies compounded of them; even so very hard as never to wear or break in pieces; no ordinary power being able to divide what God himself made one in the first Creation.

This is the meaning of "mechanical philosophy"; it is both a metaphysical principle and a program for research. As a program for research it suggests that explanation consists of reducing phenomena to the action of hard particles upon one another—a reduction later carried out with great success in the case, particularly, of heat. Early efforts, however, were not all so fortunate. Typical of the enthusiast, for example, is the Boston scientist Thomas Robie's effort to explain the workings of rattlesnake poison in these terms.[37]

With such impressive predictive and explanatory power, it is no wonder that Newton was hailed as a kind of demigod of the age, and his method was enthusiastically adopted—at least verbally—by the English-speaking learned world. Thomas Godfrey, the Pennsylvania glazier who learned Latin in order to read Newton, is a good example of the reverent enthusiasm generated by the achievement of Newton. "But to proceed upon the Copernican and Newtonian scheme of astronomy," rapturously proclaimed an almanac of 1761, "what wonders of wisdom, power, and grandeur open to our view."

Although the first known copy of the *Principia* to arrive in the colonies was one obtained by James Logan in 1708, Newtonian ideas, as demonstrated by the cases particularly of Charles Morton and Cotton Mather, long antedated the arrival of the text. The *Principia* simply gave formal expression to a point of view that had been developing during the past several decades throughout the learned world.

Into this rationalistic framework dominated by Newton— amounting almost to an idolatry of order—the work of the other great scientist of the eighteenth century fit beautifully. Carl Linnaeus, the Swedish systematist who classified all of nature and, like Newton, discovered an unchanging order in nature which he and his followers believed to be instituted by God himself, had a lasting effect on science, theology, and philosophy despite all subsequent developments. Even though Linnaeus's intellect was of a distinctly lower order than that of Newton, and although he never made a single important discovery in science, he must still be ranked next to Newton in terms of his impact on eighteenth-century thought.

Whatever lasting scientific value the work of Linnaeus has had is based on two points. First, he was successful in carrying out the binary nomenclature in connection with his careful and methodical study on the distinguishing of genera and species. The basic idea was that two names, a generic and a specific, would be assigned to each plant, and that these names would be sufficient for any student to identify the plant being discussed by simply referring to a catalogue. The old effort to include a description of the plant in its name, a goal that re- sulted in sentence-long names and no agreement among au- thorities, was abandoned. In connection with his nomenclature, Linnaeus laid down in great detail several rules which must be observed in establishing species, genera, orders, and classes, and thereby introduced a new spirit of order and clarity into the science of botany. Secondly, Linnaeus recognized that the chief task of the botanist was to discover a natural system of clas- sification which would demonstrate the true affinities existing in nature. His fragment of a natural system, in which he adopted sixty-seven "families" (orders), was used as a basis by later founders of a natural system, especially A. L. de Jussieu.

And yet, these were not the achievements that grasped the imaginations of Enlightened thinkers. The one achieve- ment for which Linnaeus was most celebrated in his own time was his admittedly artificial system that, except as it guided one toward the natural system, was simply a convenience for locating and identifying individual plants.

Taking as a guide for his artificial system the number and position of the sexual parts of plants, Linnaeus was able to divide the entire plant kingdom into twenty-four classes to one of which any plant could easily be referred merely by close observation. Although it amounted to arrangement only, the achievement of Linnaeus was an impressive one that promised

to be extremely valuable to the botanists as they discovered
and tried to assimilate the new plants which were daily being
brought in from the frontiers. Appealing as it did to the Enlightenment spirit of system, and to the eighteenth-century
naturalist's assumption that the highest and most worthy task
of a botanist is to know all species of the vegetable kingdom
exactly by name, the Linnaean system found almost immediate
acceptance. Beyond this, Linnaeus's power of framing precise
and striking descriptions of species and genera in the animal
and vegetable kingdoms by means of a few marks contained
in the smallest possible number of words made him a model of
excellence for naturalists of his time. Indeed, in this respect,
he remains a model to the present day.

The popularity of the Linnaean system with amateurs,
of which there were many among American naturalists, was
immense. Many were for the first time able to refer a flower
to a systematic catalogue; in an area where a great many new
things were being discovered, this was especially important.
Hence, it is not surprising to find that even before Linnaean
botany was being taught in England, many American botanists
were testifying to their delight in Linnaeus's system and to
their indebtedness to him. Alexander Garden, for example,
wrote to Linnaeus in 1755 that he had discovered the *Funda-
menta Botanica* a year earlier and since that time had made
greater progress than in the three preceding years when he
had been following the older and less elegant system of Tournefort. Garden reported that during the preceding summer scarcely
a week passed without his rereading the work. "Such neatness!
Such regularity!" he exclaimed rapturously. "So clear and su-
premely ingenious a system, undoubtedly never appeared before
in the Botanical World."[38]

Cadwallader Colden, who had discovered Linnaeus's
work several years earlier, reported that when he had first
arrived in America he found so much difficulty in dealing with
the many unknown plants that he was quite discouraged and
gave up all efforts for nearly thirty years, until he had "casually
met with" the writings of Linnaeus. After this testimony, Colden
characteristically offered Linnaeus a philosophical discussion of
the problem of genera.[39]

The basic natural history concept of the eighteenth century, which Linnaeus and all of his followers tended to take

as axiomatic at the very time they were violating it, was that
of the chain of being, with its assumptions of plenitude and
continuity, which had been formulated by the Greeks and fused
together by Enlightenment thought. According to the concept,
every possible creature that could exist had been struck off at
the Creation (plenitude), and organisms were separated by
the smallest possible gradation (continuity). It was these concepts that guided Linnaeus in his incomplete efforts to derive a
natural system:

> There is, as it were, a certain chain of created beings, ac-
> cording to which they seem all to have been formed, and
> one thing differs so little from some other, that if we hit
> upon the right method we shall scarcely find any limits be-
> tween them. . . . Does not everyone perceive that there is
> a vast difference between a stone and a monkey? But if all
> the intermediate beings were set to view in order, it would
> be difficult to find the limits between them. The polypus
> [hydra] and the sea-moss join the vegetable and the animal
> kingdom together, for the plants called Confervae [algae] and
> the animals called Sertulonia [marine polyzoa and hydroids]
> are not easy to distinguish. The corals connect the animal,
> vegetable, and fossil world.[40]

And yet, despite his acceptance of the scale of beings, Linnaeus, like virtually every other naturalist of the time, insisted
that species were distinct from the Creation, and that one *could*
set limits between them. "We reckon so many species as there
were distinct forms created in the beginning," wrote Linnaeus,
denying that new species could arise by natural processes, or
that species could ever become extinct.* The necessity of a
new Creation would have been an admission of the Creator's
inefficiency; extinction of a once-created form would have been
an admission of the Creator's error—neither concept was ad-
missible in Puritan-Enlightenment thought.

Although the chain of being was a static concept, like
the Newtonian one in this respect, it did furnish a program of
research for Enlightened biologists. The main task of the nat-

---

*Linnaeus sharply modified his views on the constancy of species
later in his life, at the end believing that God had first created
classes, and that the orders, genera, and species were derived by
hybridization. I can, however, find little evidence to suggest that
this change was even known to his American followers. For a discussion of Linnaeus's developing ideas on species, see James L. Larson:
"The Species Concept of Linnaeus," *Isis*, LIX (1968), 291–9.

uralist, after arranging his creatures on Linnaeus's scale, was
to search for "missing links." In this respect, the concept functioned quite like the table of atomic weights at the end of the
nineteenth century. Garden, for example, was excited at the
discovery of a toadfish, which he thought appeared to be a
middle link between the reptiles and the fish. At another time
his curiosity was aroused by a copper plate sent to him by
John Ellis, which showed the mode of reproduction of the
marine polyp. Since the method of alternation of generations
was shared by some vegetables, Garden thought that it "opens
quite a new field," and "shows us a much higher link between
the animal and vegetable World than any yet known."[41]

In this conception, exactly like the Newtonian one, God
is seen as essentially an architect and builder who had made
the universe in a single act, according to the manner described
in the book of Genesis. This single creation was no longer
insisted upon from any Biblicism, but on grounds that anything
else would have been inefficient and wasteful. The analogy with
Newton's universe was perfect. At the heart of Linnaeus's conception of nature was the idea of a rationally ordained system
of means and ends. In the biological world as in the physical
universe, the Creation was its own end: self-justifying, efficient,
coherent, and intricate motion with no waste. The earth, with
its incredible variety of climate and topography, was populated
with an equally varied assemblage of living beings, yet each
creature was perfectly adapted to the region in which it lived.
Each species obeyed the command to increase and multiply,
like producing like, yet the different plants and animals were
so related to each other and to their environments that a balance of nature was preserved and no species was ever destroyed.

There is, of course, a logical difficulty from the point
of view of the systematist who adopts the concept of the chain
of being. For the concept of continuity, pushed to its logical limit,
assumes a tremendously complicated nature in which all ab-
stractions, all general terms, are artificial, since nature exhibits nothing but particulars. Hence, it might be said that no
system could enclose nature. Despite their adherence to par-
ticularity, however, Enlightened thinkers believed that there
was an absolute order in the universe which the mind of man
could find. It was not, therefore, too important where one drew
the line between species or between genera; all that was neces-

sary was to discover a consistent framework within which known species could be arranged, and within which new species could fall into their proper places. Linnaeus had discovered one such system and his was the first to combine all the admirable qualities that it possessed. Except in the minds of a few naturalists, most of whom were in France, the fact that Linnaeus's system was artificial was unimportant;* for, from this point of view, *all* systems were equally artificial. As Jefferson said of David Rittenhouse, one might say of Linnaeus: "He has not indeed made a world, but by imitation he has approached nearer to its Creator than any other man from the Creation to this day."

Newton and Linnaeus, the one a mathematical genius who reduced the movements of the celestial bodies to a formula having predictive power, the other an ardent classifier of nature, together managed to capture exactly the spirit of the eighteenth-century Enlightenment. Both gave testimony to the power of the human mind in unraveling the intricacies of nature, both postulated an essentially static universe which continued to call upon God the Supreme Architect as First Cause but to relegate Him to an insignificant place otherwise, and both appealed notably to the love of order and harmonious regularity of the century. The same spirit that adopted the rhyming couplet as its characteristic poetic form and the formal lines of classical architecture as its ideal of building art, enthusiastically welcomed the clockwork universe and the rigorously ordered biological world. American intellectuals, sharing with their English and continental brethren the acceptance of these concepts, naturally looked to the two giants of the period for their inspiration and their guide in matters philosophical. By the middle of the eighteenth century they had become full-fledged, if still junior, members of an Atlantic intellectual community.

---

* Despite his testimonial to the Linnaean system, Cadwallader Colden did criticize all artificial systems and argue strongly for a natural system based on the scale of beings, in a letter to J. F. Gronovius, December 1744, in New-York Historical Society *Collections*, LIV (1920), No. 3, 89–92.

V

~~~~~~

Science: Handmaiden
of Enlightenment

On his second voyage to England, Benjamin Franklin barely escaped shipwreck off the coast of Falmouth. Commenting to his wife about the near-disaster, he said: "Perhaps I should on this occasion vow to build a chapel to some saint; but . . . if I were to vow at all, it should be to build a better lighthouse."

This remark illustrates the well-known rationalist bent of Franklin's mind with its thrust toward the immediately practical. America's greatest scientist of the eighteenth century, he was also one of the most outspoken proponents of the application of science to the problems of this world. "What signifies philosophy that does not apply to some use?" he asked in 1761. Franklin had an unshakable conviction that genuine science would ultimately yield useful results and he thought that the philosopher should dedicate himself to the service of mankind. Franklin carried the belief so far that he even felt apologetic about the time he had spent on the construction of magic squares—a mathematical game in which he developed great skill, but for which he could envisage no practical applicability. He was even dissatisfied with his electrical experiments until he was able to find something "of use to mankind," in his researches. There is no doubt that in Franklin's own mind, as in the minds of most of his countrymen, the lightning rod was much more important than his theoretical contributions.*

---

*I. Bernard Cohen, in "How Practical was Benjamin Franklin's Science?" *Pennsylvania Magazine of History and Biography*, LXIX

It would be a mistake to attribute any "frontier" or other uniquely "American" origins to Franklin's practicality. In truth, he was simply echoing sentiments that had been widely shared among the philosophers of Western Europe since the time of Francis Bacon. Science before Bacon had been essentially theoretical rather than practical, an adornment of life concerned only with the search for natural law and order or the determination of the limits of natural law. After Bacon's influential work, it came to be regarded more and more as an instrument to achieve mastery over nature. Observing, for example, that agricultural philosophers could neither kill weeds nor enrich the land, English agriculturist John Dove, a contemporary of Franklin, wrote that "a philosophy deficient in both these ought to be hissed out of the land."[1] This utilitarian attitude, along with the delight in the harmonious regularities of nature for their religious value, had been transported to America by the Puritans as a part of their original intellectual baggage, and developed to an unprecedented height during the following century on both sides of the Atlantic. Robert Boyle's comment, in the year of the founding of the Royal Society, is typical. He had been studying natural philosophy, mechanics, and husbandry, he said, "according to the principles of our new philosophical college, that values no knowledge, but as it hath a tendency to use." Three-quarters of a century later, the Finnish botanist Peter Kalm exclaimed that if he could follow the call of his heart and at the same time be sure of his daily bread, his life's efforts would be directed toward the application of botany to the economy of the country and to collecting "all sorts of things connected with natural history and especially to find out what use they could be put to."[2]

It was squarely in this well-established tradition that Thomas Jefferson, on receiving what he termed a "charming treatise on manures," wrote to the sender that "science never appears so beautiful as when applied to the uses of human life." At another time he emphasized "the importance of turning a knowledge of chemistry to household purposes," and he ridiculed those scientists who "seem to write only for one another." As a model to the chemists, then (c. 1820) going

---

(1945), 284–93, details these instances of Franklin's practicality, but he nevertheless concludes that Franklin was really motivated only by his curiosity and interest.

through the early phases of professionalization and therefore losing their direct interest in practicality, he emphasized "the just esteem which attached itself to Dr. Franklin's science, because he always endeavored to direct it to something useful in private life."[3] That Jefferson found it necessary to administer a rebuke to early nineteenth-century chemists for neglecting practical applications of their work was a sure indication the Enlightenment had ended and that new values were growing within the scientific community.

Also typical of the post-Baconian philosopher was the unlimited progress that Franklin foresaw as being the lot of man by virtue of science:

> The rapid progress *true* science now makes, occasions my regretting sometimes that I was born too soon. It is impossible to imagine the height to which may be carried, in a thousand years, the power of man over matter. We may perhaps learn to deprive large masses of their gravity, and give them absolute levity, for the sake of easy transport. Agriculture may diminish its labour and double its produce, all diseases may by sure means be prevented or cured, not excepting even that of Old Age, and our lives lengthened at pleasure even beyond the antediluvian standard.[4]

Even though the historical record did not give a clear indication that science had produced utilitarian results, there remained ample faith among the philosophers that it would do so. The truth was that science in the eighteenth century had not yet reached a stage at which it could offer much help to the farmer, the merchant, or the artisan in the ordinary pursuit of business. Advances in technology that did prove useful still came through trial-and-error methods more or less independent of theoretical science. Franklin's own useful inventions are indicative of this. His bifocals, his four-pane lamp for street lighting, his Pennsylvania stove—all involved only the most elementary theoretical conceptions and were simply products of ingenuity. Even his most famous invention, the lightning rod, did not depend upon his great electrical discoveries. The lightning rod is based on the idea that electricity—or lightning —will choose a good conductor in preference to a poor when a good one is available, an idea that does not necessarily assume the identity of lightning and electricity but that was one of the points of analogy suggesting such identity. The discovery had, in fact, come through actual observation of the behavior of lightning many years before Franklin's work—not from controlled electrical experiment.

The century did produce an impressive number of technological innovations by the trial-and-error method characteristic of pre-scientific technology. The power loom, the water frame, the spinning jenny, and most of the other apparatus upon which the industrial revolution was based are only the most obvious examples. Before the century's end, the essentials of the automated assembly line had even been developed and applied with great success by the American mechanic Oliver Evans; and French balloonists had taken the first long step toward conquering space. But the wedding of theory and practice, the constant interaction between science and technology that is taken for granted in the mid-twentieth century, occurred only rarely before the Enlightenment passed away.

Philosophers of the present day have considerable difficulty distinguishing between science and technology, although it is generally agreed that they are different and that the one is of a distinctly higher order than the other. The difficulty today is that a complexity of continuous interactions between the two tends to produce a blur at the edges—it is the boundaries of each that are uncertain. In the eighteenth century, the invidious distinction between the two was not made. The legacy of Bacon had resulted in dismissing the speculative works of the scholastics—the "pure scientists" of the sixteenth century—and in fusing what was left of science, along with technology and even mechanical ingenuity, under the rubric "useful knowledge." This term included the Newtonian laws that allowed one to predict eclipses, explain the tides, and trace the course of a comet; and it included the ability to construct a thermometer, a clock, or an automated factory. A number of important consequences flowed from this confusion.

First, and most evidently, it allowed philosophers to point with pride to the many real evidences of mechanical ingenuity and claim them all for "science." Thus Thomas Jefferson could rank David Rittenhouse as an astronomer second only to Newton, simply because he had constructed the world's finest "orrery": a clockwork representation of the solar system which accurately demonstrated the positions of the heavenly bodies backward or forward in time by the simple turning of a crank. Second, the confusion meant that genuinely talented scientists—as Rittenhouse was, although present-day judgment is that he did not quite deserve Jefferson's praise—would be asked, with no sense of impropriety, to perform tasks far below their levels of competence. Rittenhouse, known to be a scientist, was typically asked to survey boundaries, advise on canal routes, tend the clock in Philadelphia's state house, and construct a copper standard measure for the colony of Pennsylvania. Later he became a virtual handyman of the Revolution. Whether Rittenhouse could have become a great astronomer had he not spent so much of his time and energy on routine mechanical tasks is a question beyond the competence of any historian; at any rate, the philosophy of the Enlightenment conspired with his own location in history to deprive him of any real opportunity.[5]

The final result of the identification was a more happy one. In a day when no distinction was made between mechanical ingenuity and science, philosophers, feeling that all knowledge was of a piece, were sure that their discoveries would ultimately prove equally useful. "Of what use is a new-born babe?" Franklin is said to have retorted to a critic who questioned the value of one of his researches. Surely, he thought, it was only a matter of time before any new fact would prove itself of value. It was this unshakable faith that preserved Franklin, like other Enlightenment philosophers, from narrow philistinism and allowed him to pursue his theoretical researches with a good conscience.

American philosophers were thus able to engage even in such scientific projects of purely theoretical interest as the observation of the Transit of Venus which occurred in 1769. The transit, which would not occur again for another 105 years, was eagerly anticipated by the learned of Europe and America, for coordinated observations provided the best measure of the sun's distance from earth, consequently giving a better knowledge of the dimensions of the solar system. The American Philosophical Society, like the Royal Society, consistently spoke of its plans for observing the transit in terms demonstrating its appreciation of the general expectation that science ought to be useful. The Royal Society emphasized the possible benefit to navigation, the Philosophical Society spoke specifically of the increased possibility for determining longitudes of the principal American cities. Yet there is no evidence that the British observations aided navigation in any way; as for the American claims, the longitude of important mercantile cities was already well known and the observations were not conducted in locations where they could aid this practical interest. In both cases, the philosophers simply concentrated upon answering the scientific questions at stake, confident that practical results would inevitably follow. The results brought acclaim to American science, and increased understanding of the solar system, but any benefits to commerce were not evident.

Genuine philistinism can only arise—as it did in the following century—after distinctions are clearly drawn among various types of knowledge. Thomas Sprat, in his *History of the Royal Society*, expressed the great faith of the Enlightenment in the following manner:

> There is nothing of all the Works of Nature so inconsiderable, so remote, or so fully known; but by being made to reflect on other Things, it will at once enlighten them, and shew itself the clearer. Such is the Dependence amongst all the Orders of Creatures, the inanimate, the sensitive, the rational, the natural, the artificial; that the Apprehension of one of them is a good Step towards the understanding of the rest. And this is the highest Pitch of *human Reason;* to follow all the Links of this Chain, till all their Secrets are open to our Minds, and their Works advanced, or initiated by our Hands.

Carrying out this program, Sprat concluded, was truly to command the world.

So confident were they that science could serve man, and not being content to wait for the slow working of natural processes, the men of the Enlightenment founded a large number of scientific societies in Europe and in America which were primarily dedicated to hastening the day when every scientific fact would find its destined useful role. This orientation was even reflected in the very names of most of the early societies. "The American Philosophical Society, Held at Philadelphia, for Promoting Useful Knowledge"—a cumbersome name still carried by the most distinguished of the American societies—was in part the outcome of a political compromise in the founding of the society, but it accurately reflected the attitude of Enlightenment philosophers. All such societies had as their aim the program so admirably expressed by Franklin at the time of the founding of the original Philosophical Society in 1743. It would pursue all experiments that "let light into the nature of things and tend to increase the power of man over matter, and multiply the conveniences or pleasures of life." The program of the societies, however, rarely amounted to the direct application of theoretical science to practical ends. For the most part, they confined themselves to promoting study of the most successful practices found in any part of the world and to experimenting with new methods as freely and as frequently as possible. The pervasive tendency to take natural history observation as the model of science, despite the obeisance paid to Newton, is nowhere more clear than in the activities of the societies. This approach, contrary to those who find in it an example of American uniqueness, was common to societies on both sides of the Atlantic. The philosophers would study the practical work of the world with the detachment that had been so useful in the study of science and attempt to disseminate the knowledge thus acquired. Mankind, gaining power over nature, would be elevated to unprecedented heights of riches, health, comfort, and morality as a necessary side effect. Nature was benevolent and would unfold its blessings to those who diligently pursued them.

The careful study of existing practice in order to identify the best method was the approach that in the eighteenth century was being used in England, particularly in agriculture and in the mechanic arts, in both of which significant innovations had been introduced since the beginning of the century. By mid-century, England was in the full tide of an agricultural revolution and the accelerating enclosure movement was encouraging more efficient methods of farming. Manuring, crop rotation, and intensive cultivation were being practiced by English farmers with great success. Having only limited acreage available for agriculture and an abundant labor supply, such methods were only good economic sense in England. American philosophers were particularly anxious to emulate their English brethren in the case of agriculture, for agriculture was not only the chief support of the American people, but it was demonstrably in need of reform. The pervasive Enlightenment belief that farming was the pursuit most worthy of man and most pleasing to his Creator gave added impetus to the drive for agricultural improvement. The Jeffersonian conception that farmers were the only viable basis for a republic made the American need seem particularly urgent. "The American planters and farmers are in general the greatest slovens in christendom," concluded the anonymous author of *American Husbandry* (1775), the best surviving eighteenth-century survey of Colonial agriculture.

European observers in general agreed with the author; indeed, accustomed as they were to the tidy English and Continental farms, most of them were positively shocked by the wasteful methods being used in America. As early as 1758, the *American Magazine* had suggested that "township companies or societies" be formed to discuss agricultural improvement and to make new experiments. The editor of the magazine reported a sad story of wasteful practices on every hand: there was practically no crop rotation; no purposeful breeding, despite the early experiments on heredity that had been made in this country; no intelligent use of fertilizers. Such conditions, the natural outgrowth of abundance of land combined with a serious labor shortage, were bound to raise the ire of Enlightenment philosophers. Consistent environmentalists that they usually were, they still made a fetish of efficiency and they considered the wasteful use of nature's bounty to be essentially a blasphemy upon the Creator.*

Leadership in the organized effort to seek out improvements in Colonial agriculture was early taken by societies in England, primarily the Society Established at London for the Encouragement of Arts, Manufacturers, and Commerce—an organization chartered in 1754 which was better known as the Society of Arts, or the Premium Society. Its primary function, mainly directed toward the colonies, was to award premiums for the production of certain favored commodities and for the discovery of new techniques and inventions. Each year it offered premiums for the production of designated items—all of which were expected to supplement the economy of England rather than compete with it. In one year, for example, the society offered premiums ranging from £10 to £100 for the production of logwood, olive trees, potash, safflower, and wine. According to one contemporary, the society was responsible for the establishment of potash and pearl ash manufacture and for the cultivation of grapes in America;[6] in both cases, however, independent efforts had been made previously.

Adopting the well-tried model of English societies, several

---

*For an interesting description of eighteenth-century agriculture in America, see O. R. Bausman and J. A. Munroe: "James Tilton's Notes on the Agriculture of Delaware in 1788," *Agricultural History*, XX (1946), 176–87. Tilton's figures indicate that productivity in Delaware was very close to that of the present, and that "Hitherto we have depended chiefly on the freshness & richness of our soil." He indicated, however, that manure was coming into use and that farmers were beginning to practice crop rotation.

American groups began offering awards for improvement in agriculture—either in method, useful implement inventions, or in crop innovations. The first of these, the New York Society of Arts, founded in 1764 as a response to the Stamp Act crisis and the economic depression occurring at the same time, offered premiums on hemp, mules, and sturgeon, along with several manufactured items—the latter class of goods certainly being less pleasing to the parent society than the former. Similar societies were founded in other colonies. In 1773, for example, the Virginian Society for Promoting Usefull Knowledge awarded a Gold Medal to John Hobday for his invention of a threshing machine, a "cheap and simple" device which was credited with the capacity of beating out 120 bushels of wheat in three days. Some studies were made of crop pests and published by the societies, Thomas Gilpin's studies of the seventeen-year locust and Landon Carter's studies of the wheat fly being the most notable of these. Some efforts were even made to experiment with English methods of intensive farming, but given the abundance of land and the shortage of labor in America this would have meant financial suicide and the practical farmers generally recognized that this was the case. Much heated enthusiasm was generated among the philosophers and careful instructions were printed in newspapers, transactions of societies, and pamphlets, but almost no trials were reported.

　　To most of those concerned, agricultural improvement meant introducing new crops that offered economic advantage and finding indigenous substitutes for items that were customarily imported. There was, therefore, a continuity between the preoccupations of the age of exploration and those of the Enlightenment. "The greatest service which can be rendered any country," Jefferson observed, "is to add an useful plant to its culture; especially a bread grain; next in value to bread is oil."[7] Early issues of the *Philosophical Transactions* contained many articles on both these themes. One paper discussed the distillation of persimmons and recommended the resultant hard liquor as a substitute for the West Indian rum then in such demand. American sources of table oil were also investigated in the *Transactions:* sunflower seed, ground nuts, and walnuts were at one time or another suggested as possibilities. Jefferson served "various companies" a salad made with benne oil instead of olive oil, reporting such a favorable reaction that he resolved to grow benne himself.[8]

　　But by far the greatest interest was in foreign plants that

might be profitably introduced into the Colonies. Once again, Jefferson was the most assiduous collector. Every diplomatic trip was an occasion for studying the produce of the countries he visited and collecting seed for introduction into America. From southern France, he sent seeds of malta grass, acorns from the cork oak, and several species of olives. Crossing the Alps into northern Italy, he tabulated the plants he encountered according to their powers of resisting cold. In northern Italy he collected a supply of rice grains, smuggling them out of the country in his pockets when he found that export was prohibited by law. New strains of rice were of special interest to Jefferson, for he hoped to introduce a variety that could be grown under more healthful conditions than those required for the strain then used in Georgia and South Carolina. In typical Enlightenment fashion, the humanitarian impulse was intimately connected with Jefferson's interest in agricultural improvement.

Initial impetus for this activity, however, came from British authorities working within the framework of the mercantile theory. They wished to promote the culture of indigo, hemp, flax, vegetable dyes, and other products needed by the Empire which, at that time, were being imported. There is, for example, a long history of abortive efforts to introduce silk culture into North America, beginning in 1621 when Governor Wyatt arrived in Virginia with instructions to plant mulberry trees as food for silkworms. A law of 1658 required every Virginia landholder to plant ten mulberry trees for every hundred acres of land he held, and in 1661 a bounty of fifty pounds of tobacco was offered for every pound of silk produced in the colony. All of the southern and central colonies at one time or another made some effort to encourage silk production. An Italian family was induced to come to Georgia in 1733 and other Italian workers were brought in later; raising silkworms was taken up as a fad by the ladies of some of the wealthy planters of South Carolina; and, as late as 1770, a public filature was established in Philadelphia.

Benjamin Franklin became interested in the Philadelphia project. The Pennsylvania Assembly granted £1000 toward the filature, the American Philosophical Society contributed more, and much enthusiasm was generated. Led by William Smith, a group of Philadelphia men organized, signed a formal agreement among themselves, and contributed £1400 to the project.[9] But the effort, like others of its kind, soon died, primarily for lack of a trained labor supply but also because the cost of production could not be lowered to competitive limits. The greatest measure of success was met in Georgia, where a maximum production of 1,084 pounds was attained in 1766. In no case, however, did the value of the silk match the cost of production. In 1767, the Royal Society of Arts, having paid out a total of £1370 in silk premiums, became discouraged with the results and abandoned its efforts to stimulate silk production in America. Native efforts, however, continued through the post-Revolutionary period with such people as Ezra Stiles, President of Yale College, who participated in efforts to revive silk culture in Connecticut during the 1780's.*

The story was similar with efforts to encourage the production of wine and hemp, articles that Great Britain also had to import. These products, like silk, promised both to complement the self-sufficiency of the Empire and to increase the riches of America. Mercantilism and local self-interest here joined in happy combination. Americans of all shades of opinion could therefore join in supporting the cultivation of such products, convinced— no matter where their primary loyalties—that they were acting in the best interests of patriotism.

The culture of grapes for the purpose of making wine had been advocated from an early time in the hope of decreasing the large importation of foreign wine into Britain and her colonies. The reports of early travelers of the abundance of wild grapes growing everywhere along the Atlantic coast raised the hopes in England of an unlimited supply of wine which would make her independent of her rivals Spain and France in the production of this beverage. Some early local success was attained, particularly by the Huguenots, who had been settled for that purpose in South Carolina by the British government. After the French and Indian War, efforts were stepped up; a bounty was offered by the British government and a premium, promptly claimed by Edward Antill of New Jersey, was announced by the Society of Arts. In 1769, a vineyard was established by the Province of Virginia. Although some temporary successes were recorded, in general the results were disappointing. The American wine was judged to lack the fine quality of the familiar French brands, and export served

---

*Efforts to introduce silk culture into the New World date from the early sixteenth century, when the Spanish brought mulberry trees and silkworms.

mostly to stimulate ridicule of America and American products among the fastidious wine drinkers of England. Even Thomas Jefferson, that most enthusiastic advocate of American improvement, continued to import his Madeira, although he tried out a native wine on his visitors at Monticello—which he ranked with the Crûmartin of Burgundy—and reported that he was gratified when his guests could not tell the difference.[10]

Another crop long advocated for mercantilist reasons was hemp, used for cordage, bags, and sails. A Virginia law of 1682 offered bounties for hemp and flax, and it also ordered that every tithable person should produce two pounds of dressed hemp or flax annually. From that time on considerable hemp and flax were raised in Virginia, but the bulk of the crop was used at home. After 1750, when the British government began giving liberal bounties for the export of these products to England, some began to be shipped overseas. Similar efforts were made in Massachusetts, Maryland, and Connecticut, in the last of which a law of 1640 ordered that every family of the plantations should procure at least a spoonful of English hempseed and sow it "in some fruitful soyle, at least a foote distant betwixt every seed, and the same so planted shall preserve and keepe in husbanly manner for supply of seed for another yeare."[11] As early as 1735, an Irish pamphlet on flax and hemp was reprinted in Boston, but it was not until the pre-Revolutionary tensions of the 1760's that general interest rose. Parliament established a bounty on hemp in 1765 and Virginia and North Carolina soon followed suit with bounties of their own. Hemp was widely attempted, but although its virtues were generously extolled, it did not supplant flax, which the hemp advocates considered much less generally useful, and production continued on a limited basis. Madder and tea, which also received encouragement, failed altogether.

Implicit in the English and Colonial efforts to improve agriculture, especially after the turn of the eighteenth century, was the belief that agriculture was a science, and, like other sciences, could profit from reliable data derived from carefully conducted experiments. Agriculture, wrote Thomas Jefferson, was "a science of the very first order." The list of agricultural subjects he wrote on is very long indeed, but the writings are dominated by his overriding belief in the necessity of a more scientific approach to agricultural problems and by his belief in the vast benefits to be derived by his country from the application

of agricultural science. He studied diversified farming, soil erosion, conservation and soil building, the use of gypsum, green-
dressing versus fallows, manuring, Lord Kames's proposal for
an essence of dung; he became an expert on crop rotation, con-
tour-plowing, deep-plowing and harvesting methods—the list
is practically endless.[12] And in his preoccupation, he reflected
typical Enlightenment views. Plattes, Hartlib, and Cowley in the
middle of the seventeenth century had emphasized the need of
agricultural experiments and instruction in the handling of soils,
plant culture, and animal husbandry at the higher institutions of
learning. Francis Home, a London physician, gave a particularly
clear statement of this viewpoint, imbued with the Enlightenment optimism and faith in science characteristic of the mid-
eighteenth century:

> Agriculture does not take its rise originally from reason,
> but from fact and experience. It is a branch of natural phi-
> losophy, and can only be improved from the knowledge of
> facts, as they happen in nature. It is by attending to these
> facts that the other branches of natural philosophy have been
> so much advanced during these two last ages. Medicine has
> attained its present perfection only from the history of diseases
> and cases delivered down. Chymistry is now reduced to a
> regular system, by the means of experiments made either by
> chance or design. But where are the experiments in agriculture
> to answer this purpose? When I look around for such, I can
> find few or none. There then, lies the impediment in the
> way of Agriculture. Books in that art we are not deficient
> in; but the book which we want is a book of experiments. . . .
> in time this plan may afford fund [of information] sufficient
> for some future comprehensive genius, who, laying the differ-
> ent, and often seemingly opposite experiments together, and
> codifing all their concomitant circumstances, may be able
> to reduce the practice to fixed and permanent rules.[13]

Some experimental work on a modest basis had been conducted in America as early as 1669, when Joseph West established
an experimental garden on the Ashley River settlement in South
Carolina. The Lords Proprietors had ordered West to stop at the
Barbados and procure a supply of cotton seed, indigo seed, ginger
roots, sugar canes, and olive trees. At the experimental garden,
West was to take special care to discover the soil to which each
species was best adapted and the season of the year most favorable for planting, as well as to provide seeds and cuttings for the
use of the plantation. The experimental work was to be done with

a "man or two" and the rest of the people were to be employed
in growing provisions.[14] A similar experimental garden was established in Georgia, east of the town of Savannah, shortly after
1733. Francis Moore, who visited Georgia in 1736, enumerates
some of the plants found in the ten-acre garden:

> Fruit trees usual in England, such as Apples, Pears, etc. In
> another Quarter are Olives, Figs, Vines, Pomegranates and
> such Fruits as are natural to the warmest Parts of Europe.
> At the bottom of the Hill, well sheltered from the North-
> wind, and in the warmest part of the Garden, there was a
> collection of West India Plants, and Trees, some Coffee, some
> Cocoa-Nuts, Cotton, Palma-Christi, and several West-Indian
> physical plants, some sent up by Mr. Eveleigh a publick
> spirited Merchant at Charles Town, and some by Dr. Houstoun,
> from the Spanish West Indies, where he was sent at the
> Expense of a Collection raised by that curious Physician Sir
> Hans Sloan, for to collect and send them to Georgia, where
> the Climate was capable of making a Garden which might
> contain all kinds of Plants; to which Design his Grace the
> Duke of Richmond, the Earl of Derby, the Lord Peters, and
> the Apothecary's company contributed very generously; as did
> Sir Hans himself. . . . There is a Plant of Bamboo cane brought
> from the East Indies, and sent over by Mr. Towers which
> thrives well, there was also some Tea-seeds, which came from
> the same place; but the latter, though great care was taken, did
> not grow.[15]

The cotton referred to in this passage was the perennial variety
which proved unsuitable to American conditions. It was at this
garden that the first cotton in America was grown, from seed
sent by Philip Miller, superintendent of the Chelsea garden.[16]

The early interest, however, did not last: the public-spirited
men apparently ceased to donate after 1738; soil, climate, and
uncertain knowledge combined to produce many failures, and
within a very few years the garden perished from neglect. The
chief benefit from the garden was probably in demonstrating the
negative conclusion that oranges and tropical plants in general
could not survive under Georgia conditions. During the latter
years of its existence, it was used merely as a nursery to grow
mulberry trees in an effort to aid the silk venture.

All of these failures point up the fact that by the mid-eight-
eenth century American farmers, with abundant land and suitable
money crops already tested and proven, but with a labor shortage,
were singularly uninterested in experimental agriculture of any
kind. The idea of an experimental garden, however, was one that

the philosophers never abandoned. Thus, in 1759 Alexander Garden of Charleston suggested that his correspondent John Ellis ask the Royal Society to propose the establishment of a provincial garden. With undue optimism he suggested that "If the Society would take the trouble, in a few sentences, to point out a few of the advantages flowing from it, they would at once, America like, be all on fire, and stick at no expense to promote the plan."

Garden should have known better, for two years earlier, Charles Whitworth, a member of the London Society of Arts, had suggested a more limited project, which Garden had enthusiastically backed. The idea was to establish a society dedicated to a study of vegetable dyes with a corresponding unit in London and an experimental unit in the Colonies. The aim would be to seek plants which would yield commercially profitable dyes. Garden had conferred with "some of the most public spirited gentlemen" in the Colonies, put an advertisement in the paper, and tried to encourage planters to make experiments with vegetable dyes. At the same time, he had tried to sell Whitworth on a much broader vision, suggesting that societies be established in Georgia, South Carolina, Philadelphia, and New York in order to point out "proper articles for your Premium Society to exercise their benevolence on" and to enrich "Natural History and Philosophy" in general. Whitworth never replied to the suggestion for a broader scheme, for even the more limited one he suggested failed entirely. There was not, Garden had to report, a single trial.[17]

Improvements in agricultural techniques were, of course, made in Colonial America. One introduction, cotton, laid the foundation for a major enterprise during the next century; rice, introduced into the Colonies in 1647, eventually became one of the major crops of South Carolina; improvements in tobacco culture led to significant increases in quality and quantity. Skillful breeding techniques had even produced the Conestoga horse, a powerful and efficient draft animal which, hitched to the Conestoga wagons, became the familar symbol of America's inland commerce until the coming of the railroads; and it had also produced the Narragansett pacers of Rhode Island, in great demand for their gait and speed during the latter part of the eighteenth century. Selective breeding with a few choice rams had resulted in a breed of sheep that compared favorably with the English variety in producing wool. Such improvements, however, were all made without benefit of knowledge of the science involved.

The case was practically the same with agricultural technology: the pre-Revolutionary Carey plow with its plated wooden moldboard and wrought-iron share; the "cradle," an improvement of the scythe, which tripled the amount of grain a man could harvest in a day; several improvements to the harrow; the wonderfully efficient American chopping ax, wedge-shaped, perfectly balanced, the handle especially designed for the individual user. These, again, were products of experience rather than theoretical science. Thomas Jefferson's further improvement of the moldboard in the 1790's was, indeed, worked out on theoretical principles, and some mathematics was involved, but this seems to have been the only exception. In this one instance, the exceptional nature of which was highlighted by the very importance assigned to it, Jefferson fulfilled the Enlightenment ideal of making "the combination of a *theory* which may satisfy the learned, with the practice intelligible to the most unlettered laborer," and he received a gold medal for his invention from the principal agricultural society of France.[18] Unlike the innovations that implied an alteration of basic patterns of agriculture, Jefferson's moldboard improvement, and all the others that were adopted, helped American farmers do better what they already regarded as their main business: laying as much land under the plow as possible as rapidly as possible, and with the least possible requirement of labor.

The failure of science to provide help for native manufactures was a counterpart of the failure to alter patterns of agriculture. As in agriculture, there were two distinct phases in the encouragement of manufactures: the imperial phase, which aimed to produce pig iron, bar iron, potash, pearl ash, and other products needed to supplement the economy of the British Empire; and the American phase, marked by attempts to rediscover the techniques and processes that would permit Colonials to compete with the mother country. In the first phase, the British government actively cooperated by heavily taxing the needed items when imported from foreign sources, while admitting those from the Colonies, duty free. The latter phase was marked by determined efforts particularly to stimulate the manufacture of textiles, but such efforts spread to virtually every type of consumer goods with the rise of imperial tensions after 1765.

The technique was the same as in the promotion of agriculture—the offering of premiums and awards for successful performance. In 1771, for example, William Henry Stiegel was awarded £150 by the Pennsylvania legislature for his flint glass specimens which had previously received the endorsement of the American Philosophical Society. Annual exhibits were held at Philadelphia, at which prizes were given for the most excellent products of home industry: woolen and cotton cloth, stockings, leather, shoes, whiskey, ironware, and paper hangings were among the articles receiving this distinction. Linen manufactures were encouraged by the private societies at New York and Philadelphia. Sometimes, patents and limited-period monopolies were awarded as inducements to enterprising industrialists. As early as 1639, for example, the builder of a glass plant in Massachusetts was voted a generous grant of land as a subsidy; two years later, the company was still foundering and a money grant of £30 was voted, with the understanding that the amount would be paid back when the company was in a position to do so. In 1646, "master mechanic" Joseph Jenks successfully petitioned the Massachusetts General Court for a fourteen-year monopoly in return for erecting an ironworks for making scythes and "divers sorts of edge tools." With legislative support, Jenks's business flourished; a later effort by Jenks to introduce the wire drawing industry into the Colonies was, however, not successful.[19]

The same Joseph Jenks had earlier been engaged by John Winthrop, Jr., to build the ironworks on the Saugus, for which Winthrop had been able to secure extremely favorable terms. The General Court had offered to prohibit the establishment of any competing enterprise for a period of twenty-one years, and had also given tax exemption, land grants, and freedom from militia duty for stockholders and employees. But the magistrates had stipulated that these advantages would be consequent upon establishment of complete facilities—furnaces, forges, and mills —and furthermore that none of the company's output should be shipped outside the colony until local needs were met. Specifically, it was to supply colonists with bar iron at no more than £20 a ton. Moreover, the court indicated its hope that control of the company might someday pass out of the hands of the London stockholders, to whom Winthrop had been forced to go for financing, into those of Massachusetts residents. The company, like other Massachusetts businesses during that period of money shortage, was required to accept payment in kind for its product.

Although the requirements were seldom so carefully spelled

out, the charter of Winthrop's ironworks reflected a typical—and serious—divergence of viewpoint between investors, necessarily from outside, and colonials. Investors thought in terms of an exportable product, salable in England for a quick return on their money; thinking in terms of an English market, they were mercantilists almost by instinct. Colonials, on the other hand, thought in terms of the domestic supply of a necessary commodity, which they could not pay for in England; again almost by instinct, they were anti-mercantilists. Both groups were bitterly disappointed when two years passed and £15,000 had been spent before one sow of iron had been poured. Early in 1645, shortly after operations began, the financial backers began to protest restrictions, insisting that it was unreasonable for them to be asked to accept payment in kind for the produce of the ironworks, especially since they had invested so heavily in it. They argued that, since Massachusetts residents could not pay, the company should be allowed to export. New England, they insisted, would still profit from wages earned at the furnaces and forges and by large purchases of provisions that the company was obliged to make. The court, however, would not retreat, and the controversy continued. By 1648, Saugus was producing a ton of iron a day; nevertheless, the company went bankrupt in 1652.[20]

The failure of the Saugus Ironworks was in part attributable to the cross-purposes of investors and Colonial officials, but not entirely. By the early 1670's Winthrop had shed his Old World wishful thinking about the New World to the extent that he concluded that New England had slight promise for early mineral industries of any kind. Apparently an effort to mine graphite had cured Winthrop of his dreams, for he discovered that, with the primitive transportation facilities available, it cost more to mine the graphite and move it to the coast than it would bring in London. His belief that graphite was necessarily associated with silver also turned out to have been mistaken.[21] New England, in fact, was deficient in several important minerals, notably chalk, flints, and lime, and it had not so far developed the transportation facilities necessary for assembling the factors of production economically and for distributing the product. Until these conditions were met, efforts failed mostly because of their prematurity or because of their overly grandiose design.

Although this divergence of interests remained as serious as the better-known one between Parliament and Colonial govern-

ments, outside investors continued to be necessary for any large-
scale enterprise, and colonials continued to be dissatisfied with
the bargain they made. It was perhaps the best method available
at the time, and all of the British colonies at one time or another
awarded monopolies on this same pattern. In 1696, for example,
New York granted two expert tanners a monopoly as an inducement for them to settle there and erect a tannery. In 1712, New
Jersey authorities granted Ebenezer Fitch the sole right to erect
a slitting mill to make nails; in the following year the same colony
gave Edward Hinman a monopoly to make molasses, with the
understanding that he would make it from cornstalks—thus bene-
fitting local agriculture—and that it would be "as good and cheap
as the West Indies Variety."

The height of the American phase came after 1764 when a
series of non-importation agreements cut down the supply of
British manufactures. The Sugar Act, the Stamp Act, the Townshend Acts, and the Coercive Acts led to a succession of such agreements which did notably stimulate household production of a
pre-industrial nature. The philosophers, however, and the societies
they had founded, continued to preach their grander vision.
American scientific organizations in several colonies offered
premiums for needed inventions, various Colonial governments
offered premiums, and there was a great deal of discussion. In
Philadelphia, annual exhibits were held, at which prizes were
given for the most excellent products of home industry. Liberal
rewards were offered from the Carolinas to New England for
such needed items as iron, leather, woolen goods, linen, potash,
and firearms.

    Now it was increasingly portrayed as a patriotic duty to
support American manufactures and American produce in general: "A people who are dependent upon foreigners for food or
clothes," wrote Benjamin Rush in 1775, "must always be subject
to them." Rush, then a professor at the University of Pennsylvania,
had taken up the study of chemistry at Edinburgh primarily because he thought he could qualify himself to help in establishing
new industries in Pennsylvania. At the 1770 commencement,
Harvard students demonstrated their patriotism by appearing in
clothes entirely of American manufacture. In New York, the collection of rags for the local paper mill was publicized as a patriotic
duty. Benjamin Franklin, ever ready with a wise word, wrote a

fervent appeal in a letter to Humphrey Marshall, in which he
argued that the support of native manufactures was intimately
connected with agricultural prosperity:

> If our Country People would well consider, that all they save
> in refusing to purchase foreign Gewgaws, and in making
> their own Apparel, being applied to the Improvement of their
> Plantations, would render those more profitable, so yielding
> a greater Produce, I should hope they would persist resolutely
> in their present commendable Industry and Frugality. . . . The
> Colonies that produce Provisions grow very fast. But of the
> Countries that take off those provisions, some do not increase
> at all, as the European Nations; and others, as the West
> India Colonies, not in the same proportion.
>   So that tho the demand at present may be sufficient, it
> cannot long continue so. Every manufacturer encouraged in
> our Country, makes part of a Market for Provisions within
> ourselves, and saves so much money to the Country as must
> otherwise be exported to pay for the Manufactures he sup-
> plies. Now in England it is well known and understood, that
> where-ever a Manufacture is established which employs a
> number of hands, it raises the Value of the Lands in the
> neighboring Country all around it; partly by the greater De-
> mand near at hand for the Produce of the land; and partly
> from the plenty of Money drawn by the Manufactures to that
> part of the country. It seems therefore the Interest of all our
> Farmers and Owners of Lands to encourage young Manu-
> factures in preference to foreign ones imported among us
> from distant Countries.[22]

In Franklin's argument, there is a glimpse of the later protectionist
doctrine, which, indeed, had the same end in view.

The American Philosophical Society, as the largest scientific body in America, from its central location in Philadelphia
tried to serve as a coordinating agent in the stimulation of American manufactures. The society offered premiums for the collection of rags to be used for paper-making, tried to collect clays
from all over the country so that the deposits most suitable for
pottery could be located, and gave to William Henry Stiegel a
scientific stamp of approval for the flint glass he manufactured.
Stiegel used this to advantage in his advertising and in his successful appeal for an assembly grant. This type of activity was not
very different from that undertaken from the imperial point of
view earlier. In some respects, American and British imperial
interest in Colonial manufactures were not in conflict. In the
1750's, for example, Alexander Garden of Charleston had sent

clay samples to London in the same hope that motivated the American Philosophical Society later. In one shipment he sent specimens of seven different strata of Savannah Bluff, asking his correspondent to show them to the people who made English china in order to get their opinion of their potential.[23]

In other areas, however, the interests were in direct conflict. Iron production is an example. This, too, had been welcomed by the British to a degree. A blast furnace had been built by the Virginia Company as early as 1619 at Falling Creek, Virgina. Its failure is apparently directly attributable to the Indian massacre of 1622.[24] Its prospects, however, had never been good—the mechanics of smelting and rolling are such that it is not economical to produce on a small scale. On the other hand, more ambitious Colonial efforts at iron manufacture ran into direct conflict with English mercantilist doctrine. The imperial authorities had encouraged production of pig and bar iron, but had tried to prevent the erection of plants for finished iron products. When Jared Eliot, for example, found a way to extract iron from "black sea sand," his discovery was welcomed with a gold medal by the London Society of Arts; even though the method later proved to be of practically no importance, it was aimed at the kind of activity thought to be appropriate for Colonials. But when Connecticut granted legislative subsidies to sustain steel production, that was quite another matter, for it conflicted with English manufacturing interests. The same cross-purposes between British and American needs that earlier had plagued the efforts of John Winthrop, Jr., to establish an iron industry also debilitated later efforts.

The typical European belief in El Dorado—the easy accessibility of precious minerals—had led early colonizing expeditions to Virginia invariably to include a "mineral man" in the hope that he would discover readily exploitable wealth. Likewise, the founders of the Bay Colony, while still in England, had ordered their advance agents to be on the lookout for mines of any sort. More than a century later, the same hope moved John Bartram to propose a geological survey of the Colonies which, by borings, would uncover the mineral wealth of the country.[25] Still later, in the nineteenth century, state legislators put Bartram's program into effect and were bitterly disappointed when no easy wealth resulted. The belief in El Dorado died hard.

Unfortunately for Old World expectations, both Virginia and Massachusetts turned out to be notably deficient in precious minerals of all kinds. Such industries that were successful had far more humble origins and supplied more mundane needs. A lime industry, for example, prospered by utilizing the huge beds of fossilized oyster shells near Jamestown. Also enjoying modest success at several locations were brick-making and textile technology, the latter based on the importation of skilled workers bringing knowledge of the latest English techniques with them. In all of these and other areas where relative success was attained, small-scale production for local consumption was both desirable and economically feasible.

The war interrupted scientific activity and the conclusion of the war gave it new direction. Despite the common belief that "the sciences were never at war" and despite the best efforts of military commanders on both sides to implement this belief, the arts and sciences tended to be silenced by the hostile clash of arms. With the outbreak of hostilities, for example, the project to establish a Colonial observatory in Philadelphia, with David Rittenhouse public astronomer by royal appointment, came to naught. Rittenhouse himself was deeply involved in political matters, and did not return to science, and then only temporarily, until eight days after the British evacuated Philadelphia, when he observed an eclipse of the sun on June 24, 1778. Benjamin Franklin was deeply immersed in political matters; John Winthrop spent much of his time inspecting powder mills, advising on the production of sulphur, and trying to stimulate saltpeter production. The note that Provost Smith penned in his copy of the first volume of the Philosophical Society *Transactions* aptly summarized the period:

> Astronomical observations, 1776: This year exhibiting little else but scenes of confusion and distress amidst the calamities of an unhappy war, scarce any attention was paid by members of the Philosophical Society to astronomical or any other literary subjects.[26]

Indeed, not until after the British withdrawal from the city in 1778 were meetings of the society resumed; and attendance did not climb back to normal until 1785.

Quite as serious as the diversion of thought and energy from scientific questions was the interruption of communication with England that the war entailed. After 1775, the flow of scientific communications from England to America dried to the merest trickle, even though individuals made heroic efforts to preserve it. Franklin's diplomatic post at Paris did permit him to serve as one continuing link between American and English science, relaying letters from Americans such as John Winthrop at Cambridge to English friends. The circuitous and difficult route, however, was no real substitute for the direct flow of peacetime traffic. Not even the new intellectual contacts with France quite compensated for the loss. One of the severest penalties paid during the war by the medical profession, Dr. Rush told his old teacher Cullen on the re-establishment of peace, "was occasioned by the want of a regular supply of books from Europe, by which means we are eight years behind you in everything."[27]

On the home front, the habit of dependence upon England had left the Colonists poorly equipped for fighting a war on their own. Despite the boasts of technological self-sufficiency of people like Tom Paine, the industrial resources of the Colonies were clearly inadequate. Chemical technology was at a low level. Practically no gunpowder was being made in America, and little was known of the manufacture of saltpeter, an essential ingredient in gunpowder. The country lacked trained engineers for constructing defense bulwarks and encampments; transportation facilities for troops and supplies were inadequate. Moreover, the civilian needs of six million dollars' worth of commodities annually imported from England had to be satisfied somehow. The efforts of the philosophers were absorbed in these prosaic matters; their eventual success is a matter of record, but science neither supplied much help nor profited from the diversion.

On the whole, the philosophers did not make a very good record in their efforts to apply science to the betterment of the human condition. It was not really the fault of the well-intentioned scientists—except perhaps for their naïve optimism that made the failure seem all the worse.

Political considerations often interfered. Inter-Colonial rivalries and jealousies often wrecked planned projects; imperial restrictions and efforts to promote impossible projects, such as silk culture, often interfered. From the time of Raleigh's colonization scheme at the end of the sixteenth century, to the Revolutionary period, the New World had been viewed as a source of imperial necessities. It was to free England from the exorbitant prices it had to pay to the Baltic states for potash and naval stores, and from dependence on the faraway East for essential dyes, spices, and niter. England, needing desperately to become self-sufficient in these commodities, tended to overemphasize them at the expense of what the Colonists deemed their own best interest. Colonials, for their part, all too often tried to undertake projects that were not feasible in an unsettled society lacking adequate transportation and marketing facilities. Above all, they displayed the common frontier exaggeration of the efficacy of legislation in creating what they desired in the face of an inhospitable environment.

It is a truism that a modern industry of the type being advocated by the eighteenth-century philosophers can flourish only where its products can be marketed. This implies a relatively large population with purchasing power and good transportation. On this truism had foundered schemes ranging from John Winthrop, Jr.'s, seventeenth-century ironworks and graphite mine through Baron Hasenclever's large-scale enterprise in 1762. Despite the fact that Hasenclever had made notable technological innovations in his ironworks, he was sufficiently chastened by this and a later experience in textile manufacture to write in 1768 that "the present zeal to establish manufactures is premature" because labor was too expensive and there was too much land to be settled.[28]

One also observes a lack of an integrated program of activity, reflecting the absence of a true sense of inter-colonial solidarity that has been often noted in other areas. The failure of a scheme to unite the Delaware and the Chesapeake is symptomatic. The American Philosophical Society appointed a committee to survey several possible routes, made its recommendations, and published a summary, in Volume I of the *Transactions*, containing a map of the routes surveyed. "Public spirited" merchants in Philadelphia contributed £200 to further the plan, and a great deal of support was gathered in that city during 1769–70. Marylanders, however, were singularly unenthusiastic, for it seemed obvious to them that the purpose of the plan was to take trade from Maryland to Philadelphia. After all, Philadelphia, so the American Philosophical Society had said, would gain by drawing to itself "the Produce of the rich and growing settlement on the Susquehanna and its Branches." The Philadelphia merchants, so Marylanders reasoned, had not acted entirely out of disinterested "public spirit." This kind of jealousy was a major cause of the failure of canal building plans until the 1820's, when the first successful one of any length was constructed entirely within the limits of a single state with liberal state financing.

In manufacturing, the low state both of scientific knowledge and of manufacturing made the philosophers' exuberant hopes unrealistic. The abundance of land and the high cost of labor in America—exactly opposite from the conditions that obtained in England—made hopes of reforming agriculture along the British model particularly chimerical. Success, when it did come, would be attained by a wholly different breed of men in both areas; and it would come by operating within the framework of peculiarly American opportunities and limitations. The eighteenth-century philosophers, by trying to force historical development into a rationally determined pattern of their own, only achieved frustration. The environment was against them, and the times were against them—in marked contrast to their efforts in the realm of politics.

VI

Science in a Free Society

ℭ Even though there was wide belief in a "Republic of Science" —which transcended national ties and was immune to such things as wars, diplomacy, and other disruptions in the affairs of nations—the breaking of colonial ties with England was a shattering event in the life of science in the United States. Formal institutional supports such as those provided by the Anglican Church, the Crown, and governmentally subsidized societies were dissolved, and no new ones appeared immediately to take their place.

Compounding the inevitable losses was the popular cultural nationalism of the time, which scientists shared with their fellow-citizens. Going abroad to get an education or sending a scientific report to a society in London did not seem at all undesirable when it could be regarded as "within the Empire." Americans, after all, a short time before the war had been loyal British subjects, and they could feel a sense of shared pride in the Royal Society, the London Society of Arts, and other proud symbols of Britain's scientific eminence. They could turn, without a feeling of inferiority, to English instrument makers, libraries, museums, herbaria, publication opportunities, and patronage—all indispensable rarities in America. After independence, all the assumptions changed; what was previously a source of imperial pride became at once a symbol of national shame. What was once provided by others, American scientists found they must provide for themselves.[1]

There was abundant faith that such problems would be

easily and promptly solved by the new republic. As the republican form of government was best in politics, so it was in science. Americans who had come out of the Revolution with an intense sense of nationality felt it was their solemn duty to illustrate the superiority of their polity in every respect. To fail in this sacred obligation would be to betray the idealism of the Founding Fathers, who had fought and died for the right to establish a free society. Joseph Priestley, arriving from England in 1794, wrote that the United States government by "encouraging all kinds of talents is far more favorable toward the sciences and the arts than any monarchial government has ever been. A free people will in due time produce anything useful to mankind." There could be no question about the ultimate success of science in America; as one journal said in 1812: "Standing, as it were upon the shoulders of our transatlantic rivals, we may hope to catch new views of the prospect before us, which will enable us to shorten the road to ultimate perfection."[2] Joel Barlow, issuing at Washington in 1806 a Prospectus for a national institution, gave voice to the great Jeffersonian dream then being shared by so many. "What a range," he exclaimed, "is open in this country for mineralogy and botany!" And the same was true for the other sciences; the prospect was overwhelming in its grandeur:

> Could the genius of Bacon place itself on the high ground of all the sciences in their present state of advancement, and marshal them before him in so great a country as this, and under a government like ours, he would point out their objects, foretell their successes, and move them on their march, in a manner that should animate their votaries and greatly accelerate their progress.[3]

Even while the Revolution was under way, orators and writers emerged in every state to expound upon the theme that the freedom and liberty associated with independence would be beneficial to the growth of science. Dr. David Ramsey, Charleston patriot, speaking on the "Advantages of American Independence" on the second anniversary of the Declaration, found numerous reasons why independence would encourage the growth of science. The sciences, he said, "require a fresh soil, and always flourish most in new countries. . . . Large empires are less favourable to true philosophy, than small independent states." And, "May we not hope," Ramsey asked in his emotion-laden peroration, "as soon as this contest is ended, that the exalted spirits of our politicians and warriors will engage in the enlargement of public happiness, by cultivating the arts of peace, and promoting useful knowledge, with an ardor equal to that which first roused them to bleed in the cause of liberty and their country?" He predicted that the "arts and sciences, which languished under the low prospects of subjection, will . . . raise their drooping heads. . . . Even now, amidst the tumults of war, literary institutions are forming all over the continent, which must light up such a blaze of knowledge, as cannot fail to burn, and catch, and spread, until it has finally illuminated with the rays of science the most distant retreats of ignorance and barbarity."[4]

After the Revolution, the faith remained undiminished. The Reverend Dr. Samuel Cooper, of Massachusetts, insisted that as the arts and sciences "delight in liberty, they are particularly friendly to free States." John Gardiner, son of a Loyalist, pointed out even more explicitly that the "introduction and progress of *freedom* have generally attended the introduction and progress of *letters* and *science*. In despotick governments the people are mostly illiterate, rude, and uncivilized; but in states where CIVIL LIBERTY hath been cherished, the human mind hath generally proceeded in improvement,—learning and knowledge have prevailed, and the arts and sciences have flourished."[5]

Despite a faith in the successful outcome, the obstacles at the beginning were indeed formidable. Operating previously within the colonial framework, it had not been necessary for Americans to think seriously about the institutional apparatus of science. The English connection had provided it as a matter of course. The American scientist had gone to England for his education, looked to England for direction and for publication outlets, and typically sent the specimens he collected to England for analysis and, in many cases, for naming and classifying. Immediately after the Revolution, Americans found themselves unprepared to do these jobs. The basic instruments needed for research were in many cases unobtainable in America, for in the absence of an indigenous scientific culture, the supporting technicians, instrument makers, and other "footsoldiers in the scientific army" had failed to emerge.

Even those European-trained scientists who took over the housekeeping tasks for the first generation of Americans found that inadequate facilities severely restricted them. For example, John MacLean, a Glasgow-trained natural philosopher who taught chemistry and natural history at Princeton in 1795, announced that he would also lecture on comparative anatomy if a class could be obtained to encourage the necessary effort. Whether or not he actually taught the course that year is not known, but at any rate the cabinet of specimens that he had available was graphically, if somewhat disparagingly, described by Isaac Weld, who visited Princeton while MacLean was there. "[There] are two small cupboards, which are shewn as the museum," Weld reported. "These contain a couple of small stuffed aligators, and a few singular fishes, in a miserable state of preservation, the skins of them being tattered in innumerable places, from their being repeatedly tossed about."[6] What kind of course he could have taught with these deplorable demonstration specimens is a matter for the imagination; nevertheless, virtually every American science professor found himself in a similar situation. So limited was the equipment available to American colleges, and so difficult was it to replace lost or broken apparatus, that professors tended to guard jealously what they had. The idea of allowing students free use of laboratory materials could not be entertained seriously. An agreement, drawn up in 1816, between J. F. Dana and Professor John Gorham of Harvard is illustrative of the value placed on the scarce supplies. Dana, as laboratory assistant to the professor, was to prepare and demonstrate the experiments, then to clean and dry the glassware and return it to the shelves. In return, Dana was to receive a salary of $400 per year and be allowed the use for his private experiments of "such vessels, materials, and instruments as, if broken or exhausted, may be replaced from Boston." But even he was not allowed to use "the more valuable apparatus, tests, etc., which can be procured only in Europe."

Amos Eaton, who wandered through the New England states and New York between 1817 and 1824 lecturing on several branches of science to all who would pay a small fee to listen, has left a vivid description of his chemical apparatus in a letter to John Torrey:

> A pewter sucking bottle is my fluoric gas bottle, a stone pig and a tin tube my earthen retort, a teakettle with the cover luted on is my iron retort, etc. I have a complete pneumatic cistern with several of my own improvements. My glass retorts etc. are also regular. But much of my apparatus is sui generis . . . I illustrate the most abstruse parts by a dish kettle, a warming pan, a bread-tray, a teapot, a soap bowl or a cheese press.[7]

An extraordinarily gifted teacher, as Eaton certainly was, could accomplish a great deal with such crude apparatus, but it must have imposed severe limitations even upon him.

The situation was no better in the natural history disciplines, where one would have expected Americans to be better provided. At the turn of the century, Charles Wilson Peale's museum in Philadelphia, conducted as a private enterprise, was the only large collection of American natural history specimens available to students, and much of the material on display was not properly classified for scientific purposes. Not until 1821 did the museum take on a scientific staff and take itself seriously as an educational institution. In 1805, the entire Yale collection of minerals—and that recently acquired—was carried in a single candle box to Philadelphia for classification. Harvard's collection, although somewhat larger, had only recently been brought together by gifts from the French Republic and from Dr. John Coakley Lettsom of London. Early in the century, according to Benjamin Silliman—first professor of chemistry, mineralogy, and natural history at Yale—it was difficult to obtain laboratory identifications of even the most common minerals, such as quartz, feldspar, or hornblende, or even granite, prophyry, and trap among the rocks.[8] With such limitations, only the most common specimens could even be examined by students. After 1810, the situation at Yale was substantially improved when Colonel George Gibbes, a wealthy resident of Newport, was persuaded to place on public exhibition at the college the extensive collection he had made in Europe. The collection, numbering about 10,000 specimens, remained as a loan exhibit until 1825, when Silliman managed to raise $20,000 by public subscription to have the college purchase it.

Even as late as 1816, when Parker Cleaveland wrote his *Elementary Treatise on Mineralogy and Geology,* American sources were severely limited, and Cleaveland found that he had to rely on English, French, and German writers who had no experience in American mineralogy. The only specifically American materials he could find were a few issues of Archibald Bruce's *Mineralogical Journal,* a single paper in the *Memoirs* of the American Academy of Arts and Sciences, and another in the *Medical Museum.*[9] Excellent as Cleaveland's *Treatise* was, American students could learn little of the mineralogy of their own country from it.

Neither did the American botanist have readily accessible botanical gardens or herbaria to assist him even in the simplest taxonomic matters. Harvard's botanical garden, the first in New England, was not established until 1805. Its herbarium remained insignificant until Asa Gray began building it up in the 1840's. The most extensive American herbarium was obtained, oddly enough, by the New-York Historical Society when David Hosack placed his own collection, including duplicates he had been given in England by James Edward Smith from the Linnaean collection, on deposit in one of the organization's rooms in 1817.[10] These, along with private collections in Philadelphia, that of John Torrey in New York, and a few other scattered collections in private hands were all that American botanists had to work with.

While it may seem ironic that, through the first two decades of the nineteenth century, better collections of American specimens were to be found in England than in the country of their origin, the fact is easily explained by the former relationship between American collectors and their European mentors. Specimens were characteristically sent to Europe as rapidly as they could be collected, where, in accordance with the Baconian-inspired program, they were to be assimilated with collections coming from other parts of the world. The need to keep specimens of what was readily available in America rarely occurred to those scientists working under the old assumptions. Consequently, during the first few decades of independence efforts were made to keep up the old correspondence with Europeans, and a *de facto* colonial status persisted. As the Boston physician and botanist Jacob Bigelow was to explain to James Edward Smith in 1816, when he sought to open a correspondence:

> In the present imperfect state of science in this country, I am every moment sensible of the necessity of an European correspondence, to supply the defect of books and advantages which it is impossible here to command. . . . In the New England states we have a country as yet little explored by European botanists, and having in some degree a botany peculiar to itself. Many of our species resemble those of Europe, but at the same time differ in small particulars, so as to leave it doubtful whether they are species or varieties . . . On these plants it is difficult to decide without a comparison of actual specimens.[11]

The Reverend Henry Muhlenberg of Lancaster, Pennsylvania, although situated only sixty miles away from the gardens and libraries of Philadelphia, nevertheless felt the need for access to the herbaria of Europe. In the midst of his work on a catalogue of American plants, he appealed to James Edward Smith, explaining that he needed "a friend, who would kindly assist me to find out which plants are already described by Linnaeus, and which are nondescripts." Some of his doubts, Muhlenberg explained, had been cleared up by his friend Dr. Schreber, the editor of the eighth edition of the *Genera Plantarum*—"but very many remain."[12] Smith, the owner of the Linnaean herbarium, replied promptly and an extensive correspondence resulted, presumably saving Muhlenberg a great deal of useless duplication of effort. This kind of exchange was welcomed by most British naturalists, for it could still be as valuable to them as to their American partners. Even though primary interest in England was shifting to those areas of the world within the Empire, there remained a constant need for information about the new discoveries made in America.

Thus, the familiar working relationships between American scientists and those in the mother country continued long after formal independence had been achieved. In addition to the continuing transatlantic relations, American science profited from the fact that a small group of British scientists who had remained friends of America decided to transplant themselves to the United States at the end of the war. Among these were John Vaughn, a promoter of scientific inquiry who became a constant support of the Philadelphia community; the Reverend Charles Nisbet, who brought the learning of the Scottish Enlightenment to the newly founded Dickinson College; and Walter Minto, who brought the study of advanced mathematics to the College of New Jersey.

Despite the development of medical schools at the College of Philadelphia, at Kings College in New York, and at other places, American physicians still continued to seek training abroad. It was not merely snobbery that accounted for the superior distinction accorded by a European degree—the leaders of the medical world were at European institutions and it would be many years before an equivalent education could be obtained in America. Samuel L. Mitchill in 1802 found cause for pride in the fact that among the graduates of Edinburgh that year "there was not a single one from the United States." To Mitchill, this constituted positive evidence that American medical schools had come into their own. During the winter of 1805–6, however, Benjamin Silliman found more than thirty Americans studying at Edinburgh.[13] It would be many years before Mitchill's fond hope became a reality.

Even though the old state of dependence persisted long into the nineteenth century, it was no longer accepted gracefully; it was, in fact, considered by a growing number of native scientists to be a blemish upon the good name of the republic. The new spirit of nationalism after the Revolution affected every branch of the arts and sciences, and it reinvigorated the philosophers in their old hopes to make science truly useful. Nicholas Collin argued that it was the duty of philosophers "to cultivate with peculiar attention those parts of science, which are most beneficial to that country in which Providence has appointed their earthly stations," and Benjamin Rush expected much learning and many improvements to flow from the "harmony between the sciences and government." Nationalism and local utility joined to reject technical and classical terms in science as "the remaining pendantic formality and fustian of dark antiquity, ill suiting the plain masculine independent genius of enlightened Americans." After 1790, medical theses at the College of Philadelphia might be written either in Latin, as hitherto, or in "the language of the United States," the latter being defended by Rush as the likeliest means of diffusing medical knowledge.[14]

Even some foreigners joined the chorus of voices calling for a more concerted effort by Americans to describe their own country scientifically. Writing to Dr. Rush in 1784, Dr. John Coakley Lettsom, London physician and warm friend of American science, expressed the hope that America would now cultivate the arts of peace and he promised that this new land had "such an extent of novelty in . . . Soil, produce, improvement &c. as must stimulate genius, and gratify invention and study." The following year, he insisted: "Set your men of Science upon studying your own Country, its native and improvable productions —Your resources would influence Europe—your reflections would instruct her."[15] Friendly Europeans also continued to offer aid or encouragement to the emerging institutional framework of American science.[16]

In the minds of Americans one can detect the growth of an anxious concern that the discovery and description of the natural-history wealth of America should be made known to the world by Americans, publishing preferably on American presses. According to the editors of the *New England Journal of Medicine and Surgery,* all "true Americans" even considered yellow fever to be national property, and they resented European attempts to study the disease.[17] American scientists also grew sensitive about the natural-history wealth of North America being exploited by foreigners. As the entomologist T. W. Harris explained to John LeConte in 1830:

> . . . I expect to send to Europe, in a week or two, 1200 Coleoptera, among which some of the same insects are included, and *must not go without their names.* It is sometime since I formed and have adhered to the resolution not to send out of the country an American insect without a specific name, nor one which had not been described, except a very few which I was about to publish myself. It gives me pleasure to find this resolution adopted by several other American gentlemen.

Their reason for adopting this resolution, Harris said, was that the study of American insects was made inordinately difficult by having to consult costly, and sometimes inaccessible, books in "all the languages of Europe."[18]

Despite the best efforts of Americans, however, they were not able, during the first half-century of independence, to carry out their program. Time after time, for example, important works on American botany were published in Europe. Even presidential initiative sometimes did not suffice. Upon the return of the Lewis and Clarke expedition, Jefferson had its collected plant specimens placed under close guard in Philadelphia so that Lewis might have the credit of publishing their descriptions. But Lewis died; a visiting German botanist, Frederick Pursh, was finally hired to process the collection, and ultimately the description was published in England, where Pursh had access to the specimens collected over a period of two centuries.[19] Again, the first systematic works on American botany were published in Paris by the Michaux, father and son. Dr. Schoepf, of Erlangen, had published the most important eighteenth-century work on American *materia medica* in German in 1787. And so it went, through the list of sciences. Europeans with their superior facilities and training, time after time reaped the honor and glory, while American naturalists could only protest, formulate plans, and call upon their countrymen for aid.

But where achievement was lacking, scientists substituted hopes and fervent appeals to national pride. Science was recommended to all who would hear as the highest example of useful knowledge. It was widely proclaimed that it would not only enrich and elevate the individual, but would make the nation great in both peace and war and advance mankind in general on the road to that happiness for which it was destined.

The cultural patriotism invoked in connection with the utilitarian appeal actually antedated the forming of the nation. It seems to have grown during the colonial period along with the general feeling of separateness that was being developed. Many examples are in particular evidence following the deterioration of Empire–Colonial relations during the mid-1760's. The reinvigoration of the American Philosophical Society in 1769 and the support given to the Transit of Venus observations during the following year were in part a product of this desire to assert American cultural equality. A particularly forthright statement was that made by Ezra Stiles, in the year of the Stamp Act, in connection with his effort to organize an American Academy of Science. In defining qualifications for membership, Stiles said that in addition to scientific eminence:

> They shall ever have Two Thirds of the Associates Presbyterians or Congregationalists. The number is thus limited to defeat episcopal intrigue by which this Institution would be surreptitiously caught into an anti-American Interest. It being designed that this shall be for the Honor of American Literature, contemned by Europeans, Therefore let the Associates be all Americans & if born in America, & be 21 year, it shall be indifferent whether of English, Scotch, Irish, French or German Blood all these distinctions being lost in American Birth.

Here was patriotism, religious orthodoxy, the promotion of science and the melting pot concept all wrapped in one neat package! Despite Stiles's zeal, however, and despite another effort the following year, in which he actually appointed the officers and drew up a membership list, nothing came of his plan until after the Revolution.* It was indicative, however, of a growing movement toward independence in all areas.

---

*Statements of August 15, 1765, and November 24, 1776, Library of Congress. The feeling that they were "contemned by Europeans" was one shared by Latin American colonials; e.g., "In some parts of Europe, especially in the north, through being more remote, they

After the Revolution, scientists continued their practical appeals, except that now they were even more careful to combine them with the growing nationalistic interest. "What a field have we at our doors to signalize ourselves in," Thomas Jefferson exclaimed in a letter to Joseph Willard in 1789. "The botany of America is far from exhausted, its mineralogy is untouched, and its natural history or Zoology totally mistaken or represented." Provost Smith, of the University of Pennsylvania, while announcing the professorship of natural history and botany in its Medical Department, promised that "the philosophical parts of Agriculture, as they regard a science peculiarly interesting to these United States, will be . . . attended to."[20] "The man who will discover a method of preventing the fly from destroying turnips," the public was assured, "or who will point out a new and profitable article of agriculture and commerce, will deserve more from his fellow citizens and from heaven, than all the Latin and Greek scholars, or all the teachers of technical learning, that ever existed, in any age or country."[21] Humphrey Marshall, Bartram's successor as the leading American seedsman, accepted this estimation of values. In the preface of his *American Grove* (1785) he justified the study of botany on the ground that the importation and cultivation of foreign plants or the discovery of native productions of general usefulness would relieve the United States of dependence on foreign nations. Tea might be grown in the southern states, rhubarb deserved to be cultivated, while the discovery of a plant as useful as tobacco or the potato or of a substitute for coffee or quinine would be of inestimable advantage. "The Science of Botany," Marshall went on in a paper which he read to the Philadelphia Society for Promoting Agriculture in 1786, "certainly holds its most dignified station when subservient to Medicine; but its utility does not terminate in this alone, though it has too long

---

think that not only the original Indian inhabitants of these countries [New World] but also those of us who were, by chance, born in them of Spanish parents either walk on two legs by divine dispensation or that, even by making use of English microscopes, they are hardly able to discover anything rational in us." Sigüenza, in his *Libra Astronomica* (1690), as quoted in Irving A. Leonard: *Don Carlos de Sigüenza y Góngora, A Mexican Savant of the Seventeenth Century* (Berkeley, Cal., 1929), pp. 62-3. Nearly one hundred years later, Sr. Velásques de León commented: "The humility, fear, and difficulty which the Spanish Mexicans regularly have in producing their ideas is great, and much greater is the preoccupation of the Europeans with our barbarism. Why should they seek data from men whom they still visualized with bows and feather plumage as they depict us on their maps." Ibid., p. 63.

been considered as having no other connexion."[22] And he then showed the importance of botanical science to agriculture as it was practiced in the United States.

In medicine the new nationalism of the postwar period encouraged a great deal of thought and speculation. Invoking the widely held belief, which dates back to earliest Colonial times, that for every disease peculiar to America there must be an American remedy, the search for native drugs was undertaken with great fervor. As early as 1789 the College of Physicians of Philadelphia circularized "the most respectable characters in the United States," asking their opinion of the need for an American dispensatory and particularly requesting information of all American remedies discovered anywhere in the country. "Who knows," asked Benjamin Rush, "but that, at the foot of the Allegany mountain there blooms a flower, that is an infallible cure for the epilepsy?" It might be reserved for America, he thought, to furnish cures for many diseases which then eluded the power of medicine. Rush at one point allowed nationalistic ardor so much to overcome his medical judgment that he was accused of refusing to prescribe anything *except* American drugs for his patients.[23]

The persistence of the belief in a wisely balanced nature is indicated by a review published in the *American Medical Recorder* in 1818. The reviewer hailed Nathaniel Chapman's *Discourses on the Elements of Therapeutics and Materia Medica*, published the previous year, in the following terms:

> Living in a country, as we do, of widely extended territory—
> of various soil and climate, we cannot but believe, that there
> are many of our indigenous products whose medicinal virtues
> are of real efficacy. Many of our native plants have already
> been found to possess the most valuable sanative qualities.
> We have our indigenous cathartics, emetics, tonics, astrin-
> gents, diuretics, diaphoretics, and in short, one or more articles
> in almost every class of remedies. It is time, then, that they
> should take their proper station in the body of some classical
> book on the Materia Medica.

Chapman's introduction of many of our own medicinal articles into his book, the reviewer explained, was "a circumstance which must make it the more valuable to the Medical gentlemen of the United States." Essentially the same point was made by the reviewer of W. P. C. Barton's *Vegetable Materia Medica of the United States*, another 1817 publication, in the same volume of the *Medical Recorder*.

Moreover, direct connections were often suggested between

the practice of medicine and the political system of the United States. Dentistry was the republican art; it must be especially attended to in America, since premature loss of teeth was "a particular disadvantage in a great republic, where so many citizens are public speakers." At the end of the Revolution, Jean-François Coste, a physician with the French forces in America, told students at the College of William and Mary that since the United States was now independent, medicine in the country must forever be free from subjection to authority, even that of the most celebrated masters.[24] Rush and others argued the influence of political forms and events on the public health, now adducing evidence that republicanism and health are concomitant blessings; now, that the strain of self-government produces various mental derangements. Rush's new theory of therapeutics, with its emphasis on bleeding and heavy dosages, owed something to the prevailing nationalistic spirit. Diseases in America, Rush argued, were of a higher order than in Europe; in this strong, fresh new America sick men were sicker and needed more heroic treatment than did the enervated subjects of the worn-out monarchies of Europe.[25] Many years after the war, Rush declared that it had been the Revolution that led him to discard his old medical system and try to build an American system. Medicine, like government, had now been emancipated from the tyranny of special and esoteric interest and restored to the people. Toward that end, Rush's republican system of medicine was based upon just a few, easily comprehensible "essential principles," which could easily be taught to any American.

The simplicity in pathology and treatment that Rush preached derived from his belief that there was only one source of disease: a convulsive tension and action in the blood vessels, which was to be reduced by the debilitating use of purging and blood letting. The simplicity, Rush believed, was a truth perfectly related to the truth of republicanism, and like it the product of centuries of historical development. History over the years had evolved republicanism and along with it a new system of medicine accommodated to republican Americans and their environment. An American medical education was, therefore, the best preparation that an American physician could have, and all instruction should be in the language of the country; Rush's own University of Pennsylvania medical school was, of course, to be the revolutionary center of training for these new physicians.[26]

· · ·

After the close of the War of 1812, which was widely regarded as the final achievement of American independence, the patriotic appeals became even more direct. One of the major results of that war, according to New York zoologist James E. DeKay, was that a "proper feeling of nationality" had been widely diffused among American naturalists. This feeling had "impelled them to study and examine for themselves," he continued, "instead of blindly using the eyes of foreign naturalists or bowing implicitly to the decisions of a foreign bar of criticism."

Another writer, celebrating the natural history exploration of the American continent since the signing of the Treaty of Ghent—which he also regarded as the key event in recent history —said that one of America's chief national blessings which had resulted from the peaceful condition since that time was "that we have been enabled and induced to turn our thoughts *inward*." The spirit of investigation which had developed during that period should be pleasing to the political economist, the moralist, and the patriot alike. "Whether we look to its effects on the wealth or on the happiness of our community, we are sure that to culti-vate and to cherish it must be regarded as a sacred duty."[27]

The Philadelphia Journal of the Medical and Physical Sciences was founded by Nathaniel Chapman in 1820 as an effort to answer the attack on American culture and science made by Sidney Smith in the *Edinburgh Review*. Smith had written a gen-eral indictment of American civilization—of its art, literature, and drama, as well as of its science—but Chapman was particu-larly incensed by the question "What does the world yet owe to American Physicians or Surgeons?" This quotation was carried on the title page of Chapman's journal for almost five years as a deliberate method of urging upon Americans the patriotic duty of supporting an indigenous scientific culture. In the prospectus published in the first number, Chapman pointed out that, in regard to medicine, Smith's charge was as unjust as it was in-sulting, for the current medical literature of Europe was describ-ing doctrines and modes of practice that had been adopted there after long usage in America. As a matter of editorial policy, Chapman declared that his leading aims would be

> To trace the progress of medicine in the United States, to vindicate our claims to certain improvements, to preserve these, as well as what may hereafter be done, from foreign usurpation, and lastly, to evalue, and stimulate the genius of the country to invigorated efforts.

Chapman's cultural patriotism was also excited by the appointment in 1824 of an Englishman (Robley Dunglison) to the chair of the Institutes and Practice of Medicine at the University of Virginia. Although he was ready to concede that in many instances a man might be imported who was at that moment better qualified than any available American, Chapman saw no reason either for despair or for the wholesale importation of foreigners. There was no doubt, he thought, that the apparent superiority of foreigners was merely a temporary condition arising out of the better opportunities for leisure in Europe. If an American scientist were given such a position, where he would have the opportunity to develop his native talents, he said, echoing a question asked many times by his contemporaries, "shall we not soon make a man superior to any one who could be brought from abroad?" Chapman found it hard to believe that Thomas Jefferson, the father of the University of Virginia, could have had anything to do with Dunglison's appointment, for it was inconceivable that he would have chosen a foreigner in preference to providing opportunity for an American. In fact, according to Chapman, a favorite aphorism of Jefferson was that "the failure of almost all the great scientific or literary undertakings of Americans, is to be attributed to their employment of foreigners, instead of calling into exercise the talents of their own citizens."[28]

Views such as these were expressed by scores of scientists and writers of reviews in the early part of the nineteenth century. J. G. Cogswell, reviewing John W. Webster's geological studies of St. Michael and the Azores, had his pleasure increased by the fact that it was the production of "an author of our own." His national pride, he said, was gratified in seeing such an encroachment upon the domains of the Old World, although he did regret that Webster had been forced to go abroad to get his advanced education. Cogswell, who had himself studied in England and Germany, considered the necessity for going abroad to get an education not only belittling to the homeland but positively dangerous from the national viewpoint. "Genuine, glowing patriotism proceeds in part from prejudices in favor of one's country," he warned, "and these prejudices are very liable to be scattered by the comparisons, which travellers are sometimes obliged to make in foreign lands, to the disadvantage of their own." Even though Cogswell grudgingly admitted that European science was more highly developed than American, and he did inject the warning note, the dominant tone of his article was still one of pride and high hope for the future.[29]

Whenever a reason was given for the backward state of American science, a major role was assigned to the necessity of working for a living in a democratic society and the lack of either private or public patronage for scientific research. The Federal Government, in particular, came in for criticism because of its niggardliness, and patriotic appeals were often linked with invidious comparisons of American and British governmental support of science—no doubt as an effort to awaken a sense of shame in the public mind. It was "mortifying," said John D. Godman, well-known physiologist and propagandist for science, for an American to witness the "magnanimous policy" of England toward its scientists, and then to contemplate the pitiful contrast exhibited by the United States Government. Britain, Godman warned, was even then sending its explorers to the borders of the United States, while the American government was sitting idly by and taking little interest in the work of its scientists. This "magnanimous policy" of England was also an example of enlightened self-interest, Godman was careful to point out, for it was contributing more than anything else to the wealth and strength of the nation.

Godman was particularly incensed at Congressional efforts at economy in regard to the second expedition of Major Long to the Rocky Mountains. The expedition had been poorly financed in the first place, he objected, and adequate publication of the findings had not been provided. Such a policy was both a waste of scientific resources and a national disgrace.[30]

Such invidious comparisons of American with British support of science remained quite common throughout the early nineteenth century. In 1840 an anonymous reviewer in the *New York Review* compared the fortunes of the American W. C. Redfield, who had provided the best support to date for the rotary theory of storms, and his British follower, Colonel Reid. In the first place, the reviewer maintained, it seemed unjust that the author of such an important theory should be compelled to print his researches at his own expense, or else be confined to the limited space available in the monthly miscellanies. Colonel Reid, who was at the most a "faithful and intelligent follower of Redfield," had been able to publish all his work in "the most splendid style" by the aid of his government. But there was yet another way, the reviewer pointed out, in which the British government had demonstrated its high regard for Reid's scientific ability. While Redfield's labors had brought him neither reward nor appreciation from his government, Colonel Reid had been appointed to a sinecure in Bermuda, an ideal place to pursue his researches. Such a contrast, the reviewer said indignantly, was not "creditable to us as a nation."[31]

In addition to pointing out that Redfield had personally been treated unfairly, the reviewer attempted to appeal to the sense of national pride by suggesting that such treatment threatened to discredit America in the eyes of the world—to deprive it of its rightful place as a great nation. Such an effort to shame Americans into more active support of scientific research was evident in every field of science, but it took a particularly prominent place in the continuing pleas for a national observatory. American astronomical work had typically been confined to the limited facilities available in a few universities and small private observatories, most of which had been hastily erected as temporary quarters for the observation of some particular astronomical event—an eclipse or a transit of Venus or Mercury—or as temporary stations used in the determination of boundaries. Having served their purpose, they were generally abandoned. But by the 1820's most scientists understood that the time had passed when either isolated individuals or sporadic group efforts could contribute significantly to astronomical knowledge. Because of the need for ever-larger and more complicated instruments, and for large-scale coordination of efforts on a continuing basis, it was thought to be especially fitting for the government to enter this area of science. One writer, illustrating a favorite technique of argument, mentioned the "mortifying fact" that every "petty German principality" had been operating an observatory for years and that there had for some years been one in "the convict land of Botany Bay." Some English critic, he warned, would very soon be using these facts as the basis for another attack on the scientific intelligence of the United States; and he closed with an eloquent plea for "those who guide the councils of the nation, to deliver us from the reproach."[32]

The idea of a declaration of scientific independence from Europe remained the most common plea of those who were urging government support of science throughout the period. One of the earliest and most persistent of the many movements for establishment of a national observatory was connected with the need for a prime meridian in the United States. It was widely held that rather than follow the common custom of reckoning from either Greenwich or Paris as zero degrees in determining longitude, a sovereign nation should extricate itself from the "degrading and unnecessary dependence on a foreign nation" by laying a foundation for fixing its own first meridian.* William Lambert, a government clerk and one of the organizers of the abortive Columbian Institute for the Promotion of Arts and Sciences, concluded that continuous astronomical observations were necessary to provide accurate bases for the required fundamental surveying. During the first decade of the century he launched a campaign to secure governmental support for such a project. In 1827, he was still arguing on the same grounds for an observatory where "we might observe and compute for ourselves . . . and . . . prepare and publish an Astronomical Ephemeris, independent of the aid of European calculations."[33] Speaking on the same subject, a Whig writer in 1845 argued that an astronomical observatory would enable American navigators to keep their course through the seas "in patriotic reliance upon the calculations of their countrymen," rather than upon those of some foreigner who, presumably, could not be trusted with the welfare of American ships.[34]

The triumph of nationalism over the scientific cosmopolitanism of the Enlightenment could not be better illustrated than it was in this article and in the many others like it that the period produced. The idea that "science knows no distinctions of country" was still voiced by American spokesmen for science, but they became more inclined to qualify it by adding that the scholar was most likely to prove a useful citizen of the "great republic of letters and science" by making himself a good citizen of his own country. What was good for America, some went so far as to argue, was good for science; for was not America the destined home of science?[35] Thus spoke Charles H. Davis before his colleagues in the American Association for the Advancement of Science, as he argued the need for an American prime meridian.

It was in this same spirit that Benjamin Silliman, in a "note" to an article on the Coast Survey, said that the suspension of its operations was a great blow to science, but it should also be con-

*American State Papers, Misc., 2 vols. (Washington, D.C., 1834), II, 753. Around the mid-century there was a great deal of discussion of this issue at meetings of the American Association for the Advancement of Science. See, for example, AAAS Proceedings, II (1849), 381–3; IV (1851), 155–7.

sidered a national misfortune. This was so, he continued, not simply because of the loss of previous expenditures or of the delay in practical benefits—both of which he thought significant arguments—but it was especially unfortunate because the principles which Superintendent Hassler used were in advance of the science of Europe at that period and, consequently, their application to practice originally in the United States would have "redounded to the national honor."[36] It made no difference to Silliman that Hassler was a Swiss who sometimes had difficulty with English, that he had brought his own set of standard weights and measures with him from Europe, and that he had been forced to go to England to get his instruments made. The application here, even of borrowed technology, would have been a credit to the government and a confirmation of the vision of the Revolution.

Accompanying this cultural patriotism was an extraordinary, sometimes undoubtedly stultifying, sensitivity to the very possibility of foreign criticism. Thus in 1829, Amos Eaton in a letter to Benjamin Silliman regretted that specialists had not been put in charge of each department of science handled in the *American Journal of Science*. If this had been done, he said, "Pieces as that Dutch translation of last number, which had been fully discussed and passed by for half a century, would not have drawn a sneer upon American science as being just found out."[37] The point becomes even more clear in this 1848 comment made in a letter from James Dwight Dana to S. S. Haldeman:

> It is to be regarded, as a national calamity, we might almost say, that Prof. Peirce [Benjamin Peirce, a Harvard mathematician] should have been so hasty in his conclusions, or rather in his publication of them. In his prominent position, and in a subject of so much general as well as scientific interest— the planet Neptune, it is bad to have to renounce such grand discoveries as error, especially as he had made himself a critic upon European astronomy. He is undoubtedly a man of great ability, but is too "quick upon the trigger."[38]

Edward Everett, who was present at the meeting of the American Academy when Peirce committed his indiscretion by attacking La Verrier's calculations, is reported to have addressed the academy, begging that so utterly improbable a declaration might not go out to the world with the academy's sanction.[39]

As these comments indicate, Americans were protesting too much in declaring their independence of the Old World in matters scientific. The tendency to look to Europe for approval, and the tendency to stand in awe of European scientists, persisted beneath the surface all during this period of national affirmation. It could hardly have been otherwise in a period when virtually all of the common textbooks were simply pirated "American editions" of English works or, with increasing frequency after the first decade of the nineteenth century, translations of French works. Instruction in science necessarily reinforced the idea that important work came from Europe and that Americans were still in a borrowing stage.

American scientists did somewhat better in their efforts to provide the other essentials of science, such as organizations, media, and training facilities. The great hope of American scientists seemed to lie in the societies rather than in more formal educational institutions. These would prove the grand instruments of a democratic culture that would both satisfy provincial ambition and provide recognition of the cosmopolitan fraternity of science. Through them Americans would claim their proper place in the world. Scientific societies, so the American Academy of Arts and Sciences pronounced in its organizational statement of 1785, were to bring together persons ready to supply each other with hints of progress, to publish their discussions, and to excite a spirit of emulation that would enkindle the sparks of genius which otherwise in this America "might forever have been concealed." Through these associations, continued the statement of the academy, "knowledge of various kinds, and greatly useful to mankind, has taken the place of the dry and uninteresting speculations of schoolmen." Through them, "solid learning and philosophy have more increased, than they had done for many centuries before."

The American Philosophical Society, as the first learned society in America, was looked to as the keystone of a grand edifice of science in America. As early as 1780, Jeremy Belknap was looking from New England to Philadelphia for direction. Some means should be devised, he wrote Ebeneezer Hazard, whereby properly qualified persons might engage in scientific activities. The Philosophical Society, he suggested, might do something in this way:

> Or, if there were inferior societies, or boards of correspondents, in the several states, connected with the principal one at Philadelphia, and united in the same views, there might by such means be some valuable things brought to light, which,

if discovered by individuals, are but imperfectly known, or
neglected, or undervalued, or perhaps concealed, so as to be
of no general use. Why may not a *Republic of Letters* be realized in America as well as a Republican Government? Why
may there not be a Congress of Philosophers as well as of
Statesmen? And why may there not be subordinate philosophical bodies connected with a principal one, as well as separate
legislatures, acting in concert by a common assembly? I am
so far an enthusiast in the cause of America as to wish she
may shine Mistress of the Sciences, as well as the Asylum of
Liberty.[40]

A proposal that the members of the Philosophical Society in
each state be formed into local committees to forward the society's
business was actually presented by Lewis Nicola to the American
Philosophical Society in the same year, but nothing came of it.
Still, although no federation of learned societies was effected in
eighteenth-century America with its seat at Philadelphia, the
scientific, medical, and agricultural societies of the Quaker City
were actually looked to by those in New York, Charleston, Providence, and other places for advice and assistance.[41] Advertised
as engines of social betterment and blessed by the Philadelphia
and Boston precedents—the former a realization of the vision of
Benjamin Franklin—the institutes, societies, and academies of
innumerable localities sought to supply the democracy with lan-
terns of learning appropriate to a free people.

    Despite the grandiloquent plans, organizations were des-
tined to function on a local scale for quite some time to come. As
George Washington had accurately observed, "We must walk as
other countries have before we can run. Smaller societies must
prepare the way for the greater." Although he thought it would
be some time before a national society was feasible, Washington
did express his hope that the United States would not be "so
slow in maturation" as the older countries had been.[42] Because
of inadequate transportation and communication facilities, even
those that were nominally national were, in effect, local societies
dependent upon a large local scientific community for support.
This was true of the American Philosophical Society (founded
1769), the vast majority of whose active members were Philadelphia residents, and of the American Academy of Arts and Sciences
(founded 1784), which was an organization confined mostly to
the Boston area. In both cities, an intellectual community of adequate size had developed to support such enterprises.

The local orientation of the two major societies set the pattern that dominated American science for the first half-century of independence. Thus the Connecticut Academy of Arts and Sciences (founded 1799), was largely an appendage of Yale College; the Massachusetts Historical Society, which included a major emphasis on natural history, along with political and ecclesiastical history, in its definition of functions, was closely related to Harvard. The latter society combined an educational function with its collecting one; appended to a circular letter sent out at the founding of the society in 1792 were a series of articles giving directions for preserving animals, skins of birds, collecting and preserving vegetables, taking impressions of vegetable leaves, preserving "marine productions," and collecting mineral and fossil substances. Indicating the broad ambitions of the society, the letter was addressed to "every gentleman of Science in the Continent and Islands of America," and it stated that the aim of the society was "to collect, preserve, and communicate materials for a complete history of this country."

These and a multitude of other societies tried to fill the gap that had been created by the loss of formal English ties. The natural history orientation, created during Colonial times, continued to be evident in virtually all of them. The East India Marine Society, for example, founded at Salem, Massachusetts, under the leadership of Nathaniel Bowditch in 1799 and incorporated in 1801, was instituted for the purpose of investigating and recording facts relative to the natural history of the ocean. Its procedure was to furnish to every member when bound to sea a blank journal in which he was to enter his observations of the variation of the compass, temperatures, winds, currents, tides, and any other potentially useful information. Sixty-seven such journals had been collected and a museum of natural history formed by 1821, when the society published a catalogue.

The American Botanical Society, Philadelphia, founded in 1806, and later called the Linnaean Society, was created by a group of young men "desirous to promote a knowledge of the vegetable kingdom; and assured of the advantages to be derived from it, in a medical and agricultural point of view." The group immediately elected a number of corresponding members, from whom they hoped to receive specimens, drawings, and descriptions of plants, and they issued a public statement of objectives of the society and the services it would provide. Their aim was to obtain a full knowledge of the medicinal and dyeing drugs of the country, to expedite the discovery of useful metals, to aid native manufactures, and "to remove the inconveniences and disadvantages of individuals not possessing an acquaintance with natural knowledge." Pursuant to this last aim, the society offered to examine any plant or mineral specimen sent to it and return the sender information relative to the nature and uses of the specimen. The society would, of course, retain the specimen for its collection or for trading purposes; the practical and the scientific aspects were thus neatly combined. Nevertheless, the society gave way during the decade 1810–20 to the Academy of Natural Sciences of Philadelphia, which had broader natural history interests than botany and mineralogy and was therefore able to appeal to a broader audience.

By the beginning of the 1820's, societies analogous to these were in operation in every state. There were five natural history societies incorporated in the state of New York alone between 1815 and 1819. In 1819 the legislature of New York provided state funds for the partial support of county agricultural societies, which would have as one of their purposes the general promotion of science. The New York law granted $10,000 annually to be distributed among the different counties in proportion to their population; the proviso was that an agricultural society be formed in each county, the members of which should raise by voluntary subscription an amount equal to the sum apportioned. A further sum of $1,000 per annum was granted to a central board for distribution of seed and for printing transactions. New York's early effort at "matching funds" was apparently quite successful, for within one year after passage of the law twenty-six such societies had qualified for their share of funds. This was a modest beginning, but nevertheless it placed New York in the forefront of a trend that was nation-wide. The rapid multiplication of new societies continued unabated until 1825, after which time it slowed down greatly but never died out. Including all organizations which had as one of their aims the general promotion of science, there were more than three times as many in 1825 as had existed in 1815. The largest increase, reflecting both improved transportation facilities and the aid that had been granted by New York, was in the Middle Atlantic states, where the number quadrupled.[43]

Societies dedicated to the physical sciences did not achieve even the modest degree of success of the natural history societies.

Between 1792 and 1821, three separate efforts were made to found chemical societies, two in Philadelphia and one at Delhi, New York. Each of these promptly failed, primarily because of inadequate support. Where the natural history societies could depend upon a great deal of amateur enthusiasm and support—natural history was still essentially an amateur's pursuit—the physical sciences had already passed into professionalism. And no professional community of adequate size yet existed in America. It was, in fact, not until 1876, after the nation had been crisscrossed with both rail and telegraph lines, that a successful chemical society was founded.[44]

The societies, in addition to providing a meeting place for interested scientists, functioned in a number of ways to aid the growth of science in America. Major functions were to act as "committees of correspondence," to coordinate work especially in natural history, to facilitate the interstate trading of specimens, and to give isolated naturalists the feeling that they were participating in an important national enterprise. Sometimes the more wealthy of the societies actually financed explorations; for example, Samuel L. Mitchill's geological trip into western New York was financed by the New York Lyceum of Natural History. Finally, and perhaps most importantly, the societies' collections served an important educational function: despite the miscellaneous nature of their collections, and the confused order in which they were generally kept, they did freely open their collections to students and there was no other place that young naturalists could go to compare specimens or even to learn the names of many plant, animal, and mineral species. In a real sense, such societies served as the universities of their time. Some, in fact, maintained a small professional staff and provided free public lectures and demonstrations.

During the same period which saw the proliferation of scientific societies, scientific journals began to flourish, for there was a state of mutual dependence between society and journal —even when they were not officially connected with each other. As an inducement for members to participate, the societies needed organs for the publication of papers read at meetings, and the journals needed a ready source of material which, in the absence of more formal research institutions, only the scientific societies could supply. The earliest scientific publications quite frequently served as quasi-official organs for a number of societies. H. Bigelow's *American Monthly Magazine and Critical Review*, for example, regularly printed between 1817 and 1819 the transactions of the Literary and Philosophical Society of New York, the New York Medical Society, the New-York Historical Society, and the New York Lyceum of Natural History. For many years, the only exclusively scientific publication was Samuel L. Mitchill's *Medical Repository*, founded in 1797. A more common pattern was for general "literary and philosophical" publications to devote some space to science along with other matters. The number of periodicals publishing scientific material on a regular basis more than doubled between 1815, when there were eleven other than the transactions of medical societies, and 1825, when twenty-four such journals were in publication. The centers of publication were New York, Pennsylvania, and to a lesser extent, Massachusetts and Ohio. However, before 1850 twenty-three states had been the home of at least one scientific periodical.

Even though more specialized publications began as early as 1804 with George Baron's *Mathematical Correspondent*, the medical journals continued to play an important role throughout the first half-century. The *Medical Repository*, which under the editorship of Mitchill had brought Lavoisier's chemical revolution to America and had led in the fight against Priestley's efforts to save the phlogiston theory, stopped publication in 1824, but more than forty new medical journals had been founded during the preceding decade to take its place, including the *American Medical Review and Journal*, 1824, forerunner of the still-existing *Journal of the American Medical Association*. Along with the descriptions of cases and reviews of books, which were their main functions, these journals carried articles on natural history, chemistry, and mechanics, as well as on the expected anatomy and physiology.

The many special-interest journals that were founded were, without exception, unable to survive. Archibald Bruce's *Mineralogical Journal*, founded in 1810, could not make it through the first volume, and even in 1842 Benjamin Peirce's *Cambridge Miscellany of Mathematics, Physics, and Astronomy* lasted only one year. Similarly, Robert Adrian tried on five different occasions to found a mathematical journal, but only one, the *Mathematical Diary*, published from 1825 to 1832, was in existence for more than a few issues. Specialist publications, suffering from the same limitations as specialist societies, could not find a large enough market to make them economically self-supporting, and this was before the day of university subsidization.

All of these journals played some part in the growth of

scientific activity in America, but the crowning achievement of the
first two decades of the century, in terms of publication, was the
founding of the *American Journal of Science and Arts* by Benjamin Silliman in 1818. The first scientific journal in America to survive the century, Silliman's quarterly could very well be credited
with having been the greatest single influence in the development
of an American scientific community. Nothing better illustrates
the contemporary status of Silliman's journal than the remark
made by Nathaniel Hawthorne upon seeing an issue on one of
the desks in Radcliffe Library at Oxford. It was, he said, "the only
trace of American science, or American learning or ability in any
department, which I discovered in the University of Oxford."

Silliman's periodical, like all the others, passed through
a number of difficulties during its early years, but, through Silliman's efforts and the support of the Connecticut Academy of
Science, it managed to weather each storm. At various times, he
had to contend with rivals who sought to lure his supporters away,
and he also had difficulty with supporters who wanted him to
make the journal "more miscellaneous," but even though the "*and
Arts*" was not dropped from the title during his lifetime, Silliman
was resolved to keep it as much as possible a purely scientific
journal. He also had the difficulty, usual for periodicals of the
time, of collecting for subscriptions; and much to the embarrassment of Yale College, which offered only moral support for his
publication venture, he was forced several times to appeal to the
public for aid. By 1829, however, the fortunes of the journal had so
much improved that editor Silliman was able to announce that he
had begun paying for original contributions. Those who were still
willing to write for the journal "from other motives" were requested to do so.[45]

With the success of Silliman's journal a new era can be said
to have dawned in American science. Increasingly since the end
of the War of 1812, Americans had sought their primary identifications with their own societies, science was being taught to some
extent in institutions all over the more settled areas of the country,
scientific societies had developed collections which made it
possible for American natural history productions to be classified
in America as well as in Europe, a vigorous nationalism made
Americans wary of placing themselves in positions of dependence
upon European scientists, and finally, a truly national scientific
journal had by the late 1820's proved its ability to survive. American science, as one historian has recently said, had come of age.[46]

VII

The Democratic Age in
American Science

℀ On February 11, 1785, a number of public-spirited gentlemen,
most of them residents of Philadelphia, met to found the
Philadelphia Society for Promoting Agriculture. Leaders of the
group had been members of the respected American Philosophical
Society, which, to men of their particular cast of mind, had become much too esoteric in its concerns since the days of Franklin's
active leadership. The new group, with its single-minded attention
to the dominant sector of the American economy, would be a
much more congenial instrument for the distribution of certain
types of practical scientific knowledge than the organization from
which they withdrew.

In Boston, a group of men with similar interests began to
criticize America's other learned society, the American Academy
of Arts and Sciences, on the grounds that it could not give the
required attention to agriculture because its aim was "universal
investigation." One dissident member proposed that in its stead
there be founded a specialized society dedicated wholly to the promotion of agriculture and financed by a large proportion of the
dirt farmers in Massachusetts.[1] Although even it did not have the
grassroots support that had been envisioned, an agricultural
society was founded in Massachusetts in 1792.

It had been the hope of the founders of America's two
learned societies that those organizations would be able to lead
the way in the application of science to the affairs of this world.
Agriculture, as the chief support of the American people, had
been a particular concern. Agricultural improvement had been

high on the list of priorities drawn up by Benjamin Franklin when
he originally founded the old Philosophical Society. The Pennsylvania Assembly was still thinking in these terms when it made
a grant of £150 to the American Philosophical Society "for the
purpose of encouraging agriculture and commerce, by enabling
that learned body to obtain such discoveries as have been made
in *Europe* and other countries." The legislature of Massachusetts
had the same end in mind when, in incorporating the American
Academy, it declared, "The Arts and Sciences are the Foundation
and support of agriculture, manufactures, and commerce."[2]

In the case of neither society were the expectations fulfilled. The American Academy did collect and publish in its first
volume of *Memoirs* (1785) a number of papers upon agricultural
subjects, including Manasseh Cutler's botanical catalogue, which
was intended as a guide for farmers. It also appointed a committee
on agriculture that sought not only to stimulate experiments and
to collect papers but to encourage better agricultural techniques
among the farmers by awarding prizes for achievements in practical farming. The academy's efforts inspired little interest among
the dirt farmers of Massachusetts, however, and agriculture remained only one among its many interests.

Despite the stated interest of the American Philosophical
Society, its first volume of *Transactions* (1771) had been dedicated to a comprehensive treatment of the Transit of Venus. Its
publication had brought the society favorably to the attention
of scientific bodies all over the world and had considerably enhanced the standing of American science, but it contained nothing
to encourage application. The second volume (1786) not only
had little of interest to practical agriculturalists in it, but according
to some critics, was wholly given over to "abstract speculation"—
certainly not a very profitable matter in the minds of farmers,
artisans, or other "practical" men.

Since both the older societies were falling short of expectations, the obvious thing to do, so it seemed, was to replace them
with organizations of more limited scope. Following the lead of
the Agricultural Society, a great many other special interest organizations were founded in the decade after 1785. Besides the
large number of agricultural societies, which were represented
in nearly all the states, these took the form of manufacturing
societies or more general "societies for the diffusion of useful
knowledge," forerunners of the nineteenth-century "mechanics

institute." All were deliberately and narrowly utilitarian societies in a sense that the more general organizations had not been. Their appearance was in part connected with the democratization of knowledge in America, as some historians have suggested, but they were also a reaction to the growing professionalism of the broader scientific bodies' failure to concentrate on the immediately practical. In other words, they are a sign that the Enlightenment faith that all knowledge was useful was breaking down into the somewhat contradictory elements that the Enlightened mind had so neatly balanced. The multiplication of the organizations after 1800 reveals a growing dissatisfaction with the broad program formulated by the Enlightenment philosophers.

Significantly, only a few members of the new societies could, even with a wide degree of toleration, be regarded as scientists. In the agricultural societies were gentlemen farmers, city dwellers with a nostalgia for the farm, and well-intentioned amateur horticulturists who felt that they had a mission to reform American agriculture along the lines laid down in the latest English treatises. The manufacturing societies for the most part attracted incipient or would-be entrepreneurs who combined to investigate and exploit industrial opportunities. The "Pennsylvania Society for the Encouragement of Manufactures and the Useful Arts," for example, founded in Philadelphia in 1787, erected a textile factory which its manufacturing committee managed for a brief period. The societies for the diffusion of knowledge, aimed at the working men of the cities, were composed of an uneasy combination of zealots for mass uplift with radical democratic social assumptions and of those who wished, in the tradition of the English societies, to use mass education as a means of social control. None had much sympathy for the more esoteric interests of the older societies. It was these groups, along with the efforts made by scientists to appeal to them, that were primarily responsible for the intense practicality scholars have generally attributed to that period in the history of American science.

Although the more advanced scientists never really lost their interest in abstract science, to become engulfed in narrow practicality, the utilitarian emphasis that scholars have placed on nineteenth-century American science is not all illusion. There had long been a tradition of practicality stemming from the Baconian assertion of the utility of science, moderated in Enlightenment times by the conviction that *all* knowledge, even the most abstruse and theoretical, was ultimately useful. This belief had been widely accepted in Europe, and it was disseminated and generally approved in America. The American social environment strongly reinforced this European idea by providing important support for scientific projects only when some clear utilitarian gain seemed likely to immediately ensue. The war, by forcing single-minded attention to mundane matters connected with the Revolutionary need for self-sufficiency, intensified this tendency in America; the separation from England further intensified it by forcing Americans to rely more on their own resources.[3] Although Thomas Jefferson regretted that Rittenhouse was throwing away the genius of a Newton on the affairs of a crown, Benjamin Franklin's remark, made long before the onset of the Revolution, was probably more to the point and better reflected the attitudes of most American philosophers. "Had Newton been Pilot but of a single common Ship," he said in a letter to Cadwallader Colden, "the finest of his Discoveries would scarce have excus'd or atton'd for his abandoning the Helm one Hour in Time of Danger; how much less if she carried the Fate of the Commonwealth."[4]

As one scholar has noted, the American Revolution appears to constitute a negative landmark in the history of American science; for it stood for a radical break with the whole idea of élitism in a form that was doubly injurious to science. It not only undermined both the institution of patronage by the rich and well-born and the disinterested pursuit of learning by these men themselves, but it also reinforced to almost a pathological degree what was already a vigorous American tradition: the work-ethic which required of everybody that he should follow an obviously useful calling. "There is a vast deal to be done," wrote Dr. James Jackson, Sr., in vetoing a career in medical research for his son, "and he who will not be doing, must be set down as a drone."[5]

Thus, the observed failure of the abstract sciences combined with American experiences to effect an invidious distinction among types of knowledge that Enlightenment philosophers would never have made. "You are not to live in the sun, nor moon, nor to ride upon the tails of a comet," college graduates were told in a magazine article of 1790. "A *few* astronomers are enough for an age."[6]

Actually, there is good evidence that this attitude had always existed; its appearance after the Revolution may only be a case of the emergence of "middle-brow" ideas in the more democratic atmosphere that prevailed. The Enlightenment was definitely a minority movement; the general failure of Enlightenment intellectuals to persuade rank-and-file Americans—artisans, farmers, and others—to join in their enthusiasm gives some indication of the limited regard in which the philosophers were held. As Whitfield Bell has shown, despite evidences of popular interest in science before the Revolution, even in Philadelphia, the majority of people looked upon natural phenomena chiefly—almost solely —as curiosities, and they continued to do so afterward.[7] They might very well attend a popular lecture on science for its entertainment value, particularly if some impressive demonstrations were promised, but it is not clear that they even understood, much less sympathized with, the philosophers' reasons for interest. Lecturers, in fact, seemed to understand the basis of their appeal, and in their advertising they invariably promised showmanship. The following announcement placed by Ebenezer Kinnersley, one of Franklin's co-experimenters, in the *Pennsylvania Gazette* of December 28, 1758, is a good example:

> For the Entertainment of the Curious It is proposed to Exhibit on Tuesday next in the Apparatus Room of the College, a Course of Experiments in Electricity to be accompanied with a lecture on the Nature and Properties of that entertaining Branch of Natural Philosophy.

To the "curious," Kinnersley offered such items as "An *artificial Spider*, animated by the *electric Fire*, so as to look like a *live one*"; electrified coins, which he said "scarce anybody will take when offered to them"; a battery of eleven guns discharged by lightning after it had passed through ten feet of water; eight musical bells played by electricity; and probably most attractive of all, "The Salute repulsed, by Fire darting from a *Lady's Lips*, so that the Lady, tho' she gives free Permission, may defy any Gentleman to salute her."

Surely Kinnersley's lecture had educational value, but one could hardly assume the purposes of those attending such lectures to be educational. The Philadelphia populace, Bell concluded from a careful study of journals and diaries, laughed at Fitch's steamboat, jested about balloons with scarcely a thought of their potentialities, and saw in electricity only a child's toy— spectacular but of little value. Telescopes and the movement of the stars in their courses were interesting, and little more.[8] With such attitudes, it would have been perfectly natural for patriotic Americans to conclude that the new Republic could no longer afford such entertainment. The insistence, which became very strong by about 1800, that scientists confine themselves to the immediately useful, naturally followed.

"We cannot avoid noticing," said a writer in the *Edinburgh Review* in 1810, "*the air of business* which seems to play about everything American." The Americans, he said, had not yet found leisure to pursue science for amusement. Chemistry, for example, "always appears among them in connexion with some useful and gainful occupation. . . . It is therefore studied only with a view to improvements in the arts of preparing cements and manures, dyeing, bleaching, distilling, purifying infected air, tanning and currying leather. These are the topics to which the chemical books, published or imported, principally refer."

The plans that laymen made for the future course of American science generally reflected this reorientation. An example of this is provided by Dr. Nicholas Collin, the rector of the Swedish Churches in Pennsylvania, who wrote the introduction to the *Transactions* of the American Philosophical Society in 1793 (Volume III). The article, entitled "On those inquiries in Natural Philosophy Which at Present are Most Beneficial to the United States of North America," was essentially a program for the rational exploitation of the natural resources of the United States. Even though philosophers are citizens of the world, said Collin, it is their duty to cultivate particularly those inquiries most beneficial to that country "in which Providence has appointed their earthly stations." This was the wisest course, he said, "because we have the best means of investigating those objects which are most interesting to us." Collin pointed out five areas to which philosophers should direct their attention:

1. Medical inquiries: All countries, he said, have peculiar diseases which derive from the climate, style of life, and occupations of the people. It is the duty of American scientists to investigate those peculiar to North America.

2. Agricultural inquiries: Collin thought that Americans would be wise to forever make agriculture the principal source of their prosperity and wealth, for anything else would be "perverting the order of nature." In line with this thought, he suggested particular needs of the farmers, which, he said, would create a great deal of business for the natural philosopher; and he also revealed a strong interest in conservation. Among the tasks he thought suitable for investigation were discovering what

trees are of the quickest growth, at what age trees increase the most, the proper planting distance between them, the best methods of pruning and promoting their growth, and so forth.

3. Physico-mathematical inquiries: Under this head he listed the need for "machines for abridging human labor," gave a list of agricultural machines needed, adding, without foreknowledge of the cotton gin, that the "humane philosopher" must especially note that such machines would in the Southern states depress the labor of slaves. He spoke also of the need for a "tolerably accurate" map of the United States, adding that for this purpose astronomical observations would have to be made for those places which are "most essential for the figure of the whole country, or for the situation of certain parts in a political or economical view"; and he suggested that a treatise on the best means of surveying private estates should be prepared.

4. Inquiries in natural history: Here Collin began with the standard poetic passage on the sublimity and beauty of the wonders of nature, particularly in America, and he charged that "Neglect of natural history under circumstances so alluring would indicate a want of rational taste." Yet, the actual program he outlined included only such things as a search for medicinals and for wild plants and fruits that could be domesticated for human consumption or other utilitarian purposes, such as making dyes or soap. In discussing the animal domain, he began with the need to study the Hessian fly, the canker worm, caterpillars, and other vermin.

5. Meteorological inquiries: His only concern here was with prediction of the weather for utilitarian purposes, such as navigation, agriculture, and the preservation of health.

The only point in Collin's entire program that would have been of interest to "pure science" came in his discussion of quadrupeds, when he suggested that studies should be made of variations in species between the Old World and the New, and the variation of species under different latitudes in the New World. These were modest aims, indeed, when compared with the consuming curiosity of the earlier natural philosophers. In Collin we have a near-perfect example of the corruption of the Enlightenment faith in the utility of science into the assumption, characteristic of early nineteenth-century American laymen, that if scientific research is not evidently useful it is then not to be taken seriously.

As the pressing needs of a new society dictated the emergence of a severance between practical and esoteric science during the final quarter of the eighteenth century, the popular American reaction to the French Revolution completed the process of separation. The association of Enlightenment ideas with France was well known; the extent of the rejection of both became clear during the election of 1800, when the Federalists linked Jefferson's scientific interests with deistic religious ideas and his partiality for the French, even going so far as to question the legitimacy of scientific attainments for a public man. "If one circumstance more than another could disqualify Mr. Jefferson from the Presidency," asserted one writer, "it would be the charge of his being a philosopher."[9] That men like Rittenhouse and Rush were leaders in Jacobin societies indicated to their opponents a sinister connection between science and dangerous ideas. The fact that Federalist propagandists should devote space to the charge of dangerous knowledge shows their low estimate of the popular appeal of science and, as A. Hunter Dupree suggests, is perhaps a key to the long record of congressional reluctance to help it throughout this period.[10]

Thus, the new democrats and Federalists alike came to regard science with suspicion if not outright hostility. To the Federalists and to the conventionally religious, science was tainted by its association with the French Enlightenment; to the radical democrats, many knowing little of the French Enlightenment, it was tainted by its élitist associations and its failure to produce practical results. In either case, the pure scientist in America would have a long, hard struggle toward respectability. "It is certainly discreet to avoid the imputation [of being a man of science]," wrote Samuel L. Mitchill regretfully but with resignation in 1814 in observing that a newly published author had expressed his apprehension "lest the character of a man of science be fastened upon him." In a society like that of the United States, Mitchill explained, "a mere suspicion of possessing eminent attainment of this kind, too often lessens the confidence reposed in an individual as a man of business."[11]

Despite the depressed fortunes of natural philosophers after the Revolutionary period, some evidence of the persistence of a certain amount of spontaneous popular interest in science can be found. The public, then as now, demonstrated its interest in dramatic events, such as the discovery of a new planet, the appearance of a spectacular comet in 1807, and the fall of meteorites in Connecticut in the same year. These and similar occurrences were noted in the public prints and occasioned considerable comment and speculation—much as an orbital flight or a nuclear explosion does today. Charles Wilson Peale managed to make his museum in Philadelphia a paying enterprise by exploiting public interest in the marvelous, and John Griscom's annual course of chemistry lectures in New York was well subscribed year after year beginning in 1800, remaining a familiar institution for well over a quarter of a century. Griscom was, however, definitely an exception during the opening decades of the century, and his popularity was undoubtedly a function of his showmanship rather than a testimony to extensive public interest in science *per se*. Most public lectures on science before 1815 were one-season performances before small audiences, often without recompense to the speaker. Some of these, however, were well attended, and they gave evidence of a lingering interest in science which would later be effectively exploited by Benjamin Silliman, Edward Hitchcock, and other giants of the age of popularization. None of the earlier lectures, however, matched the grand performances at Boston's Lowell Institute during the second quarter of the century.[12]

The main channel of popular information about science after the first quarter of the century was the lyceum, a typically American institution created in response to the growing demand for education and self-improvement—a demand fed by a variety of sources, all connected with the growth of social democracy. Josiah Holbrook, the leader of the movement, was one of the many Yale graduates who had received inspiration from Benjamin Silliman. After an early experiment in agricultural education on his farm in Derby, Connecticut, Holbrook in 1828 organized the American Lyceum, or Society for the Improvement of Schools and Diffusion of Useful Knowledge. The movement spread rapidly. In 1831, Holbrook and others organized a National Lyceum, which held annual meetings until 1839. In 1834 there were already nearly 3,000 lyceums in the United States; in 1839 Horace Mann counted 137 in Massachusetts alone. Although their character varied from place to place, they generally went in for discussion of such questions as corporal punishment, the teaching of ancient languages in schools, the necessity of teachers' training schools, the advantages of a manual-labor system in schools, and other such broad educational questions. Popular lectures in the sciences were an integral part of most of the lyceums; it was in this respect that they performed their most valuable function.

For the lyceum movement Holbrook constructed apparatus to promote the teaching of science—a relative novelty at the time, which later became widely adopted. The program of the *American Lyceum*, first published in 1829, advertised tools for instruction in geometry, arithmetic, astronomy, geology, chemistry, and "natural philosophy." The geometrical apparatus consisted of 2 sheets of diagrams, 15 geometrical cards, 4 transparent figures, 26 solids, and a book with questions; the apparatus for astronomy contained an orrery, a tide dial, and an instrument to show eclipses; that for natural philosophy had levers, pulleys, an inclined plane, wedge, screw, wheel, and axle. The chemical apparatus consisted of tubes, flasks, retorts, crucibles, and a compound blowpipe.[13]

College professors, politicians like Daniel Webster, and other public figures appeared on the lyceum platforms. Fees for a lecture ranged from $5 to $100, depending upon the location and the anticipated drawing power of the speaker. During the winter of 1837–8 twenty-six courses were delivered in Boston, each consisting of more than eight lectures and attended by an estimated number of thirteen thousand people. In that area the lecture system was substantially improved by a generous bequest from John Lowell, the frail son of the industrialist who had brought the power loom to America and helped to introduce the factory system. When John Lowell died in Bombay in 1837, he left $250,000 for the maintenance of an annual course of free lectures in Boston. The Lowell Institute opened its doors in 1839 and was an immediate, one might say spectacular, success. Early scientific lecturers included Benjamin Silliman, Thomas Nuttall, Jeffries Wyman, Asa Gray, Henry Rogers, Charles T. Jackson, and Benjamin Peirce, all distinguished American scientists. With the large fees that could be paid, foreign talent could even be attracted to explain science to the people of Boston. Sir Charles Lyell twice came from England to lecture, introducing the American public to his new uniformitarian geology; Louis Agassiz first came to America to deliver a course of Lowell lectures and decided to remain, accepting a position at Harvard.

Many lecture courses lasted several weeks; some had enormous attendance. Benjamin Silliman, already a practiced and popular lyceum lecturer, opened the institute in 1839 with a series on geology. When tickets were distributed for his second course on chemistry, the crowd is said to have filled the streets adjacent to the lecture hall to such capacity that they crushed into the windows of the Old Corner Bookstore. Often an extremely popular lecturer would attract as many as eight to ten thousand applicants—far more than the capacity of the hall. In such cases tickets were distributed by lot.

The success of the new popularization efforts was such that many felt that the promise of mass enlightenment was finally to be achieved in America. Entire educational systems were enthusiastically planned specifically for the "diffusion throughout the whole people of a knowledge of the principles of science and the application of science to the arts," as one enthusiast wrote in 1842.[14] Technical schools began to spring up and the curricula of existing institutions began to feel the pressure of the democratic movement. Harvard, for example, was accused by one Massachusetts democratic leader of being twenty-five years behind the times and failing to "answer the just expectations of the people of the state." While Harvard should be trying to make better farmers, merchants, and mechanics, it was offering instruction better suited to an aristocracy.[15] "Where is science now?" exclaimed William Ellery Channing from a Philadelphia platform:

> Locked up in a few colleges, or royal societies, or Inaccessible volumes? Are its experiments mysteries for a few privileged eyes? No, science has now left her retreats, her shades, her selected company of votaries, and with familiar tone begun the work of instructing the race. Through the press, discoveries and theories, once the monopoly of philosophers, have become the property of the multitude. Its professors, heard not long ago in the university of some narrow school, now speak in the mechanic institute. There are parts of our country in which lyceums spring up in almost every village for the purpose of mutual aid in the study of natural science. The characteristic of our age, then, is not the improvement of science, rapid as this is, so much as its extension to all men.[16]

In keeping with this "characteristic of the age," scientists were urged by lecture managers to deliver addresses "in a plain, intelligible manner, divested as far as practicable of technical phrase-ology and such terms as tend to discourage . . . a love of science."[17] Scientists of the caliber of Silliman, Agassiz, and Hitchcock complied with the request, demonstrating the "wonders" of chemistry or electricity, the "marvels" of natural history, or, perhaps, as Edward Hitchcock did, comparing the "wonders of science" with the "wonders of romance"—much to the disadvantage of the latter. "How stupid must that intellect be," Hitchcock exclaimed, "which is not roused and interested by such paradoxes" as those to be found in the sciences.*

The "science" imparted by the lecturers was seldom the most up-to-date, for there was a common tendency to give the same lecture again and again, over long periods of time that witnessed fundamental changes in the data and the theories underlying it. Moreover, most lecturers realized that popular audiences could not be expected to comprehend much above the level of superficiality, and so generally contented themselves with engendering "enthusiasm" for science by retailing "wonders." Nevertheless, by mid-century social democracy had gone so far in science as to make it possible for the common man to believe he was as capable of understanding and making decisions in that area as he was in comprehending and evaluating religion and politics. Cheering the lyceum lecturer who brought fascinating exhibits showing nature's wonders had a great deal in common with applauding the stump politician whose vision of progress one admired. One voted for a particular brand of science, or even a particular scientific doctrine, just as one did for a particular vision of the social order or a particular piece of legislation. For its support, science, in common with other areas of national culture, became subject to popular taste and approval.

Science, then, had become democratic in every sense of the word. "The language of nature is not written in Hebrew or Greek," said an early post-Revolutionary spokesman for the Democratic Age. "The understanding thereof is not involved in the contemptible quirks of logic, nor wrapt in the visionary clouds of metaphysical hypothesis. The great book of nature is open to all—all may read therein."[18] Moreover, it was not only open to all, it was good for all. It would promote the economic interests of the

---

*Edward Hitchcock: "The Wonders of Science Compared with the Wonders of Romance," in *Religious Truth, Illustrated from Science* (Boston, 1857), pp. 139–40. This was a lecture that Hitchcock had delivered many times over the past decade.

people generally—the small businessman and the mechanic as well as the rich merchant and the large farmer—and it would bring relief to the sick and the destitute. A new and far better world was not only possible, but just around the corner in America:

> It is conferring on us that dominion over earth, sea and air, which was prophesied in the first command given to man by his Maker; and this dominion is now employed, not to exalt a few, but to multiply the comforts and ornaments of life for the multitude of man.[19]

In his answer to Thomas Carlyle's attack on mechanistic civilization, an American writer suggested that because of the cultivation of science "the terms uphill and downhill are to become obsolete" and that machines would soon be performing all the drudgery of man "while he is to look on in self-complacent ease." Becoming somewhat lyrical in an introduction to a treatise on scientific method, Samuel Tyler in 1843 propoesd

> To give some account of the philosophy of utility—the philosophy of lightning rods, of steam engines, safety lamps, spinning jinnies and cotton gins—the philosophy which has covered the barren hills and the sterile rocks in verdure, and the deserts with fertility—which has clothed the naked, fed the hungry, and healed the sick—the philosophy of peace, which is converting the sword into the pruning hook, and the spear into the ploughshare.[20]

And still another commentator, writing in the *New Englander* in 1851, claimed that by virtue of scientific advances more had been done within the past fifty years than in all previous recorded history to place the comforts and "elegancies" of social life as well as the necessities within the reach of all classes.[21]

The reasons for the scientist's success in overcoming earlier hostility or indifference are clear. The growth of urbanization and industrialization, the spread of public education, and the growth of a strong desire for self-improvement among the artisan class were necessary conditions which only came into existence during the second quarter of the nineteenth century. But just as important was the technique generally adopted by spokesmen for science. In their public utterances, whatever their private inclinations, leading scientists jumped on the bandwagon of practicality, social democracy, and, as will be seen later, of religion—thus appealing to the dominant forces of their time in a way the Enlightenment philosophers had never been able to do. Superficially, the success of the new generation of scientists was impressive, and their techniques were probably necessary in that milieu. A profession must have a constituency, and for continued support it must find some method of appealing to that constituency. Failure in this respect had spelled the doom of Enlightenment plans.

But success, as well, proved to have its price. Having convinced the people that science was democratic, and having submitted it to the bar of popular approval, scientists quickly found that they had given away more than they intended or could well afford. Charles Brockden Brown's comment at a suggestion that the American government establish a society of linguists to collect and analyze Indian vocabularies was indicative of a chief problem of American science for the next century or more. "The American citizen," he said, "will smile at this proposal. The *great* importance here bestowed on the business of collecting the dialects of barbarous tribes, who are hastening to oblivion, for the sole purpose of throwing a faint light on the question whether these tribes originally came from the north of Asia, will hardly be felt by the busy merchant, artisan, or farmer, or by their public representatives."[22] If the people were truly to judge, then they must certainly determine what types of science were worthy of their support.

But there were even more serious limitations than the restriction to practicality. The democratic assumptions of Jacksonian America militated against professional expertise of every sort. If one man was truly as good as another, what need was there for special training before one could contribute his mite to science? The scientific analogue of the spoils system in politics, with its assumption that the affairs of government were so simple that any honest man could perform the necessary functions as well as another, was the practice of soliciting the aid of the untrained as scientific observers. Thus, it was suggested at one time that the Smithsonian Institution send out sets of meteorological instruments to missionary stations—the missionaries, being on the scene in widely scattered parts of the world, could send in regular observations and soon the weather systems of the world would be known.[23]

Meteorological observations, which did require minimal knowledge to make, were a favorite method of contributing to science. Navy surgeons at sea took advantage of their opportunity to send in sets of observations, which in the early days would be duly published in the *American Journal of Science and Arts*. The first holder of the post of Surgeon General of the Army, Joseph Lovell, had early begun to have his far-flung medical officers collect weather data, partly on the theory that some relation might exist between disease and weather. Lovell managed to obtain good results for several years, even without additional appropriations.[24] The Secretary of War, at Alexander Dallas Bache's request, once sent out a circular to military posts asking whether any unusual meteoric display had occurred on the night of November 13, 1834, when a huge shower had been reported by Denison Olmstead at New Haven. Enough replies were received in this case for Bache to decide that the meteors had been a merely local occurrence.[25]

The informal data collection system, however, was not restricted to meteorology. Robert W. Fox, for example, submitted "Questions relative to mineral veins . . . to practical miners";[26] William Shaler, while a U. S. consul at Algiers, improved the occasion by performing ethnographic researches on "The Language, Manners, and Customs of the Berbers, or Brebers, of Africa."[27] As eminent a professional scientist as Asa Gray at one time recommended that the Smithsonian prepare a manual of scientific observation and research "especially adapted to the use of the American inquirer." This should include directions for observing natural phenomena and collecting specimens, and it should be placed in the hands of officers of the army and navy, as well as civilian travelers. Such a practice, he suggested, "would greatly tend to develop the natural resources of our extended country, and to the general advancement of science."[28] Although his motion was adopted by the AAAS, nothing further was heard of the idea. Gray himself, however, did make effective use of amateurs for limited purposes. Taking advantage of the high accuracy of transmitting botanical information by dried specimens, he was able to organize a network of collectors who fed a steady stream of information into his herbarium at Harvard, much of which appeared in his *Manual of Botany*. When organized and directed by the professional, the system did have its virtues.*

---

* This feature was pointed out to me by A. Hunter Dupree.

But enthusiastic spokesmen for the Democratic Age did not restrict their vision to such modest, professionally directed activities. The meteorological correspondents the Smithsonian already had in the United States, for example, could be put to much broader use. "How easy to call upon the trained meteorological correspondents for information upon other subjects," wrote Spencer F. Baird to Joseph Henry:

> the distribution and local or general appearance of certain forms of animals, vegetables, or minerals, the occurrence of various diseases over the entire country; the spread and rate of progress of a pestilence as small pox, yellow fever, or cholera through the land; the range of action of noxious insects, as the Hessian fly, the cotton or tobacco worm, etc. with an infinity of others. I have long dreamed of some central association or influence which might call for such information, digest it, and then publish it in practical form to the world, and I see that my dream is not far from realization.[29]

As Baird's comment indicates, there were those who believed that in the Smithsonian, America had the one scientific institution that a democratic society would need to take its place in the world of science. Operating on limited funds, drawing upon existing manpower, controlled by public officials, including a museum for public entertainment, and its works open to public scrutiny, it was the ideal institution for an essentially anti-institutional society.

It is ironic that the Democratic Age should have reached its height at precisely the time when the growth of the sciences was demanding an ever greater specialization of interests and when, consequently, the amateur was becoming largely irrelevant as a genuine contributor to science. There were still some fields, such as meteorology, which remained in a primitive descriptive state during the early nineteenth century, to which the unskilled observer could contribute, but these fields were becoming rare. Astronomy had long since passed beyond the common man's comprehension, although he could still admire its "wonders," and if properly organized and directed, he could still contribute limited amounts of information. By the first quarter of the nineteenth century, chemistry, paleontology, geology, botany, and zoology had all adopted theoretical structures of such complexity that only the specially trained intellect could adequately comprehend them. Given this divergence between the ideals of the society, and the requirements of a developing science, complaints were to be expected, and mutual misunderstanding was inevitable. As Chester Dewey, one of the important opponents of the natural system of classification, wrote to Asa Gray in 1842, "The natural method takes botany from the multitude and confines it to the learned." Earlier, Thomas Jefferson, on much the same grounds, had rejected the natural system and expressed his preference for the Linnaean, even though he was not completely satisfied with the latter. The difficulty with Linnaeus's system was that some determinations required anatomical dissection. It would certainly be better, he thought, to base a system upon "such exterior and visible characters as every traveller is competent to observe, to ascertain and to relate."[30]

In this environment, the changing point of view of professional scientists could not be expected to find much public sympathy. The public suspicion of expertise was, in fact, something that the leading professionals were aware of as a major problem to them in their professional aspirations. Alexander Dallas Bache, for example, in an 1851 presidential address before the American Association for the Advancement of Science, spoke of the "modified charlatanism" that he saw as the real danger the association would have to face:

> This form of pretension leads men to appeal to tribunals for the decision of scientific questions, which are in no way competent to consider them; or to appeal to the general public voice from the decisions of scientific men or scientific tribunals, in matters which, as they only are in possession of the knowledge necessary to make a right decision, so they only can give one which is valid.[31]

Bache's fears were shared by many leading professional scientists during the first part of the century. They were, in fact, generally wary of forming associations beyond the local level, where they could be assured of control. As many dangers as advantages were recognized in large-scale associations. There was, of course, the ever-present fear that the charlatanism which Bache attacked so vigorously would actually gain control and subsequently "true" science would be subordinated to extra-scientific considerations and perhaps misdirected. There is clear evidence that the failure of professional scientists to support the National Institute for the Promotion of Science in 1844 was a major cause of its collapse—even though the institute had circulated the false notion that it had the support of the other scientific societies and of the scientific community in general. Dallas Bache, to be sure, had attended the national meeting, but only so he could be "on the ground to direct . . . the host of pseudo-savants . . . into a proper course," as he explained to his friend and fellow-professional Joseph Henry. Henry's own long-felt conviction that a "promiscuous assembly of those who call themselves men of science in this country would only end in our disgrace" was a characteristic professional reaction. The heavily political, almost wholly amateur, membership of the institute was sufficient reason for its rejection.[32]

Bache, speaking in 1851, explained that the "most progressive" had opposed the early founding of a general scientific association, because they did not have enough strength to control it. It was wise, he said, that the first organization had been left to the geologists, who were working on common problems. Among them, he said, "to be heard, a man must have *done* something."[33]

As Bache suggested, it was a small group of working geologists who, in 1840, had founded the first truly professional association in America. It had expanded its membership slowly until 1847, when the name was changed to the American Association for the Advancement of Science. By organizing in this fashion, a continuity of leadership could be maintained and the association could be held more closely to professional lines. Leadership was exercised by a standing committee which gained an increasing amount of power as the years went by. On the first standing committee were three men who had been founding members of the parent association. During the first fourteen years, Bache served on the committee eleven times, James Dwight Dana seven times, Benjamin Peirce and the younger Benjamin Silliman six times each, and Joseph Henry five times. By 1856, when a new constitution was adopted, this committee had gained the power to exclude papers from presentation, to decide which of the papers or other proceedings should be published, to suggest topics for reports, and "to manage any other general business of the Association" between sessions. Papers judged unworthy of publication by the standing committee were not even listed by title.

The association performed its duty as arbiter of scientific matters in various other ways, assigning research projects to members, appointing investigating committees, reporting on controversial papers, and at times even ruling on questions of priority among members. This last was a questionable matter, Bache thought, but it was probably a lesser evil than avoiding a decision, for if such questions were excluded, the association would, in effect, be driving its members to appeal to the public.

Naturally, the growing role of the association as a scientific tribunal was a source of contention, especially since it was perceived that a small group of men exercised great power. Marginal members of the profession tended to view the leaders as a clique, banded together for their private interest. There were bitter protests, in public and in private, against the "reign of scientific despotism," as one member of the out-group put it. The censorship function of the association was often misunderstood and bitterly resented; in the case of Daniel Vaughn, who had three papers refused at the 1856 meeting, the disappointed author privately distributed a paper to the members "because original inquirers meet with so little liberality from our American Association."[34] One would-be contributor had created such a storm that a committee of the AAAS had been appointed to investigate. The committee report, signed by Dana and Peirce, was a resounding affirmation of the standing committee's decision. "The character of the paper was such—its conclusions so erroneous, and its reasoning so false—that any other action would have been wanting in fidelity to the interests of the Association and the science of the country."[35]

Despite the protests and charges of conspiracy from outsiders, the more professional, for their part, were never satisfied with the association because it was not entirely in the hands of professionals. Thus, in 1860, Silliman, Jr., spoke of his amazement that "that crazy man from New York" had been allowed to read his "foolish speculations on the Atomic theory." Two or three such blunders would be enough to taint a whole meeting, "as a single dead rat a house."[36] The "foolish speculations," however, were not listed in the *Proceedings*, so it was not a total catastrophe.

By the same token, the unwillingness of professional scientists to have the public judge scientific matters was not accepted gracefully by everyone. Why could not the common man, having the evidence presented to him, judge of scientific matters for himself? He had proved himself in politics; why should it not be so in science? To deny the common man his due in this respect was to deny the democratic dream. John Warner, an amateur mathematician involved in a long plagiarism controversy with Benjamin Peirce, thinking in terms of common democratic assumptions of his day, continually demanded that Peirce publish his explanation: "The public could and would form a just opinion." For his part, Peirce was willing to have the matter adjudicated by the AAAS, but would not submit it to an "incompetent tribunal." He would not, he said, "be drawn into such a filthy court for trial."[37] A number of newspapers, including *The New York Times*, the Springfield *Republican* and the Albany *Argus*, in editorial comments jumped to the unwarranted conclusion that Peirce was guilty, and his contemporary reputation suffered greatly for it. Fellow-professionals, however, generally defended his refusal to submit to public judgment. John L. LeConte spoke for the professional community in general when he published an essay on the impropriety of carrying such arguments into the public press. The partisans of Warner, convinced that he could not receive justice before a scientific body, continued their one-sided appeal to the public.

The hostility of democracy to professional expertise was perhaps understood better by Oliver Wendell Holmes than by any other man of the time. A physician in a society where quackery and home cures flourished as in no other civilized community, Holmes had special reasons for understanding the phenomena. In 1844 he had a sober warning for the graduating class at the Harvard Medical School. They were about to enter their professional careers at a peculiar time in history, Holmes told them:

> Society is congratulating itself, in all its orations and its periodicals, that the spirit of inquiry has become universal, and will not be repressed; that all things are summoned before its tribunal for judgment. No authority is allowed to pass current, no opinion to remain unassailed, no profession to be the best judge of its own men and doctrines. The ultra-radical version of the axiom that all men are born free and equal . . . has invaded the regions of science. The dogmas of the learned have lost their usurped authority, but the dogmas of the ignorant rise in luxuriant and ever-renewing growths to take their place. The conceit of philosophy, which at least knew something of its subjects, has found its substitute in the conceit of the sterile hybrids who question all they choose to doubt in their capacity of levellers, and believe all that strikes their fancy in their character of reverential mystics.

This was the spirit with which the young medical graduates would meet daily, Holmes said, with more than a trace of bitterness in his expression. The spirit might be reduced to the following formula:

> A question involving the health and lives of mankind has been investigated by many generations of men, prepared by deep study and long experience, in trials that have lasted for years, and in thousands upon thousands of cases; the collected results of their investigations are within my reach; I who, have neither sought after, reflected upon, nor tested these results, declare them false and dangerous, and zealously maintain and publish that a certain new method, which I have seen employed once, twice, or several times, in a disease, of the ordinary history, progress, duration, and fatality of which I am profoundly ignorant, with a success which I (not knowing anything about the matter) affirm to be truly surprising, is to be substituted for the arrogant notions of a set of obsolete dogmatists, heretofore received as medical authorities.

As Holmes correctly perceived, it was not simply that a few erroneous opinions had gained currency and could be laid to rest by exposure. The roots were deeper, drawing upon the very fabric of democratic society. It was, in his terms, "Folly who is masking under the liberty-cap of Free Inquiry."[38] The anti-authoritarianism of the democracy had been extended to a hatred of *all* authority; the anti-monopolistic rhetoric formulated for politico-economic purposes spread rapidly into the field of learning. Thus, homeopaths and practitioners of botanic medicine, sincere but misguided mathematical quacks like John Warner—or, in a later day, patent-medicine pushers and cancer-cure quacks—have been assured of a sympathetic hearing when they cast their appeal in terms of these ingrained democratic prejudices. In such a climate of opinion, the public sympathy for Warner and the hostility to Peirce was predictable.

Even though the professionals had proceeded cautiously in the formation of the association and had maintained as much professional leadership as possible, a general scientific organization still could not completely escape problems inhering in the differential rate of development of the sciences. Practitioners of the less specialized sciences, not being faced personally with professional problems, generally took a public point of view. Periodically, suggestions were made that sessions be made public, a step that would have oriented the association more toward the diffusion of science rather than its advancement, which was thought by the leaders to be the proper function. Members were reminded in AAAS presidential addresses, "It is in the hope of communicating new truths and of adding something to the common stock of knowledge that we have associated ourselves."

An insistence upon this distinction, which is basic to modern professional institutions, first appeared in American science in the 1840's. Earlier, scientific journals had been conceived primarily for the "conveyance and diffusion of knowledge." Even Silliman's *Journal* contained entirely too much popular matter to suit the most advanced professionals. It was not until 1849, when the *Astronomical Journal* appeared, that a journal unequivocally dedicated to "not the diffusion, but the advancement" of scientific knowledge, existed in America. Similarly, the emphasis upon the distinction by the leadership of the AAAS was so new that many members did not even fully comprehend it.[39] As Joseph Henry noted in his *Report* of 1859, few people realized that "the advance of science or the discovery of new truths, irrespective of their immediate applications, is also a matter of great importance, and eminently worthy of patronage and support."[40] And it would be nearly a century before democratic America recognized that principle—even then, recognition was by no means universal.

VIII

~~~~~~

Geophysics and Politics

The rise of the common man and the increasing power of the electorate in general were not entirely without blessings for science. During the first few decades of the nineteenth century, an expansive democracy spread westward into the interior of the continent and, at the same time, opened up the interior portions of the older, more settled states. Innovations in transportation and communications—this was the era of turnpikes, canals, steamboats, and railroads—revealed a nation incredibly wealthy in untapped natural resources. The expanding commerce of the United States caused Americans to look outward as well, to develop a new interest in Latin America, Asia, and the exotic lands of the Pacific. Through universal white male suffrage, manufacturers and businessmen as well as Western farmers and city laborers had new power to make government do their bidding. In fields where they could see some practical advantage from science, common men began to press legislatures to use this tool. States, as usual before the twentieth century, taking the initiative ahead of the Federal Government, began to authorize surveys of natural resources. The hope of unsuspected wealth in minerals, in timber, or in new agricultural uses—a hope carefully nurtured by the scientists—spurred on these efforts.

The early nineteenth century was a period of much lengthy debate on the general question of "internal improvements," a term that included government-supported scientific activity as well as the more conventional roads, canals, and railroads.

Under this heading were debated such topics as the propriety of establishing a national university or a government-financed observatory, erecting a telegraph line, or attaching civilian scientists to a military expedition. Those who took a narrow view of governmental power in general—the "strict constructionists" —were inclined to raise objections on constitutional grounds to each of these projects. To be sure, such constitutional arguments were often mere masks for sectional interests; Congressmen tended toward "loose constructionism" in proportion to the expected direct benefits to their own states of any proposed improvement.

But even those who opposed specific internal improvement did not direct their argument against science itself. The debate, in a word, did not reveal any attitude, favorable or unfavorable, toward science. The main driving force of American politics at that time was *local interest*, and within this constricting framework scientific activity fared precisely as other kinds of activity.

Despite a great parade of constitutional scruples, successive chief executives actually approved grants-in-aid to build specific improvements, and they increasingly financed certain types of specific scientific activity. The average annual appropriation of the Federal Government for internal improvements increased with each administration from Jefferson through Jackson. Although John Quincy Adams has been termed the great champion of internal improvements, appropriations for all types of internal improvements averaged only $702,000 annually while he was President; under Andrew Jackson, well-known for his veto of one Federal highway project and his hostility to government in business, the average increased to $1,323,000. An exploring expedition to the Pacific Northwest, defeated when the scientific-minded John Quincy Adams proposed it, turned into a four-year expedition ultimately costing over $1,000,000 and producing several volumes of important scientific reports as authorized by the allegedly anti-intellectual Andrew Jackson. A national observatory, the most hated and ridiculed of Adams's proposals, became a modest reality under the Jackson administration and expanded into a more elaborate enterprise under Jackson's chosen successor, Martin Van Buren.[1]

As even these evidences of growth indicate, federal aid to science in that period was limited, as was the activity of the Federal Government in general. Despite the obvious trend toward greater and greater involvement, a trend which seemed to be independent of the opinions of chief executives, people at that time still tended to think in terms of limitations of the Federal Government rather than in terms of what it could do. General acceptance of the idea of a positive government was still far in the future. Although specific arguments more often than not masked sectional or local interests, there were many, including some leading scientists, who were positively hostile to extensive governmental activity on the grounds that government was inherently inefficient. Government patronage, so Louis Agassiz told the members of the newly formed American Association for the Advancement of Science in 1849, was "not worth the time consumed in imploring it." All the governments on earth, he said, would not develop as much scientific knowledge "as one humble individual, whose lot is cast in a garret." These individualistic sentiments were prompted by a committee resolution calling upon the government to add qualified scientists to each public expedition that might be organized. Despite Agassiz's impassioned speech, those who proposed the resolution carried the day.[2] Most scientists, like most other people, vote in terms of what they perceive to be their interests.

In such a climate, inevitably the most significant kind of governmental activity turned out to be that which was most evidently in tune with the main preoccupation of the American people at that time—namely, the conquest of a continent and the exploitation of its resources, or, in other words, the types of activities most clearly connected with the general drive for "internal improvements." Little thought was given to the idea of public support for science *per se;* it was only specific scientific enterprises that could be viewed as necessary instruments for securing some public purpose which the democratic society could properly patronize. In such an intellectual environment, geographical research naturally came to dominate the sciences in America, as geographical extension dominated the imagination of the public and the rhetoric of statesmen in that age of Manifest Destiny. Under the banner of geography in the service of Manifest Destiny, a kind of unity of the sciences emerged whereby large numbers of investigators enlisted in the practical task of describing the physical and natural characteristics of the nation, and of other portions of the earth which were of interest to Americans.

The geographical science of that era can most usefully be defined as composed of two great divisions: natural history and geophysics. The first consists of the non-experimental parts of the life sciences and geology, especially emphasizing taxonomy. Geophysics is the study of the physical properties of the earth, including its dimensions, surface features, oceans, inner structure, and the surrounding atmosphere of the planet. Gravity, terrestrial magnetism, and meteorology are in its purview, and astronomy is its handmaiden.[3]

In the age of the common man, it was the practical aspects of geographical science that helped loosen public purse strings. Legislators, and the public they represented, remained interested in using the established sciences in fairly conventional ways to achieve specific ends. They wished to be informed of possible hidden mineral wealth, of agricultural possibilities, of the existence of better channels into harbors; or they wished to have land surveyed for quick disposal and exploitation by their eager constituencies. The scientists involved, while they certainly appealed to such practical interests and no doubt shared them, invariably took their researches beyond the level of immediate usefulness to the level of possible contribution to the growth of the theoretical structure of their sciences. Often, in fact, they seemed to be concerned with theory in a manner that interfered with the practical results the public so confidently expected. The legislators from the mining states, like the men in the mining camps, had very little use for a geologic map; nor did they care for complex analyses of stratigraphy or paleontology. The background approach of the scientist, with his insistence upon thoroughness and his interest in historical time sequences for their own sake, seemed especially irrelevant to the harsh necessities of frontier life. And it was on the frontier where geographical science flourished. This was the basic conflict: scientists were placed in the position of serving two masters, the one offering funds for research, the other offering recognition by their peers for contributions to science. The results of this tension were often unsatisfactory to both interests involved.

The dual aims that caused this tension were evident in our first federal-sponsored expedition. In the fall of 1802, Thomas Jefferson asked the Spanish minister "in a frank and confidential tone" whether Spain would object to an exploration of the Missouri River. The American explorers, he said, would really "have no other view than the advancement of geography," but he would "give it the denomination of mercantile, inasmuch as only in this way would Congress have the power of voting the necessary funds." It was not possible, he explained, for the government to appropriate funds for a "purely literary expedition," since the Constitution had given Congress no such power. But when Jefferson sent a secret message to Congress in January 1803 to request the required funds, his chief selling point was that the expedition would pave the way for wresting from the British the Indian trade in the upper Missouri. He reassured Congress that Spain and France would consider the expedition a mere "literary pursuit."[4]

There is no reason for assuming that Jefferson entirely misrepresented his motives either to the Spanish minister or to Congress. Actually, the expedition would serve both purposes, and Jefferson undoubtedly thought them both important. In view of his constitutional scruples, however, he had no choice but to convince himself (and Congress) that the expedition could be justified by reference to some clause in the Constitution. Jefferson demonstrated his personal interest in the scientific aspects of the expedition by the care with which he had Meriwether Lewis trained as a naturalist before dispatching him into the interior: sending him to Philadelphia for six weeks of intensive instruction by members of the American Philosophical Society. One important reason for Jefferson's own interest in exploring the continent was his belief in the fullness of the Creation and the consequent impossibility of any created species becoming extinct. In the hitherto unexplored areas of the continent he hoped to find living specimens of the fossil forms that had been discovered.[5]

The story of the background of the Lewis and Clark expedition is revealing in that it makes clear the ambivalent justifications underlying American exploration in the nineteenth century. The public justifications for such expeditions were nearly all utilitarian—they would serve military, commercial, industrial, agricultural, or mining interests. Exploitation and Manifest Destiny were the keynotes. Nevertheless, all government-financed explorers were carefully instructed to bring back botanical and zoological specimens, as well as the obviously useful mineralogical ones—and they were further instructed to bring back not only military and commercial information regarding the Indians, but also to study their languages, their religion, and their manners. All aided the cause of science in some way. The Long expedition, which left Pittsburgh in the spring of 1819 to explore the region as far west as the Rocky Mountains, was provided with a botanist, a zoologist, a geologist, a "naturalist," and a cartographer. And it was the plant specimens collected on this expedition by Edwin James that John Torrey used as a basis for introducing the natural system of classification to America.[6]

If nothing else, the expeditions aided science simply by providing a relatively safe way for a scientist to get into an unexplored area. In the early days, the civilian scientists accompanying military expeditions were not paid, but they did get transportation, rations, and protection from hostile Indians. Explorations were an area of activity in which the government found a steady scientific occupation to which its participation was indispensible. Reflecting the needs of the basic forces of commerce and westward expansion that dominated the period, explorations made up the great bulk of the scientific research paid for by the government before the Civil War. The government's interest thus reinforced the geographical emphasis in American science at the time. Sciences in which laboratory work played a large part found no place within this framework.[7]

The railroad surveys and public lands surveys later were likewise of immense value to science. For many years, after about 1820, when Sylvanus Thayer strengthened the science and engineering departments at West Point, the military academy had practically to itself the field of engineering education in America. Thayer, who had studied in Europe and become acquainted with the methods of the École Polytechnique in Paris—a school famous for its great teachers, outstanding textbooks and equipment, and its many world-renowned graduates—began to introduce French methods, textbooks, and even instructors when he was appointed Superintendent of West Point in 1817. Consequently, Army engineers were among the best-trained in the country and, in accordance with governmental policy, they often accepted supervising positions in strictly civilian enterprises. Railroad surveying, almost always done under the supervision of West Point graduates, was among the most important of these non-military occupations. Between 1824 and 1838, sixty-one railroad surveys were made at gov-

ernment expense, using military engineers. The total cost of
the surveys was $75,000—an appreciable sum in terms of the
then-current level of governmental expenditures. But the value of
the governmental contribution was enhanced even more by the
scarcity of trained engineers in the United States.[8]

Surveying in the early nineteenth century was considered
to be a genuinely scientific occupation because of its connection
with astronomy. One had to know how to compute latitude and
longitude, and the fact that there were few known reference
points necessitated extensive astronomical observations. Moreover, the West Point-trained surveyor was far more than an
engineer. Under the leadership of Thayer, there was assembled
at the Military Academy some of the best scientific talent in the
country to instruct cadets in a variety of scientific knowledge.
William H. C. Bartlett, for example, taught natural philosophy;
John Torrey, the foremost botanist of his day, taught chemistry
and mineralogy; and Jacob W. Bailey, a specialist in the study of
fresh-water algae and the techniques of microscopy, provided
further grounding in natural science.

With this scientific background, the surveyor was generally
a man of broad scientific interests whose primary identification
was with a scientific community extending far beyond the Army.
A member of the broad scientific fraternity, he could therefore
be counted on to be friendly to other scientists, often allowing
them to attach themselves to a survey crew, sometimes in the
capacity of employee, sometimes simply as a kind of "camp
follower."

During this early period, certain common features of governmental explorations were evident. In the first place, they were
almost always military in organization and command. This was a
reflection both of the rationale behind the explorations and of
the notion, inspired by the desire for frugality in government so
important at the time, that military officers should be usefully
employed during times of peace. With the heavily scientific curriculum at West Point, such employment seemed natural for its
graduates during peacetime, both from the point of view of the
government and of the scientific community. Secondly, the expeditions almost always depended upon civilians either at the
Smithsonian or elsewhere for working up and publishing the
results. Finally, although they showed some continuity in personnel and purpose, they were usually ad hoc missions, flourishing

awhile and then vanishing. The idea of a permanent scientific institution was one that was still bitterly resisted. In 1838, a somewhat more formal arrangement appeared when a separate Corps of Topographical Engineers, consisting of thirty-six officers, was created. From that date, this small, select group operated both as the chief agents of Manifest Destiny and as the leading collectors of scientific information.

The officers of the Topographical Corps, although primarily conceived of as surveyors, were in reality, and finally in intent, a great deal more. From the time of the Mexican Boundary Survey through the great Pacific Railroad Surveys of the 1850's, the aim of the topographers was to produce a scientific description of the trans-Mississippi West, beginning with an accurate outline map to which would be added the findings of scientists in all the other disciplines, including ethnology. Their aim was only partly utilitarian, since much of the data accumulated had no direct bearing on any of the practical problems at hand.

Major William H. Emory, the head of the Boundary Survey, made most of the thousands of astronomical observations from which were derived 208 separate points of latitude and longitude from the Gulf to the Pacific. He also personally supervised the making of a barometric profile showing variations in altitude all along the line. To this profile, the field geologist added data showing the various geological strata. Several pages of the *Report* were devoted to the determination of eight points of declination, dip, and horizontal intensity of the field of terrestrial magnetism.

But despite Major Emory's work, and the work of other scientists brought in by the field commanders, the most significant scientific contributions were made by the scientists who examined the collections in detail back in the Eastern centers of learning. The basic division of labor between field collectors and museum or university specialists was beginning to appear in American science, and the new opportunities brought by the surveys hastened the transformation already begun.

Emory's geological report was of particular importance, for with the help of Eastern scholars it became the first attempt to construct an over-all interpretation of trans-Mississippi geology. Going far beyond the mere drawing of a map, the three geologists primarily concerned with the report, Parry, Schott, and New York State geologist James Hall, attempted to derive causal principles from the mass of observed data and use these principles

to reconstruct the geologic history of the region. Their recon-
structions, necessarily general and often superficial, nevertheless
included an acceptance of surprisingly advanced theories, so that
even today they are for the most part valid as far as they go.
Schott attempted to reconstruct the periods of inundation by the
great coastal seas, noting various cretaceous deposits which indicated the ebb and flow of a series of inundations. James Hall,
using the report in an effort to correlate various formations, found
evidence of still another southwestern sea during the later coal
period. To their observations should also be added the remark of
Lieutenant Michler, who, when observing the arroyos and canyons
draining laterally into the Rio Grande from the Llano Estacado,
concluded, "They are but miniature creations of the same power
which forced a passage for the Rio Grande"—thus implicitly
accepting the doctrine of uniformitarianism as opposed to catastrophism.[9]

Other sciences also benefitted from the Boundary Commission's work. John Torrey, on the basis of 2,648 species brought
back by the field collectors, produced an immense list that systematized not only all of the new and old species collected on the
Boundary Survey, but also those previously assembled by the free-
lance collectors in the region. The zoological specimens were
classified by Spencer F. Baird and Charles F. Girard of the Smithsonian.

As impressive as the early work was, the Pacific Railroad
surveys dwarfed all the earlier efforts of the Topographical Corps.
Conceived as a way to settle a divisive sectional argument concerning the route of the proposed transcontinental railroad, the
surveys were charged with finding the "most practicable and economical route for a railroad from the Mississippi to the Pacific
Ocean." The operation was not to consist of minute projections
of actual right of ways; instead, there were to be general topographical surveys aimed at determining the relative merits of the
competing routes. For this purpose, a wide variety of data was
required. The primary task was that of locating and describing
the geological factors that would be relevant to the construction
of the railroad: the relative hardness of the rock where tunnels
were to be constructed, the strength of roadbeds whether through
canyons or along the sides of mountains, and so forth. They must
report on the exact elevation and grades of the mountain passes,
and on the climate, resources, and economic possibilities of the

regions through which the proposed routes would pass. This latter class of information was needed in order to ascertain whether a population to support the railroad could be sustained along the route. Precious and semi-precious stones and building stones were to be important objects of search for the four parties.

Because the task called for such a broad collection of information, each of the field parties included a contingent of expert scientists who were to compile reports on the various aspects of natural history that were in any way relevant to the construction of the railroad. The role of science was defined as that of providing the backlog of basic knowledge which legislators would need in order to make the most practical decision. Here, as never before, was a chance to compile a great scientific inventory on all levels and at the same time to make that data relevant to the national problem at hand. The opportunity for science to serve as a positive instrument of public policy had never been greater.

Unfortunately, the political result of this ambitious effort to apply science was the direct opposite of that intended. Far from producing a conclusive report on the "most practicable and economical route," the surveyors discovered that several extremely practical routes existed—more, in fact, than the originally proposed four—and that it was merely a matter of preference which one was exploited. Sectional competition was thus intensified and the Topographical Corps tended to be discredited as an instrument of public policy. Thus, ironically, the surveys became at once the final stroke of doom for any transcontinental railroad before the Civil War and the beginning of the end for the Topographical Corps itself. But even though they were a political failure, the scientific results of the surveys were most impressive.

Perhaps the best geologic work by Survey men was their description of the forces at work upon the land. No place in the world offered a better opportunity to observe the work of erosion, abrasion, weathering, volcanic action, and other natural forces than did the barren, naked contours of the Basin and Range region, where a great deal of the work was carried out. "It is a remarkable feature in the character of the country between the Rocky Mountains and the Sierra Nevadas, that whole formations disappear, as it were, before our eyes," Schiel had remarked upon returning from his first vew of the region. "The wearing and washing away of mountains takes place here on an immense scale, and is the more easily observed, as no vegetation of any account covers the country, hiding the destruction from the eye."[10] Survey scientists noted how the prevailing westerly winds were deprived of their moisture by the Sierras before they reached the Basin, thus causing its dryness. One geologist in the group made a comprehensive study of the southwestern deserts, noting how the wind caused sand to pile up around solid objects to form dunes, and how blown sand acted as an abrasive to groove and polish the rock formations. His paper, read before the American Association for the Advancement of Science in 1855, was an early recognition of the fact that water and ice were not the only agents of erosion, as had heretofore been assumed. Another geologist on the expedition pointed out the volcanic and igneous nature of much of the southwestern country; the repeated action of earthquakes was noted in California; and evidences of glacial action were found in the Cascades Mountains. In making it clear that there were many processes at work simultaneously shaping the earth's surface, the results of the railroad surveys were a powerful argument for uniformitarianism.

When one adds to this basic geological work the wealth of botanical, zoological and mineral specimens, the meteorological reports, and the information about the Indians brought back by Survey scientists, the result adds up to an impressive scientific enterprise. By the time Torrey and Gray completed their work on the botanical specimens brought back from the West, the main lines for future exploration had been laid down and a great deal of the basic descriptive classification of American flora had been accomplished.[11] Three separate volumes, all prepared under the supervision of Spencer F. Baird, were devoted to the zoological reports. Of these, the reports by Baird on mammals and birds were of greatest importance in the history of American science. Going far beyond the compilation of Western specimens, they included notices of all specimens of North American mammals and birds. As Baird wrote in his preface, "None of the published descriptions of the old and standard species were sufficiently minute and detailed to furnish the necessary means of comparison." Once again, as in the earlier period of explorations, new knowledge of American nature had made evident the limitations of the old knowledge. Baird's collection at the Smithsonian of specimens of all known American species for the first time made it possible for American scientists to work while making constant studies at first hand rather than accepting second-hand accounts from

European museums. An important result of the surveys, therefore, was to carry American natural history a long way toward the independence from Europe that had been the dream of an earlier scientific generation.[12]

For years there had been a reciprocal arrangement between the officers of the Topographical Corps and the country's leading scientists, in which the officers received credit in the scientific world for their findings in the field while the museum men were assured of a constant flow of new specimens, the opportunity of securing employment for their young protégés, and government contracts for evaluating the scientific results of the exploration. Thus within the framework of a simple Jacksonian democracy there existed an unofficial federal patronage of science that was indissolubly linked to the practical development of the country. In all, 106 scientifically trained men, assisted by a small army of "chainbearers," "rodmen," and part-time soldier naturalists, took part in the surveys either as field collectors or as museum classifiers. The Smithsonian Institution, the country's leading scientists, and various learned organizations had an important part in determining who would go on the expeditions. Baird worked diligently to see that the surveys were supplied with the proper scientific collectors—a number of his own protégés were appointed. The Smithsonian also supplied detailed advice on collecting techniques and furnished instruments when the Army engineers found their own resources inadequate. James Hall of Albany and John Torrey of Princeton also wielded considerable influence in the selection of personnel; Torrey was personally responsible for the selection of most of the botanists. Louis Agassiz managed to get his friend Jules Marcou attached to the survey as chief geologist to the Whipple expedition. The total effect of this system of advice and recommendation was the creation of a powerful scientific lobby that not only controlled appointments but was actually instrumental in persuading Secretary Davis to embark on this ambitious scientific project to the extent that he did.[13]

The military engineers in their time were a potent force in the development of the nation. They laid out the basic lines of communication upon which ultimate economic success depended, created real estate values, and decided the location of towns and cities and capitals. They were also a primary means of attracting new settlers into a territory: the existence of a good transportation system or even the promise of one was often enough to attract the first venture capital into a wilderness area. But the very involvement of the military engineers in the economic life of the West was ultimately to be their downfall, and it limited their scientific importance while they existed. Always the immediate needs of the local frontier conflicted with and prevented the execution of any over-all plan for the Western country. Practical problems of supply and defense took immediate precedence over long-range ideas. Searching for water on the Llano Estacado of Texas, for further sources of gold in California, or engaging in other immediately practical projects dissipated their scientific energies. The system of commercial rivalries that had developed among Eastern promoters, each section having its own idea as to the ultimate destiny of the West, combined with local boosterism on the frontier itself to arouse hostilities against the topographers —exactly as it would against John Wesley Powell a generation later. Major Emory's opinion that the plains west of the 100th meridian were wholly unsuitable for agriculture was as unacceptable as was Powell's similar opinion a generation later; and Emory's suggestion that the whole legislation of Congress must be remodeled and reorganized to suit the new phase of life was ignored—much as Powell's advice was to be either ignored or reacted to with hostility.

Often the scientific interests of the topographers seemed to interfere with the rapid exploitation demanded by Westerners. One revealing incident of this nature occurred in 1856, when Senator John B. Wheeler of California introduced a petition calling upon Congress to authorize construction of a road over the mountains into northern California. The road should be constructed by civilians, Wheeler said, and no part of the responsibility should be entrusted to the War Department. When pressed for his objection to the War Department, he explained:

> Because it will be placed in the hands of topographical engineers and in my judgment, they are not as good road makers as civilians. If any Senator would go to the State which I have the honor to represent on this floor, he would find all the roads made by practical men; they are made by stage contractors, who, instead of taking instruments to ascertain the altitude of mountains take their shovels and spades and go to work and they overcome the difficulties of the mountain, while an engineer, perhaps, is surveying the altitude of a neighboring hill.[14]

It was thought by many that the public lands surveys—
the purely linear surveys involved in laying out township boundaries—could serve science in the same fashion as the railroad
surveys. Prompted by Douglass Houghton, Michigan State Geologist, the Association of American Geologists and Naturalists
began memorializing Congress in 1844 on the necessity of adding
a geological survey to the linear work of the public lands survey.
Houghton, in calling the attention of the association to this
opportunity, pointed out that the present federal survey resulted
only in the collection of geographic information and the sub-
division of public lands which were to be offered for sale, and
that, as a result, the government was carrying forward a vast
system of surveys without obtaining knowledge which would
enable her citizens to understand their country. He saw no reason
why minute geological and topographical information could not
be obtained at little additional expense by the United States surveyors. He mentioned that he had persuaded the deputy United
States surveyors in Michigan to aid him in this way and that the
system had proved successful.

The association enthusiastically adopted Houghton's resolution and appointed him chairman of a committee of six charged
with pressing the plan upon the Federal Government. The outcome, after a visit to Washington, was that Houghton himself
signed a contract with the government to run a combined survey,
but his early death prevented the fulfillment of the contract and
the idea was not generally adopted.[15] The association, however,
continued its prompting of state and federal agencies, and its successor, the American Association for the Advancement of Science,
was likewise watchful for opportunities to attach science to existing agencies, such as the Mexican Boundary Survey, or to extend the scientific work in states by urging the beginning or continuance of geological surveys.

Perhaps the most dramatic incident of that period was not, however, involved primarily with the exploitation of the interior but
with the provision of another need of America's commercial interests. This was the United States Exploring Expedition, which,
under the command of Lieutenant Charles Wilkes, spent four
years (1838–42) exploring the Antarctic, the South Pacific, and
the Pacific Northwest.

Conceived and defended primarily as an aid to whaling

and to the expanding Latin American trade, the expedition was also by design a large-scale scientific enterprise. It shared, therefore, in the dual aims common to other expeditions of the period; and, as the outcome was to indicate, it shared also in the problems inherent in such dual purposes.

The commander, by whose name the expedition was to be known, was specially chosen for his scientific competence; many other officers on the voyage had some scientific training and there was a core of professional scientists and naturalists—although not nearly enough to satisfy the scientific societies, which generally resented the part played by the Navy. These civilians included James Dwight Dana of Yale, the geologist; Timothy Pickering and Titian Peale, naturalists; Horatio E. Hale, philologist; John P. Couthouy, conchologist; William Rich, botanist; William D. Brackenridge, horticulturist; and Joseph Drayton and Alfred T. Agate, draftsmen. Together these men assembled the first of the enormous scientific collections that characterized the age in American science.

Even though dissension among the various backers of the expedition very nearly turned it into a fiasco before it sailed, the results—coming from Latin America, the Antarctic, the Central Pacific islands, and the western coast of America—were impressive. They touched the sciences of ethnology, anthropology, zoology, geology, meteorology, botany, hydrography, and physics. In addition, the surveys resulted in a large number of charts, some of them accurate enough to guide Marine divisions landing at Makin and Tarawa one hundred years later. The United States Botanic Garden, founded with the seeds and live plants brought home by Wilkes, still exists under congressional control; similarly, the United States National Museum is an outgrowth of the expedition's collections.

The story of the Wilkes expedition after its return in 1842 shows the attempt of the government to find proper means to accomplish the necessary clean-up chores of a scientific expedition. The collections were of little value until they had been integrated into the body of existing knowledge and the results made available to the scientific community. This had been the most significant failure of the Lewis and Clark expedition, the scientific results of which were not made known until years later, and then only in part, by a German botanist who published descriptions of the plants in England. The government had, in

this respect, made great progress during the forty years separating
the two expeditions. The publication program was to be administered by Wilkes acting under the direction of Congress's joint
committee on the library. Enlisting an able group of specialists
to work on the various volumes, Wilkes tied up, as one historian
has pointed out, a measurable portion of the available scientific
manpower in America for thirty years, producing twenty-four
large volumes before the work was completed. Even though a
great many wrong decisions were made—for example, limiting
the official printing to one hundred copies—and even though
Congress eventually decided that it was not the proper body to
oversee the details of scientific publication, the work would have
to be termed a reasonable success. The government had proved
that it could administer a scientific enterprise, running to well
over $1,000,000 in total cost and involving cooperation between
military officers, civilian scientists, and governmental officials.
That the cooperation was not always smooth was to be expected;
that so much of the work was accomplished at all was a testimony
to the increasing maturity of the country.[16]

The first published product of the expedition was a five-
volume *Narrative* by Charles Wilkes (1844); this was followed,
at intervals punctuated by battles with Congress for additional
appropriations, by thirteen other volumes, each by scientific
specialists, before the work was discontinued in 1874. Several
other volumes, including one on botany by Asa Gray, one on
physics by Wilkes, and two on ichthyology by Louis Agassiz, were
never printed. The published volumes, mostly distributed to foreign countries, with one copy for each state in the Union, were too
poorly circulated to have much immediate impact on American
science. The limitation in the printing called forth vigorous protests from scientists, who presented numerous memorials to
Congress regarding the matter. The government, however, refusing its blessing to the advancement of science or the education
of scientists, insisted upon the token publication and prized the
volumes only for their value in advertising the enterprise of the
United States.

Efforts to aid commercial interests of the nation were of better service to science in another institution which became a significant force during this same period. The Coast Survey, organized
in 1807 under the Swiss immigrant Ferdinand Hassler but taken

from him and allowed to languish under the Navy after 1818, was revived by Congress in 1832 when it authorized the employment of anybody considered desirable, thus paving the way for Hassler's return. The commercial interests of the country needed charts ever more urgently and the naval officers in charge of the survey had minifestly failed to provide them. Hassler and others of the scientific community had never given up hope that the survey would be returned to civilian control, and that Hassler would be allowed to continue the work.

Despite the long-awaited favorable decision, however, congressional hostility came to the surface in a specific provision that "nothing in this act, or the act hereby revived, shall be construed to authorize the construction or maintenance of a permanent astronomical observatory." An obvious slap at Adams's "lighthouses of the skies" and a prohibition of the transformation that Hassler would undoubtedly attempt to make, this provision also revealed a distinction important in the minds of Congressmen throughout the period. What they feared most of all was the creation of a permanent scientific bureaucracy which involved long-term commitment of funds. The survey of the coast was a specific task which on completion would allow the people involved to disband.[17] It was therefore—if hemmed in with the proper controls—not considered a threat. The idea that it might grow into a permanent agency, simply by increasingly refining its objectives, did not occur to the Congressmen who voted for it.

With Hassler's return to the survey came his insistence that it be a true contribution to science and not just a compiled map to serve commercial interests. As anticipated, he also brought his scheme for detailed triangulation from a few accurate base lines, the determination of which actually required astronomical observations of a high order. Hassler's method was to begin at New York and work north and south. By 1841 the survey covered eleven thousand square miles from Rhode Island to the Chesapeake Bay. Hassler, however, desiring to get as near perfection as possible, steadfastly refused to publish results as he went along, saving everything for a final report "according to good principles of science." Despite growing congressional irritation, appropriations continued year after year, always increasing. The $20,000 of 1832 became the $100,000 of 1840, and no end was in sight. One early and dramatic find which had been released despite Hassler's principles, the discovery of a new channel into New

York harbor, was probably responsible for continued congressional sufferance.

By 1841, however, it appeared that sufferance was at an end. The House of Representatives undertook a full-scale investigation, and a serious effort to cut the appropriations began. The Investigating Committee finally recommended and the Congress adopted a reorganization in which the survey kept its $100,-000 appropriation, but required that all plans be passed by a majority of a board consisting of the superintendent, his two principal assistants, two naval officers who were to be placed in charge of surveying parties, and four topographical engineers. Hassler, protesting and squirming under the yoke of governmental control, returned to the field, where he died in the summer of 1843. By the time of Hassler's death, the coast survey had been extended from New York eastward to Point Judith in Rhode Island, and southward to Point Hinlopen in Delaware. His successor, a far more politically adept man than Hassler, managed to continue the work with few of the difficulties Hassler had encountered.

Alexander Dallas Bache, the successor to Hassler, was the ideal man to represent science in government at that time. A grandson of Benjamin Franklin, he carried with him an illustrious scientific ancestry, and one close to the hearts of his countrymen. A graduate of West Point who moved easily in high political circles, he was able to wend his way largely unscathed through the intricacies of military-political attacks on the institution. And as a well-known scientist in civilian life, he had both the respect and the active support of the scientific community.

By the simple expedient of dividing the Atlantic and Gulf coasts into eight sections and placing detachments in all of them simultaneously, Bache was able to gain support from Congressmen of all the coastal states and was able to turn out immediately useful charts much more rapidly than his predecessor. Unlike Hassler, Bache was willing to compromise with the ideal perfection, although he never gave it up as a goal. Territorial expansion also aided Bache in converting the Survey into a semi-permanent institution. The acquisition of Texas, he pointed out at one time, had added two years, California and Oregon much more. Bache lost no time in getting operations underway on the Pacific Coast, producing preliminary sketches by 1850. By attaching his work to the expansive and commercial interests of the country, by playing up projects of immediate usefulness, by the skillful cultivation of legislators from the inland states, and by mobilizing scientific opinion on his behalf through the new American Association for the Advancement of Science, Bache was able to survive each attack on the Survey and, in fact, to secure annual increases in appropriations. By 1854, the appropriation was approaching the half-million-dollar mark, and still no end to the theoretically limited operation was in sight.

Besides the simple geographical extension made possible by the expansion of the country, Bache managed to move the Coast Survey far beyond its original narrow function of chart-making to a wide range of scientific inquiries. Terrestrial magnetism, which he considered "eminently practical . . . though reached by a scientific discussion which seems to pass beyond the bounds" of applied science, became a continuing study. By 1858, 103 stations were in operation in the Atlantic, Gulf, and Pacific areas. One Survey scientist, Sears C. Walker, coordinating the private efforts of several other investigators, worked out the mechanical and organizational details necessary to put the telegraph to use in accurately determining longitude—a technique which became known as the "American System." Other Survey scientists worked out the pattern of tides in the Gulf of Mexico; in the Atlantic they extended their interest sufficiently offshore to include the Gulf Stream, where they came into conflict with Navy scientist Matthew Fontaine Maury, who considered the Gulf Stream to be definitely in the Navy's sphere. Some biological work even came out of the Survey's work. Bache enlisted the best-known microscopist of the day, Professor Jacob W. Bailey of West Point, to study for organic-remains samples of sea bottom which had been collected. Louis Agassiz, under the auspices of the Survey, studied the origin, growth, character, and probable future progress of the coral formations of the Florida Keys.[18]

Despite the temporary, ad hoc, or informal nature of the Federal Government's early aid to science, it did represent a positive beginning. Federal Government involvement in science has grown at an accelerating rate ever since. Nevertheless, in the nineteenth century, states were of far greater relevance than the Federal Government generally, and this was also true of the support of science. The earliest efforts at the state level were likewise solidly in the geographical tradition, the "geological and natural history survey" being an important early development. Both the idea and the method used in the various state surveys had been conceived in Europe, especially by James Hutton and William Smith in Great Britain. Nevertheless, the formal "geological survey" came to be almost an American institution. Like the national exploring expeditions, the state geological surveys were connected with the drive for internal improvements, both in ideological terms and in terms of the stimulus such improvements could give to the survey. They also had the same basic tensions as those in the exploring expeditions.

The direct forerunner of the state survey was the systematic geological survey of Rensselaer County, New York, made by Amos Eaton in 1821 and paid for by the wealthy patroon Stephen Van Rensselaer. This survey had a definite economic purpose, for in addition to making a geological section and giving a lithological description, Eaton noted the kinds of soil, the most suitable crops, and the best methods of cultivating them—all matters of concern to a landowner. In the following year, again under Van Rensselaer's patronage, Eaton made a similar survey of the entire route of the Erie Canal. The cutting of the canal had left the strata neatly exposed, giving Eaton an unprecedented opportunity for study. His findings, published in 1824, gave a cross-section of the rock formations from Williamstown, Massachusetts, on the east, to Buffalo on the west, supplemented by a section made by Edward Hitchcock across Massachusetts. Thus, from the beginning there was the idea that the geologist, in addition to his concern about the description of rocks and minerals, their origin, and their history, could also locate and evaluate mineral and soil resources so that they might be exploited as sources of wealth to individuals and to the state. This was precisely the approach of the state legislatures when they began to become involved, during the decade of the 1830's.

Two early rudimentary "surveys" which attracted a great deal of contemporary attention had been carried out in North and South Carolina during the preceding decade. Denison Olmstead, a recent Yale graduate who was then professor of chemistry at the University of North Carolina, was responsible for the first of these. In North Carolina, as in other Eastern states, there was much concern with the problem of improving transportation routes, particularly by dredging rivers and digging canals. In 1819, the legislature created a Board of Internal Improvements to prepare plans and surveys. At the same time, the board was to make a recommendation on whether or not there would be enough economic advantage to justify the cost of an extensive canal. The problem, as it was presented to the board, was the same as that later faced by the national railroad surveys. It was on this point that Olmstead hoped to shed some light when he wrote to the board asking for an appropriation of $100 to pay his expenses while he made a geological survey during his summer vacation. Olmstead hoped to collect specimens for use in his lecture, advance his own knowledge, perhaps contribute some new geological information on this unknown area to his fellow-geologists, and as a byproduct, supply the board with information useful to it. Even this modest grant was refused to Olmstead on his first try, but in 1823 the state appropriated $250 a year for four years for a survey to be conducted under the auspices of the State Board of Agriculture.

In the following year, Lardner Vanuxem, professor of chemistry and geology at South Carolina College, received $500 for "making a geological and mineralogical tour during the recess of the college and furnishing the specimens of the same." Both surveys provided useful information for agencies of the state government, but gave little of permanent public value because the legislature did not provide money for publishing the geologists' reports in either case. Although these two were precedents and deserve comment as "firsts," they were relatively minor affairs even when compared with the state surveys of the following decade. They did have a certain amount of propaganda value. Edward Hitchcock, reviewing Olmstead's reports in the *American Journal of Science* in 1828, pointed out that they revealed the existence of mineral riches previously unknown, and he held them up as an example for other states to follow. "What an accession would be made to our resources, and to the knowledge of our country, were a thorough examination to be instituted into our mineralogical, geological and, even botanical, riches!" he exclaimed. This was to become the constantly reiterated theme of state survey promoters for the next several decades.

Following the Carolina precedent, the first full-blown state-supported geological survey was that of Massachusetts, authorized by its legislature in 1830 and begun under the direction of Edward Hitchcock, a professor at Amherst College, the following year. The frugal legislators, grossly underestimating the work involved in such a survey, directed that it be attached to the general trigonometric survey authorized earlier, and at first appropriated only $1,000 for the additional work, although this amount was later supplemented by two additional grants.[19]

Underlying the new interest of state legislators in geological surveys was the economic expansion of the country and the speculative tendencies that accompanied this expansion. The gradual opening of the West and the movement of population away from the seaboard, the growth of industry, innovations in transportation, and changing conditions in agriculture greatly increased the need for a deeper knowledge of the nature and resources of the expanding nation. New states, rapidly added to the Union, provided an ever-expanding number of governmental units to which scientists could appeal. In the spirit of frontier boosterism, residents of new states were particularly anxious to survey their economic potentialities—which they never doubted were abundant—and to take their place quickly on what they regarded as the inevitable spiral of financial prosperity. Private and public institutions were under a great deal of pressure to devote their attention to the new questions being asked about the country, and the existing institutions—the colleges, societies, and other voluntary groups—were inadequate to the demand. For the first time there was a public need which required a centrally administered scientific program directed toward specific ends. The changing society demanded a new approach to science to replace the old pattern of the lone investigator following his own inclinations.

Interest in these problems came from many sides—from people who wanted to know where to build a factory, a mine or a railroad; from a turnpike or canal company which knew that roads and canals should be located where there would be the most commodities to transport; from land speculators who wanted to know about the productivity of the soil in the unsettled areas of their states. The state governments themselves, entering for the first time into roadbuilding on a large scale, knew that, in order to justify large expenditures for roads, new needs for them should be developed and the long-range potential of the area should be assessed.

To these can be added the eager enthusiasts for education who wished to add to the educational uplift of the common man. State legislatures were often moved by such arguments, the members believing, as did most liberal men of the day, in the

Jacksonian idea that education should be available to all men. In Massachusetts, the state legislature provided funds for a library, an experiment station, and a college to develop better farming methods and to disseminate information about them. Geological surveys, which would provide information to all the people about their state, could be seen as a natural extension of this kind of education. Usually the geologist in charge of the survey was required to collect mineral, rock and ore specimens and display them in some central place; he was frequently instructed to distribute suites of specimens to academies and colleges within the state. He always had to make a formal report, which, in some cases, was printed and put into the hands of state legislators and other officers and was made available to educational and scientific organizations. In all states a major purpose was to locate, describe, and publicize such natural resources as salt and mineral springs, building stones, shales, clays, slates, coal, and ores. With this information in hand, any person, it was thought, would have a basis for judging how successfully they might be exploited. The educational and exploitative aspects of the surveys were thus intimately associated with each other.

The aims underlying governmental interest were succinctly stated by Governor Levi Lincoln in his recommendation to the Massachusetts legislature: "Much knowledge of the natural history of the country would be thus gained," he said, "especially the presence of valuable ores, the extent of quarries, and of coal and limestone, objects of inquiry so essential to internal improvements, and the advancement of domestic prosperity, would be discovered, and the possession and advantages of them given to the public."

Edward Hitchcock, appointed state geologist under the act, was a man of broad scientific interests, but he understood his instructions as directing his attention primarily to subjects of possible practical utility. He spent three years exploring Massachusetts, traveling 4,550 miles and collecting five thousand specimens of rocks. He reported that he was delighted by the way his mission was received by the people. A "Universal disposition," he wrote, "was manifested by all classes of the citizens of the Commonwealth, and in every part of it, to do all in their power to forward the objects of my commission. . . . The excursions have greatly exalted my opinion of the kindness, intelligence, and happy condition of our population, and sensibly increased my attachment to my native state." Despite Hitchcock's pleasure with his reception, he found that such popular interest did have its drawbacks. There was pressure on him to publish his results quickly, so much so that he decided to release some of his findings early in 1832. The device he adopted was one generally followed by later state geologists, torn between public and legislative pressures and their own scientific interests. Hitchcock's first report, called *Economical Geology*, was a pamphlet of seventy pages, of which six hundred copies were issued. It located and described the classes of rocks that lay under the soil, emphasizing rocks and minerals that were "useful in the arts": building stones, slates, clays, marls, peat, coal, graphite, and the ores of lead, iron, zinc, copper, manganese, silver, and gold. He even went out of his way to announce a deposit of gold in southern Vermont. Only building stones were present in valuable quantities in Massachusetts, although he said there were "many other natural products" whose value could be determined only after further prospecting.

Parts II and III, issued together in 1833, comprised the *Topographical Geology* and the *Scientific Geology* of the state. Part II, also directed to the popular taste, was really a sort of tourist guidebook. Its main purpose, he said, "will be to direct the attention of the man of taste to those places in the State, where he will find natural objects particularly calculated to gratify his love of novelty, beauty and sublimity." He gave descriptions of mountains, of autumnal scenery, of rivers, of the view from the Boston State House with moral and political contemplations. Part III was a textbook on geology with field examples. It compiled the opinions of foreign authors on land formations, applied De Beaumont's theory of mountains to Massachusetts, and struggled with the Mosaic records. In the drift of erratic blocks distributed so freely over New England, Hitchcock found evidences of the deluge; and in the fossil record he found evidence of special creation. To the scientific part of the report was appended a catalogue of animals and plants of the state, compiled from contributions by different specialists.

Fulfilling another part of his mission, although it had not been incorporated in the original resolution, Hitchcock made a collection of 1,550 specimens of rocks and minerals, which was given to the Boston Society of Natural History. In addition to this, he made for the three colleges in the state a collection of 900 specimens each.

Following the lead of Massachusetts, twenty other states established surveys in the decades of the 30's and 40's. In at least nine of them, there was a close tie between the demand for systems of internal improvements and the authorization of geological surveys. In all, there was foremost in the minds of promoters a desire to unearth the hidden treasure of their states—a desire that was effectively exploited by geologists in their petitions to legislatures. "Details and facts, belonging strictly to pure scientific geology," wrote William Mather to the New York state legislature, "will not be made public until the final report. The object of the annual reports is to give publicity to such facts and localities as may be of practical utility, so that the benefit may be derived from a knowledge of them during the progress of the survey."[20] Success in launching a survey could only come when geologists could manage to arouse the particular interest of powerful groups within each state. Thus, in Pennsylvania it was the coal interests, desiring to find new sources of the material that had already become a major product, that rallied most effectively behind the survey. As *Niles Weekly Register* reported, a few "public spirited citizens" had opened up producing coal mines which were contributing to the "prosperity of our enterprising citizens," and it was believed that there must be many unknown rich coal beds— the "hidden treasure" of the state—which "might reward the labor of a full topographical and geographical survey."[21]

In Kentucky a different group had to be appealed to. The geologist W. W. Mather, after a brief exploratory survey in 1838, reported to the legislature that it had been more than sufficient "to impress one with an assurance of the vast mineral resources of the Commonwealth . . . ready for the hand of industry and enterprise to apply them to the various useful purposes of life." The legislature, however, was not impressed, and it repeatedly denied appeals to institute a survey for the next sixteen years. Only when Robert Peter was able to get the agricultural interests behind the survey by persuading them of the potential value of the soil analysis that would be an important part of it, did a bill get through the legislature.[22]

And so it went, in state after state. The purpose of the geologists' seeking state support was generally to help them pursue their research at an accelerated pace. Paleontology, a particularly useless pursuit from the point of view of the layman, was a major interest of all geologists by the 1830's, for by this time they had generally accepted the idea of dating strata by the fossils they contained, as opposed to the older mineral classification. Economic interests—farmers, miners, industrial groups, turnpike and canal companies—desired creation of the surveys to facilitate the more rapid economic exploitation of regional resources. During the boom eras of the 1830's and 1850's, all were able to support such surveys for the attainment of their common ends: the discovery of new natural riches. The differing goals of their sponsors became increasingly clear, however, when the surveys were put in operation; the results of these differing goals, time after time, were deep rifts and strong antagonisms resulting in a partial disrupton, if not a complete defeat, of the scientific work.

In the first place, there was a general confusion of ends and a lack of any clear conception of the task at hand. Exactly what was economically useful information in the first place was never determined, either by geologist or by state legislature. An extreme example, but only a matter of degrees of confusion worse than others, was the 1847 effort to promote a Mississippi survey. It was not clear to the Mississippi promoters exactly what it would be. It was thought that it would be sufficiently agricultural to include, by counties, a statement of the number of acres in cotton, corn, oats, and potatoes, "how many slaves, how many Negro children born, proportion of sexes and ages on each plantation; number of horses, brood mares, colts reared . . ." After this miscellaneous listing the promoter concluded, "in short, every species of information."[23] Under such conditions, what the legislature accepted as being "useful" was necessarily the result of subjective interpretation of the geologist's report. If legislators saw that the geologist had located seams of coal exactly—not according to the strata of rock or the fossils with which it was associated, but in the geographical terms of the rectangular survey—and if he had determined the thickness, direction, and inclination of the vein and stated whether it was workable, and if he gave a chemical analysis and stated clearly whether the coal could be used for heating, smelting, or something else, *then* the work was "useful." Similarly, if he gave an analysis of the soil and clearly stated its deficiencies, suggested ways to improve it, and recommended the proper crops for each area, then his work was "useful." It was also difficult to gauge the educational value. The training received by those on the surveying teams probably did not count, nor did the information contained in the report have evident educational value unless the geologist prefaced his work with an elementary textbook of geology, as many of them did. But if the geologist presented an orderly display of rocks, minerals, and fossils, each specimen labeled, and the collection could be housed in some public institution, then it had educational value.

Sometimes, the scientists were themselves responsible for their problems with legislators. James Hall, for example, agreed in 1843 to do the paleontology of New York in one volume in one year. In 1847, he presented to the Governor the first of thirteen quarto volumes with hundreds of plates and thousands of pages. Two years later, in a letter to a friendly senator, Hall explained how he had violated his contract with Governor Bouck "for the better service of Science." During the following years, Hall resorted to one subterfuge after another, repeatedly went to the legislature for more funds, sometimes even used his own personal funds hoping to be reimbursed. Somehow, Hall managed to stay on the job for more than forty years, and the volumes he produced were monuments to scientific paleontology, but they did nothing to reassure legislators of the good intentions of scientific men.[24] His survey, Hall claimed near the end of his career, "has resulted in far larger and more interesting collections; and in far more interesting and valuable publications than could have been anticipated by the original promoters"—surely an understatement if Hall ever made one.[25]

Hall was a great deal less successful when he tried the same methods in Iowa, where he became superintendent of the Geological Survey in 1855. Instead of going into the coal lands, as the legislators wished, Hall had a volume published on the geology of the eastern part of the state and another expensive volume on the fossils of eastern Iowa. In order to develop a complete section of the rock strata, he had followed the Mississippi River, hoping to connect the geology of Iowa with that of Minnesota, and he had found it necessary, he said, to trace some formations far beyond the Iowa line into Missouri. Unfortunately for peace between legislators and geologists, geological formations were not always arranged along state lines. Although Hall had sound scientific reasons for proceeding as he did, the legislature was not pleased with the result of his first year's work and accordingly withheld his appropriation. Following the method that had worked so well in New York, Hall began collecting on his own for a second volume on paleontology. But the press became quite hostile, and the legislature, deciding that he was "more desirous of adding to his scientific name than of instructing the people of the state in relation to its resources," refused to continue his appropriation and furthermore refused to reimburse him for private funds that he had spent. Years later, Hall was still trying, without success, to persuade the Iowa legislature to reimburse him. Hall, never one to learn a lesson from a legislature, later tried the same technique in Wisconsin with equally unhappy results.[26]

But it was not all a matter of chicanery on the part of scientists. Even when he followed his instructions to the letter, the geologist did not always please his sponsors. Being a geologist for a state survey was a "dangerous occupation," wrote John Locke to Robert Peter as the Kentucky survey was about to be launched. "The states in ordering surveys seem to have expected impossibilities, and unless the geologist discovers a gold or silver mine on every man's plantation, they stop the survey and cry down the reputation of the surveyor."[27] Indeed, it was not always simply the unwillingness of geologists to subserve practicality that alienated the public from them. Sometimes, it was merely the esoteric form of a geological report that disturbed legislators. When the Kentucky survey report was presented in 1858, for instance, a committee of the state senate admitted that in it there was a great deal of useful information, but regretted that the language "was not better suited to the comprehension of the great portion of the people of Kentucky."[28] Sometimes, as the quotation from John Locke suggests, the geologists gave the "wrong" advice. For example, by 1839 the question of the occurrence of coal in New York had been settled as far as the geologists were concerned, but the people were incredulous. Sir Charles Lyell reported in 1841 that he heard complaints on all sides to the effect that the geologists, having been unable to find coal, had decided that no one else would ever be able to find any. Lyell himself ventured to dissuade an adventurer from sinking a costly shaft for coal but got a reply which led him to the reflection that "Every scientific man who discourages a favorite mining scheme must make up his mind to be as ill-received as the physician who gives an honest opinion that his patient's disorder is incurable."[29]

There are abundant examples of the truth of Lyell's remark. The geologist Josiah D. Whitney came under fire both in Wisconsin and in California for his unpopular advice. In Wisconsin he had been engaged by James Hall, then serving as superintendent of the state survey, to make a report on the lead region where very active operations were under way. Whitney showed clearly with his sections and his crevice maps that the lead deposits were shallow, and frankly discouraged "deep mining." The miners' open and bitter opposition to this advice was a large part of the empiric ignorance that dealt the death blow to the survey.[30] Similarly, in California Whitney came under fire for discouraging petroleum exploitation in the southern part of the state, giving his opinion that the wells would not pay. Even the scientific world was divided on this latter matter, for Benjamin Silliman, Jr., touched perhaps by the speculative spirit, had earlier claimed that oil reserves abounded in southern California and that they could be exploited easily with rather small investments. Even though Silliman's judgment was based upon a hasty examination of inadequate samplings, his was naturally the preferred opinion. "Petroleum is what killed us," Whitney later noted in a letter to his brother.[31]

Whitney, who became head of the California survey in 1860, was perhaps the most tortured geologist of his time. Not only did he call down the wrath of the speculators upon himself time after time, but his own value system was directly counter to that of most honest Westerners as well. As a devoted geologist, he considered the pursuit of pure science to be the most important part of his job. Although explicitly instructed to locate gold-bearing rocks, he wrote shortly after beginning to work: "The first thing required . . . is a knowledge of the general geological structure of the state." With this end in view, he divided the Survey into three divisions. First, he created a Topographical section charged with making maps for the various portions of the state. A second division constituted the Geological group, which was concerned with the investigation of the general geology, paleontology, and economic geology of California. The third section was the Division of Natural History, which studied California botany and zoology —then mostly unexplored. Once this ten-year general reconnaissance was completed, Whitney hoped to devote more time to the "practical" objectives of the agency. Meanwhile, as Whitney blissfully planned his long-range scientific program, the legislators anxiously awaited the practical results they expected to be imminently forthcoming. As the subcommittee on Mines and Mining Interests noted hopefully in 1863:

> We . . . expect that the reports will be of direct industrial value—particularly those portions relating to the Monte Diablo Coal Region, and the modes practiced in this state for extracting gold. The Botanist has devoted much attention to the grasses and clovers and is now engaged in extensive inquiries to ascertain what varieties are most palatable and nutritious to herbivorous animals.

Finally came the publication of the Survey's eagerly awaited first volume—an intensive study of four kinds of fossils. Anticipating the disappointment of the legislators, Whitney had remarked that "It is not the business of a geological surveying corps to act . . . as a prospecting party."

As technically correct as Whitney was, it is still easy to imagine the feelings on the part of a group of intensely practical Westerners, earnestly engaged in seeking a quick fortune for themselves and prosperity for their state, listening to a description of prehistoric reptiles in a book published at state expense. In addition, Whitney had the bad judgment to choose this as his first publication in the midst of the wild speculative boom promoted by war conditions and the Comstock Lode.[32]

In spite of the frustration of men like Whitney, it is clear that a great deal of worthwhile work was accomplished during this early period of the partnership between government and science. On both the state and the national levels, the aid given to science was significant even if it was at times grudgingly given. Before the Civil War there was very little other scientific employment available in America, and hundreds of researchers found opportunities to pursue their specialties under this American form of patronage. The expeditions and surveys also provided training for a whole generation of scientists that was available nowhere else, and, by creating circles of scientific influence which persisted long after the original reason for their existence faded away, grounds were laid for a continuing interrelation between government and science that has been characteristic of the American experience over the past century. Further, by setting men at work on common problems throughout the Union, new scientific specialties were created and a great impetus was given to the drive for scientific organization which marked the century. The American Association for the Advancement of Science could hardly have become a going concern had it not been for the opportunities for professional employment offered by governments, and it is significant that the association grew out of a society of geologists, who had been the first group to organize successfully. Working as they were on common problems, and having a common interest in such matters as standardization of nomenclature, financing, and publication, their organization was a vital one from the beginning.[33] By the same token, the National Academy of Sciences could not have come into being had it not been for the positions of power and influence that scientists had achieved within the government.

From a purely scientific point of view, the period could not be called barren either. Admittedly, a great deal of the scientists' energies were dissipated working against barely tolerable restrictions and in attempting to meet impossible expectations. The preparation of lengthy and largely irrelevant annual reports consumed time that, from a professional point of view, could have been better expended in studying the collections made during the previous season. Undoubtedly, state governments were overly parsimonious in their appropriations, and the quality of the work would have been better had more funds been available for personnel and equipment. Requirements, such as that attached to the act establishing a survey of Delaware in 1837, that an equal portion of the appropriation be expended in each county, could have served no useful purpose other than that of allaying local jealousies. The efficiency expert, sent by the state of New York to instruct its veteran paleontologist, James Hall, in the ways of cataloguing, accountability, and businesslike methods in general, was another example of an unwise interference that only succeeded in driving the almost-octogenarian paleontologist nearly to distraction and, consequently, in disrupting his work.[34]

But still a great deal was accomplished. The natural-history map of the country was virtually laid out in outline and required only the filling in of details—the major geological formations were discovered, most of the flora and fauna were classified, points of latitude and longitude were determined, and the basic meteorological and climatic data were made known. In short, in the three decades between 1830 and 1860, the preliminary work of cataloguing and labeling was completed and scientists were ready to move on into more sophisticated questions of origin, of relationship, and of explanation. In no other nation in history had the indispensable natural history survey been taken so quickly.

That scientists were thwarted in their more esoteric aims, although regrettable from one point of view, was not the great disaster which those who lived through it naturally thought it to have been. By being forced into superficiality, the scientists were able to lay out the broad outlines more quickly; esoteric speculations before the collecting and cataloguing had been done would probably have proved fruitless. It is not at all clear even from a scientific point of view, for example, that intensive work on the fossils of Iowa would at that time have been more valuable than simply digging up more fossils in some other place—the option that Hall was ultimately forced to take because of the "benightedness" of the Iowa legislature and the Iowa press. Frustration, mutual distrust, and outright hostility were quite often the results, and the legacy of those formative decades remains in the twentieth century. Nevertheless, out of it all came the best possible foundation for a scientific understanding of American nature.

One would also have to conclude that the public received a good return on its early investment in science. Scientists on the frontier often did discover water and sometimes found gold. If the petroleum deposits they found in southern California did not result immediately in economically productive wells, they did soon after the associated technology was sufficiently improved; those found by Pennsylvania geologists became immediately valuable. In addition, scientists on the frontier laid out the major transportation routes, aided in the agricultural development of the region, helped brush aside the Indian menace, and helped secure United States title to a great deal of land in the Southwest. In the Eastern and Midwestern states, they laid out canal and railroad routes, and sometimes located valuable minerals. Despite mutual dissatisfaction, on the whole it was a good bargain for all parties involved.

IX

~~~~~~

The Religion of Geology

℀ Speaking before an association of his colleagues in 1844, the geologist H. D. Rogers referred to the "truly favorable field" which the North American continent "with its wide expanse and its peculiar surface" afforded American geologists for testing the leading theories of drift then being advocated in Europe.[1] His theme, that America should serve as a testing ground for European theories, was a common one in the scientific literature of the period. Among the peculiar advantages of America, it was said that the relative isolation of the country and the absence of "unworthy national prejudices" enabled Americans to be eclectic in the best sense of the term; that is, they were able to judge European systems with impartiality and extract the best from each of them. "The remoteness of our situation supplies the time and place," said James E. DeKay in an 1826 speech surveying the possibilities in American natural history, "and we may be supposed to decide between the conflicting opinions of European naturalists, with the same justice and impartiality as if we were removed from them by intervening centuries."[2]

Such claims were made by Americans regarding every field of science from medicine to mineralogy. Americans, because of their theoretical noncommitment and the absence of unworthy prejudices, were able to judge the work of Europeans; and the peculiar features of American nature did often provide a critical test for European theories. In a way, such claims for the superiority of American nature were connected with the belief, kept alive from Revolutionary times, that America was the chosen home of the sciences. It was also a convenient method of explaining away the lack of theoretical contributions by Americans. If one's role is that of arbiter of theories, it is a bit unfair to ask him to formulate some of his own. In many ways the judge's role could be viewed as the more important.

There is a real sense, however, in which the claim was more than either nationalistic rhetoric or apologetic. America from the beginning had been viewed by Europeans as a testing ground for theories, and during the period of discovery it had played its part in the demise of some rather formidable ones. Furthermore, it was true that the natural history exploration of the American interior, which began on a grand scale in the first decade of the nineteenth century, had proved damaging to a number of exclusive theories. Classification systems were being rewritten to account for the newly described American species, many of which proved intractable for even the greatest systematists, and the geology of America by the beginning of the nineteenth century was becoming a great embarrassment to both major geological schools. By the mid-1820's, in fact, it was generally recognized that there were facts about American geology which were simply irreconcilable with either the Huttonian or the Wernerian school of thought. Even Amos Eaton, while pledging allegiance most of his life to the classification of rock formations adopted by European geologists, and never making more than a faint bow to fossils as keys to chronology, still recognized in America a form of geological succession so controlled as not to fit the European categories; and he formulated the new idea of cycles of sedimentation, or the rise and fall of the sea bottom. Eaton, to be sure, based his conclusions on such limited knowledge that he could not adequately defend them; the result was that his contemporaries scorned the conclusions and his successors largely ignored them. But Eaton's vaguely formulated idea was shown to be fundamental to interpretation of stratigraphic succession a half-century later, when John S. Newberry, who had been on the Western surveys, brought forward the conception in clearer definition and Charles Schuchert interpreted it in terms of changing continental shelves.[3]

American geologists were at a positive advantage over European geologists in many respects. The wide stretches of strata in America allowed for easier tracing than the broken character of Europe did, and, as James Dwight Dana observed,

"America has thus the simplicity of a single evolved result," which he contrasted with Europe, "a world of complexities."[4]

The first generation of American geologists, represented most notably by William McClure, Amos Eaton, Benjamin Silliman, and Edward Hitchcock, did their work at a time when the old geological framework resting upon the biblical account was being seriously questioned, and when frantic attempts were being made to reconcile the new discoveries with the Creation story in the Bible.

In the beginning, said the first chapter of Genesis, God created the heaven and the earth; on the sixth day, He created man in his own image, and on the seventh day he rested from all his work. By computing the generations of Adam, the student could readily discover that the date of Creation was about four thousand years before the Christian Era. All of the present features of the earth—the mountains, rivers, lakes and canyons, the oceans, the broad plains—had been created some six thousand years ago during the few days preceding the creation of God's final handiwork, man. Similarly, all the present species of animal and plant life had also been created during this same period. These were the presuppositions of men's thinking, the framework within which all theories of cosmogony were set before the nineteenth century.

Evidence that challenged the old way of thinking had been accumulating slowly since the sixteenth century. That fossils represent organic remains, that they have been laid down under water in sequence, that they reveal a long period of sedimentation, that the surface of the earth has been undergoing constant change, and that earth processes are still in progress; these and other related views had to struggle one at a time for acceptance against the mass of inherited belief. Only gradually were the new discoveries incorporated into men's thinking. Thomas Jefferson is only one example among many genuinely talented scientists who were never able to accept the full implications of the new discoveries; this despite the fact that the best evidence for them had come from the New World, of which he was so ardent a proponent.

By the end of the eighteenth century, however, most scientists had accepted the new ideas and two rival theories had been elaborated to unify all the new knowledge into comprehensive structures. Abraham Gottlob Werner (1750–1817) a professor

of mineralogy at Freiberg, thought that rocks, which he recognized to have been laid down in succession over time, were the
precipitates of a primeval ocean which had once covered the
globe and that they followed each other in successive deposits of
worldwide extent. Basalt and similar rocks, now known to be
of igneous origin, were believed by him to be water-formed accumulations of the same ocean. Volcanoes were regarded as ab-
normal phenomena of little consequence, probably caused by the
combustion of subterranean beds of coals.

James Hutton (1726–97) rested his own explanation of
earth history on the concept of a constantly shifting balance between earth and sea. The earth, he recognized, was subject to
constant erosion, and the material removed was the source of
sediments in the sea. Molten matter, pushed up from below served
as a counteracting force to rebuild and maintain the continents.
The present rocks on the earth's surface, then, had been formed
out of the waste of other rocks, laid down under the sea, consolidated under great pressure, and then subsequently disrupted
and upheaved by the expansive power of subterranean heat.
Hutton's theory, unlike Werner's, gave a crucial role to volcanic
action. For this reason, the followers of Werner became known
as Neptunists and Hutton's followers became known as Vul-
canists.

Although Hutton's theory is the more modern, and Werner's
superficially more easily reconcilable with the Bible, both required
profound alterations in the original biblical story. The essential
point of both is the uniform action of natural processes over time:
both required relatively long periods of time and neither left any
evident room for the purposive actions of a benevolent God.
Werner's theory was the more acceptable to the religious mind,
for he clearly postulated both a beginning and an end. For Hutton,
the processes seemed eternal; there was "no vestige of a beginning, no prospect of an end." This was uniformitarianism with a
vengeance—presumably eternal processes, constantly at work,
operating blindly to produce the earth as we now know it. There
was no room for dramatic interventions in the natural processes,
no allowance for an intelligent creative power. It was this position
that was firmly established by the publication and gradual acceptance of Charles Lyell's *Principles of Geology* in three volumes
(1830–3). Lyell buttressed the Huttonian position with the results
of recent developments in stratigraphy and argued so persuasively
that he very soon carried most geologists with him. With the

publication of Lyell's volumes, the implications of Hutton's theory struck the popular mind with full force, and for three decades the conflict between the geologists' account and the Mosaic story of Creation raged furiously, both in England and in America.

American geologists, accustomed to the regularity of strata they found in their own work, were predisposed to accept many of the new uniformitarian principles of Lyell, despite the obvious danger to the religious outlook. They had long noticed that there was a definite pattern underlying stratigraphic succession, and geologists on the Western surveys had even pioneered in tracing the operations of certain natural agencies. To Lyell, more than any other writer on geology, said Benjamin Silliman,

> . . . we owe our recovery from the illusions of dreams and visions, regarding imaginary powers supposed formerly to exist, but to have become exhausted or greatly enfeebled or even extinct, in modern times. He has proved to us, that the powers of nature are the same now that they have ever been; that except the act of creation and the first outbreak of the new-born elements and energies, there was nothing in the geological laws of former ages different from the present; and that the causes now in operation, acting with greater or less intensity, are sufficient to produce the effects of earlier epochs.[5]

Despite Silliman's apparent contempt for catastrophic geology, and despite the apparent trend of their thinking, American geologists did not in general rush headlong into the extreme uniformitarian point of view of Lyell's followers. They preferred instead to accept some parts of the "new views" and to retain elements of the old. In general, they could not envision that the origin of a large number of the earth's geological features could be explained by referring to processes and forces already in existence. To them it seemed that, no matter how important ongoing processes were in explaining minor modifications of the earth's surface, there were definite limits to what these processes could accomplish. To fully explain the revolutions in earth history, they believed it necessary to turn to forces of a different order from those operating in nature or else appeal directly to the original Creation of the world. This tendency, which placed a great amount of natural phenomena outside the realm of scientific explanation, operated with the same effect when problems of organic evolution arose. Even Silliman, in his praise of Lyell, made a careful point of the "first outbreak of the new-born ele-

ments and energies" and stressed that the causes now in operation had acted with varying intensity in the past. In short, Silliman for all his efforts to remain abreast of modern discoveries,
like most American geologists, was at heart a catastrophist.

Catastrophism, as an alternative to Lyell's concept of incredibly slow, invariant processes, dates its immediate ancestry
to an 1811 paper published by the French paleontologist Georges
Cuvier. In the following year, Cuvier set forth his completed
theory of the earth in a preface to a larger work. The preface,
under the title *Discours sur les révolutions de la surface du globe*,
went through numerous editions both in the parent work and in
separate publication. In it, he set forth the main argument of
catastrophists and indicated the main lines of evidence to support
their claims. Cuvier drew a bold sketch of earth history which he
thought accounted for the successive accumulations of fossils in
the clearly defined strata now found on the dry land areas of the
earth. All of the dry land, he said, had been laid dry and reinun-
dated several times in geological history. Areas now inhabitated
by men and land animals had already been above the surface at
least once—possibly several times—and before the appearance
of the present order of things the land had once sustained quadrupeds, birds, plants, and terrestrial productions of all types.
There had been successive uprisings and withdrawals of the sea,
and the final result had been a general subsidence of the sea to
its present level.

The really important point to notice, Cuvier said, was that
these repeated inroads and retreats were by no means gradual:

> On the contrary, the majority of the cataclysms that produced
> them were sudden. This is particularly easy to demonstrate for
> the last one which by a double movement first engulfed and
> then exposed our present continents, or at least a great part
> of the ground which forms them. It also left in northern coun-
> tries the bodies of great quadrupeds, encased in ice and pre-
> served with their skin, hair, and flesh down to our times. If
> they had not been frozen as soon as killed, putrefaction would
> have decomposed the carcasses. And, on the other hand, this
> continual frost did not previously occupy the places where the
> animals were seized by the ice, for they could not have existed
> in such a temperature. The animals were killed, therefore, at
> the same instant when glacial conditions overwhelmed the
> countries they inhabited. This development was sudden, not
> gradual, and what is so clearly demonstrable for the last
> catastrophe is not less true of those which preceded it. The
> dislocations, shiftings, and overturnings of the older strata

leave no doubt that sudden and violent causes produced the
formations we observe, and similarly the violence of the
movements which the seas went through is still attested by
the accumulations of debris and of rounded pebbles which in
many places lie between solid beds of rock. Life in those times
was often disturbed by these frightful events. Numberless
living things were victims of such catastrophes: some, inhabitants of the dry land, were engulfed in deluges; others,
living in the heart of the seas, were left stranded when the
ocean floor was suddenly raised up again; and whole races
were destroyed forever, leaving only a few relics which the
naturalist can scarcely recognize.[6]

Complementary to the catastrophic argument, which usually—although not in Cuvier's formulation—assumed the total
destruction of all living species at each catastrophe, was the evidence provided by the new science of comparative anatomy, particularly applied to the vertebrates by Cuvier and his followers.
By this method, thought Cuvier, one could with a high degree of
success investigate the past, the number and order of global revolutions, and the history of creation. The latest researches of
paleontologists, armed with the new tool of comparative anatomy,
had shown that fossils occur in strict stratigraphic successions, in
the exact order of their creation. First came the fish, then the
amphibia, followed by reptilia, and finally one finds the first
mammalia. Moreover, the older the strata, the higher the proportion of extinct species that one found. Morphological evidence
disproved the Lamarckian contention that existing forms were
derived from a gradual modification of earlier ones, for there
were no transitional forms anywhere in the fossil record. Since
no human fossils had turned up anywhere, it followed that mankind must have been created at some time between the last
catastrophe and the one preceding it; this occurred probably no
more than five or six thousand years ago.[7]

The harmony of Cuvier's catastrophism with older modes
of thought is obvious. The notion of successive catastrophes, each
one removing an older creation and preparing the way for a new
—and higher—one, harmonized with religious beliefs about the
unfolding of God's plan by direct, Providential design and apparently separated man, God's highest creation, from all earlier
creations by an unbridgeable gulf of cataclysmic proportions.
Any suggestion that man had been developed from older forms—
suggestions that had been appearing with ominous frequency

since the eighteenth century—were at once ruled out by Cuvier's concept combined with an appeal to the fossil record. Again, the nature of the last catastrophe practically invited identification with the Noachian deluge and thus managed to fit nicely with the biblical account and with the prevailing interpretations of existing earth features. The sluicelike conformation of hills and valleys; the pattern of drainage basins; the occurrence of gorges, ravines, and water gaps torn through mountain masses, and of buttes and mesas separated from the parent formation by valleys of denudation; immense deposits of gravel, carried from distant regions, on hills and slopes where no river ever could have drifted them; the nature and condition of organic remains in diluvial gravel; the consistency of similar phenomena all over the world —all this required a torrential deluge to explain the present appearance of the earth. Cuvier, by solving at once the most perplexing problems of structural geology and by appearing to support the accuracy of the biblical account, appeared to his contemporaries the scientist of the age—the man who had finally wrested from nature the secret of Creation itself and thus restored to God the supervision of nature which the Enlightenment philosophers had tried to take from Him.

"There is decisive evidence," wrote Silliman, "that not further back than a few thousand years an universal deluge swept the surface of the globe." This deluge, brought about through direct intervention of the Creator for the purpose of punishing and partially exterminating the race, had been sudden in its occurrence, short in duration, and violent in its effects. Silliman, like others of his time, went out of his way to bring his science into harmony with Genesis. In order to account for the biblical expression "the fountains of the great deep were broken up," he offered the suggestion that, contemporaneous with the forty days and nights of rain, a deluge of water burst forth from the bowels of the earth, whence it was forced by the sudden disengagement of gasses. Silliman had already accounted for the presence of subterranean water by assuming that it was derived from the primeval ocean of Wernerian geology at the time it shrank away and left the dry land. Sufficient water to cover the highest mountains, he calculated, could have been contained in a cavity the cubical contents of which was only one two hundred and sixty-fifth part of the globe.

Assuming that the antediluvian mountains were the same

as today, but somewhat higher (say, 5½ miles), and accepting
the fact that they were covered with water, Silliman proceeded to
show that, with a time limit of forty days, the water must have
risen at the rate of a foot in two minutes, or 726 feet in twenty-
four hours. The effect of even this tremendous rate would have
been increased by inequalities in the surface of the land. Silliman
was able to draw a very graphic picture of "the inconceivably
violent torrents and cataracts everywhere descending the hills
and mountains and meeting a tide rising at the rate of more than
700 feet in twenty-four hours."[8]

Ironically, the very advantages enjoyed by American geologists
predisposed them to catastrophic explanations. Thanks largely
to the state surveys and the Federal Government's exploration in
the West, by the 1850's the geological system of the United States
was thought to have been fully revealed, and it required only a
further tracing across the continent. Morton had laid open the
cretaceous, Conrad the tertiary, Hall the palaeozoic strata, and
the brothers Rogers the carboniferous, said Dana, expressing his
satisfaction with American geology.[9] Others were busy filling in
the gaps left by the work of these leaders. The publications of the
New York survey had presented to the world a rich series of strata
extending through the Cambrian, Silurian, and Devonian periods.
The thickness, vast extent, and undisturbed order of the beds
suggested the presence of tranquil and uniform conditions
throughout long epochs of time.[10] But, paradoxically, the very
completeness of American strata, which had made American geologists conscious of uniformly acting processes, tended to lead
them away from the extreme uniformitarian conclusions of Lyell's
followers. As Dana pointed out, the "abrupt limits of periods" was
also more readily observable in America than elsewhere.[11]

   "The sudden and remarkable changes in the organic contents of the strata as we pass from one formation to another,"
wrote Edward Hitchcock, "coincides exactly with the supposition
of long periods of repose, succeeded by destructive catastrophes."
One found these sudden changes even when, as was generally
true where Hitchcock had worked, none of the regular strata were
missing.[12]

   As the Hitchcock quotation indicates, the relatively smooth
nature of American strata easily led American geologists to the
conclusion that they had the complete record before them. The

abrupt limits of strata, in one case easily traceable for nearly three hundred miles along the Erie Canal route, the absence of transitional forms in the fossil records, and the apparent evidence that one eon had followed another suddenly, without warning or preparation, suggested to them that a succession of catastrophes had wiped out all life on earth many times in the past and that it had begun again by the act of a Creative power. Amos Eaton, for example, believed that the strata exposed by the Erie Canal route indicated that there had been at least five general destructions of all animal life in geological history. Each of these catastrophes had been followed by a new Creative act.* If Omnipotent Power, as Dana said, had been limited to making monads for development into higher forms,

> Many a time would the whole process have been frustrated by hot water, or by mere changes in the earth's crust, and creation would have been at the mercy of dead forces. The surface would have required again and again the sowing of monads, and there would have been a total failure of crops after all; for these exterminations continue to occur through all geological time into the Mammalian Age.[13]

In the absence of any physical mechanism to explain their findings, the evidence contributed by paleontologists and geologists actually reinforced and seemed to demand a belief in miraculous intervention. Earlier, geology had been highly suspect to the religious mind. But now, in providing what seemed to be a sound basis for catastrophism, it appeared to be the most congenial of the sciences, and was quickly pressed into theological service. How remarkable, said one lay religious thinker, that geology is now disclosing a period when the earth was inhabited by other species of animals than at present. This discovery, the writer concluded, makes absolutely necessary the admission of a Creative power, for the only two choices open to man is to assume a first cause, or to assume that the series of present races and animals had no beginning. Geological discoveries, in clearly ruling out the second alternative, had done a great service for natural theology.[14] The atheist, observed another writer on the

---

*George P. Merrill: *The First One Hundred Years of American Geology* (New Haven, Conn., 1924), p. 299. Reflecting the touch of *Naturphilosophie* in his thinking, Eaton attached a kind of mystical significance to the number five; see, for example, his article, "The Number Five, The Most Favorite Number of Nature," *American Journal of Science*, XVI (1829), 172-3.

same subject, "says all things continue as they were at the beginning"; the geologist can confute him by showing direct evidence of many changes, including the appearance of new life forms requiring Creative power—"divine interposition."[15]

Catastrophism is probably the key to the new rapport worked out in the first half of the nineteenth century between science and religion. Discontinuous progress of organic life throughout the earth's history provided a defense against the threat that science posed of representing the universe as a great cosmic machine propelled by its own blind forces and laws, unresponsive to divine control. This had been the ever-present threat lurking behind the Enlightenment concept of a watchmaker God —the Creative power that had, in the beginning, created all the laws of nature and then relaxed to watch his Creation work out its destiny without further "fiddling around" with it. If matter can sustain itself independently of God, said one concerned minister, "it is but a step farther on the same road to suppose it equally independent in its origin."[16] Francis Bowen, Harvard logician and self-appointed defender of the faith, spoke of "a set of metaphysical atheists" who were seeking to "pull down the eternal from his throne in the hearts of man, and substitute in his place a principle—an idea,—a nothing,—without consciousness, personality or intelligence."[17]

Geology, to minds that had always been uncomfortable with the Enlightenment God, came as a heaven-sent opportunity to reassert the continuing activity of the Deity. "The false wisdom of the eighteenth century," said one critic, "armed with some superficial scientific notions, pretended to destroy the edifice of religion." But recent developments in cosmogony, geology, and natural history had caused "their calumny" to fall upon the authors.[18]

To the religious mind, in a word, geology was providing incontrovertible proof of divine activity in the earth's history. The periodic catastrophes, the destruction of life attending them, and the reappearance of new forms of life demanded appeals to divine intervention; and the process seemed to indicate the continuing interest of God in his Creation:

> The fact that various systems of organic life have appeared adapted to varying conditions and that development took place on earth before life appeared and in preparation for its appearance shows that God has always exercised over the globe a superintending Providence.[19]

Geology, said a reviewer of the second edition of Hitchcock's work, "abounds with such interventions." The facts of the other sciences might be brought into ceaseless cycles, but geology showed a divine hand cutting asunder the chain of causation at intervals and beginning new series of operations.[20]

It would be difficult to overestimate the importance of this line of thinking in the pre-Darwinian era. The appeals to miracles and special providences demanded by catastrophic geology and paleontology were viewed as openings to vast new resources for the natural theologian. Previously, natural theologians had relied upon the argument from design in the universe. However, there were obvious shortcomings in this argument—especially in view of the criticisms brought against it by Hume, which neither Paley nor the other exponents of the design argument had over-come effectively. As Edward Hitchcock pointed out, Paley had not counteracted the Humean criticism that the act of Creation which the natural theologian sought to prove was a unique, isolated event, and as such it could not be established by the empirical methods of proof the human mind generally depended upon.[21] Besides, as others had observed, men really see in material nature nothing different in kind from the manifestations of design ex-hibited by the lower animals, and there is no reason to assume the existence of a higher power.[22]

The doctrine of design in its original form also possessed the weakness that it did not offer evidence of God's Providential control over nature. It was developed during a period in which a static view of nature dominated, and there was little need to ac-count for developments in nature subsequent to the original act of Creation. The "nice contrivances" which Cotton Mather and other eighteenth-century spokesmen had become ecstatic over, had been made *once*, at the Creation, and it was only in witch-craft, epidemics, or other disasters that Mather's generation could find evidence of continued divine activity. Historical geology and paleontology, however, placed a premium on demonstrating con-tinued divine activity in nature; and it suggested that this divine authority was essentially benevolent.

The argument from design was, to a certain extent, still useful for suggesting continued divine activity, and it was used vigorously throughout the nineteenth century—three books by Paley had passed through ten American editions before 1841. Its strength, however, was further weakened by uniformitarian geology, which demonstrated that apparently designed effects could be produced by the operations of natural forces well known to science. In view of these considerations there existed a pressing need for convincing evidence that the hand of God operated directly in nature. If this could not be shown, the door would remain open for theories of mechanistic materialism. The existence of general and inexorable laws certainly did not preclude the existence of a personal God; but still, so long as events moved on in their unvarying consistency, one could not infer the existence of a Being who was *above* law; who had in Himself anything higher than the laws themselves manifested. But, said Mark Hopkins in commenting on the weakness of the design argument, "could this uniformity be once broken up, could this rigid order be once infringed for a good and manifest reason, it would change the whole face of the argument."[23]

According to Hitchcock, geology provided a defense against such reasoning—a defense which had not been available to Paley and other exponents of the argument from design. The effectiveness of this defense derived from the fact that geology revealed developments throughout the earth's history which could only be attributed to God's direct action. Thus it did not matter that the theist could not prove the original creation of matter in such a way as to satisfy the Humean skeptic, for he could now take his stand on "the arrangement and metamorphoses of matter":

> We shall not contend with him [the atheist] as to the origin of matter, but challenge him to explain, if he can, without a Deity, its modifications, as taught by geology. These give theism all that is necessary for its defense.

No other science, claimed Hitchcock, "presents us with such repeated examples of special miraculous intervention in nature." Even though one might grant that the earth had evolved from a molten state to its present condition by natural means—as most American geologists after the 1830's did—this still did not account for the origin of life. Since no known scientific laws could explain the beginnings of life, it was necessary to turn to a creative act—not simply one Creative act but many of them—in order to account for the various organic populations that had inhabited the earth. The geologic record, Hitchcock asserted, reveals that again and again the earth has been peopled by creatures which have been completely erased from its surface, to be replaced by new species. Geology gives no support to theories that later races had

developed from earlier races, for there were clear breaks in the paleontological record. Here, then, was evidence of the numerous
Creative acts which had been lacking in the prior efforts of natural
theologians to counteract the doctrine of the eternity of matter.
As Hitchcock said:

> If we can prove that the power, the wisdom, and the benev-
> olence of the Deity have again and again interfered with the
> regular sequence of nature's operations, and introduced new
> conditions and new and more perfect beings, by using the
> matter already in existence, what though we cannot by the
> light of science, run back to the first production of matter
> itself?

All of the facts of geology, he concluded, made the doctrine of
the world's creation a more legitimate presupposition than the
doctrine of the eternity of matter.[24]

The final evidence presented by Hitchcock was the recent
commencement of the human race. "And who will doubt that his
creation demanded an infinite Deity?"[25] Hitchcock did not, nor did
he doubt that man's creation was the end for which the whole
system of nature was established. Geology furnished numerous
ways in which the earth had been prepared for his comfort and
well-being. There was the soil, subject to weathering and disin-
tegration, that vegetation might grow in it for man's use. The
broken and upturned condition of the strata of the earth had
made it possible for man to discover various mineral products. A
conclusive argument was based on the distribution of water
throughout the globe: "We should expect . . . that this element . . .
must be very unequally distributed, and fail entirely in many
places; and yet we find it in almost every spot where man erects
his habitation."[26] What better example of a "nice contrivance"
had the older natural theologians found than this?

Geologists differed in the exact manner in which they
sought to harmonize Genesis with the portions of the new geological ideas which they accepted; few had any doubts that
harmony could be reached somehow. Some, like Silliman, proposed that the Creation story seemed to indicate a process that
took place in stages over a period of time. The six days were
lengthened into periods or ages; then it was a fairly simple matter to achieve at least superficial harmony between science and
Genesis. Silliman, realizing that his ability to demonstrate this
harmony contributed enormously to the plausibility of his theories

and the status of his science, made the most of his opportunity by adorning his works with appendices, charts, and diagrams showing the correspondence between the two. Hitchcock, for his part, thought that Silliman's view was unsound, for he perceived that even under a loose interpretation the order of Creation in Genesis did not correspond exactly with the order of fossils in the rocks. His solution, at which he had arrived by 1823, like the earlier one of English catastrophist William Buckland, was to postulate a long period between the first Creation "in the beginning" and the six demiurgic days.[27] In that period, the processes of sedimentation were at work, and many species of plants and animals, whose remains are now found as fossils, were created by God, lived their allotted span of ages, and then passed away. The Mosaic account pays no attention to these Creations, because they are irrelevant to the main moral purpose of the Bible. The six days were the latest period of new creation of species, during which the world was fitted up for man and its other present occupants.[28] As another geologist said, the Genesis account was about the "renovation" of the surface of the earth, not its creation.[29]

There is no doubt about the sincerity of Silliman, Hitchcock, and most of the others who sought to reconcile science with the Bible. Personally, most American scientists during that period were deeply religious men, and they were genuinely concerned with the bearing of their work on their religious beliefs. Neither Silliman nor Hitchcock ever approached science on the basis of mere ulility or idle curiosity. Foremost in their consciousness remained the firm belief that they were investigating the laws by which the Creator worked, and that an important part of their task was to glorify that Creator. In delivering his introductory lecture to his class at Yale, Silliman always spoke first of religion: "It is at the head of all science; it is the only revealed one, and it is necessary, to give a proper use and direction to all the others." The belief that "we honor the Divine Author by tracing the operation of his laws" recurs far too many times in Silliman's writings for it to have been mere rationalization or self-seeking.

Silliman tried to press not merely geology, but all of the sciences, into the service of religion. Speaking of a lecture he had delivered on chemistry at the Lowell Institute in 1836, he commented that the "moral and religious bearing of the lectures was decided in illustration of the wisdom, power, and benevolence manifested equally in the mechanical and chemical constitution of our world." He lectured there again in 1841 and had the satisfaction "to contribute not only to the mental illumination of the people, but to the increase of their reverence for God."[30] Hitchcock was equally guided by religious values. His whole teaching career had as its purpose not to explain scientific facts but to illustrate by those facts the "principles of natural theology." He made no claim to give his students more instruction in the natural sciences "than is necessary to understand their religious bearing." But this, he said, is the most important use of the sciences, "as it is of all knowledge."[31]

But despite the obvious sincerity of the reconcilers and the exemplary Christian lives most of them lived, they were never able to entirely remove the suspicion in which science was held by a part of the population, and they were never able to quiet all of the critics. A part of the difficulty was in their inability to agree with each other about the exact mode of reconciliation. The history of geology, said one disillusioned critic, presents a long array of theories which have had their brief day and were successively exploded by fresh discoveries in the science. The physical facts are always the same, he said, but in deducing principles from these facts, "geologists differ as much from each other as men differ on other subjects." The writer admitted that he had once accepted the theory of a long interval between Creation and the final arrangement, a theory much like that of Hitchcock. But now that Silliman had begun to argue for indefinite periods to represent days, he did not know which authority to accept. The best solution, he thought, was to return to acceptance of Moses' literal account and the concept of the short duration.[32] Give to one of the more "modern geologists," said another, "certain millions of years, a sufficient number of earthquakes, a whole battery of volcanoes, a few dozen deluges, and the rise and fall of half a dozen continents, —and he will frame a theory off-hand, which will account for the most perplexing phenomena."[33]

No matter which form the reconciliation took, it was likely to severely challenge the literal accuracy of the Genesis account. This was threatening to men who from childhood had been accustomed to cling firmly to the Bible as the one absolutely true revelation in an otherwise perplexing world. If the account of Creation is a myth, wrote one theologian in a leading religious journal, so is the Decalogue and, for all we know, everything else in the Bible.[34]

Scientists, many critics continued to argue, should entirely avoid all such speculation concerning the Creation of the earth. It is not necessary to our happiness, said one writer, to know how the earth we inhabit was formed. It is enough for us, in the time-honored Baconian way, "to endeavor to become acquainted with the materials that compose it, their relative positions, and the laws by which they are respectively governed."[35] "Why in the name of common sense," said another in a review of Silliman's effort at reconciliation, should we "torture our brains in accounting by the known laws of nature, for an event which was confessedly the result of miraculous interposition, and of course, in direct opposition to the action of those laws?" And the reviewer penetrated straight to the heart of the difficulty when he charged: "To acknowledge the deluge to have been the result of a miracle, and in the same breath endeavor to account for it by the action of natural causes, we humbly consider to be a palpable absurdity."[36]

Despite the best efforts of American scientists, lingering doubts remained concerning the possible threat that science posed for religion. But in general most Americans were persuaded that science is to "*confirm* faith, and *proclaim* religion," as was commonly stated. And it was generally held, by scientists and laymen alike, that one of the scientist's main duties was to demonstrate the "entire harmony" between nature and revelation. Catastrophic geology offered one route to harmony, and the growing preoccupation of American geologists with paleontology, along with the regular nature of American strata, predisposed them to accept catastrophic explanations. As long as at least a superficial "fit" was easy to contrive between the facts they were uncovering and the facts important to the religious view of life, conflict generally remained below the surface. At the same time, the obvious threat of the harmonizing effort to the accuracy and dependability of the scientist's work remained only potential. But what would happen when harmony became difficult to achieve? When science reached a conclusion that was so manifestly opposed to the entire religious view of life that conflict could not be avoided? When lingering doubts about science in the minds of a few became a swelling chorus? At the very time that American geologists were rejoicing in the security that catastrophism had brought them, research was being conducted that would bring all of these conditions to pass.

X

~~~~~~

Evolution Comes to America

In July 1857, the distinguished Harvard botanist Asa Gray received what was to prove the most important letter of his life. An English correspondent with whom he often exchanged scientific information had decided to let him in on a theory he had been developing for the past nineteen years. Previously, Charles Darwin had only shared his secret speculations with a handful of English friends. But a year earlier he had opened a correspondence with the American botanist, who had generously supplied him with information crucial to the development of a startling theory. Apparently persuaded that the American botanist was both trustworthy and a man of genuine scientific merit, Darwin had decided to swear Gray to secrecy and add him to that select circle of intimates. "To be brief," wrote Darwin, "I *assume* that species arise like our domestic varieties with much extinction; and then test this hypothesis by comparison with as many general and pretty well-established propositions as I can find made out. . . . And it seems to me that, *supposing* that such hypothesis were to explain such general propositions, we ought, in accordance with the common way of following all sciences, to admit it till some better hypothesis be found out." Darwin, saying he was certain that the confession would make Gray despise him, concluded by hinting that he could explain the mechanism by which nature had created the amazing variety of species then known to naturalists.[1]

In reply, Gray presumably satisfied his English friend that he did not despise him for his heterodox ideas, for the transatlantic mails soon brought a draft outlining the theory in greater detail. And by 1859, when Darwin's famous book finally appeared, Gray had become at least a partial convert, for it was he who led the battle for American acceptance of the new theory of evolution by natural selection, favorably reviewing the book in the authoritative *American Journal of Science,* writing a series of popularized defenses for the *Atlantic Monthly,* and arranging for an American edition of *The Origin of Species.*

A product of over twenty years of thinking, observing, writing, and rewriting, Darwin's book bore a laborious Victorian title: *On the Origin of Species by Means of Natural Selection, or the Preservation of Favored Races in the Struggle for Life.* Its thesis, stated in the title, could only be described as intellectual dynamite; it is perfectly understandable that Darwin thought Gray might find it repugnant. In place of the static conception then prevailing—that species of animal life were fixed forever at their first Creation—Darwin proposed that one species had developed out of another by a series of slow changes through countless eons of time. The process of change was essentially natural, requiring no creative interventions on the part of a Deity, and no miraculous works of any kind. If there had indeed been a Creation in the theological sense—and Darwin was inclined to believe that at the beginning the Creator might have "breathed life into" four or five forms at the most—it had occurred at some time in the dim geological past and was largely irrelevant as an explanation of the current life forms.

The publication of Darwin's elaborately documented argument provoked a major intellectual battle that, in fact, outlived the century and still emerges from time to time. A common reaction was that of Adam Sedgwick, a geologist and Darwin's former teacher, who dismissed the theory as "a dish of rank materialism cleverly cooked and served up merely to make us independent of a Creator." The American theologian Charles Hodge, in answer to his own rhetorical question, "What is Darwinism?", stated flatly: "It is Atheism." As if in confirmation, others, such as Thomas Henry Huxley, who became known as "Darwin's bulldog" for his vigorous championship of the theory, hailed it as man's liberation from metaphysical and theological dogma and announced that he was ready to go to the stake for it. To John Fiske, American historian, philosopher, and scientific popularizer extraordinary, Darwinism was the crowning glory of the nineteenth century, which Fiske characterized as the "period of the decomposition of orthodoxies." Still others tried in varying ways to accommodate the new theory to old ways of thought.[2]

Actually, the idea of evolution was already very old, and for limited purposes it had even attained a certain respectability. As Asa Gray noted in one of his many reviews of Darwin's contribution:

> To us the present revival of the derivative hypothesis in a more winning shape than it ever before had, was not unexpected. We wonder that any thoughtful observer of the course of investigation and of speculation in science should not have foreseen it.[3]

In this revealing statement Gray was not referring simply to his own personal knowledge of Darwin's hypothesis over two years before publication of *The Origin of Species*. In retrospect, at least, one can see that the whole drift of thought, both scientific and non-scientific, had been leading toward more dynamic views of nature since the end of the eighteenth century. By the beginning of the nineteenth century, historians, political theorists, and philosophers were making free use of evolutionary concepts in their speculative works, astronomers were articulating an evolutionary theory to explain the formation of the universe, and geologists were employing evolution to explain the formation of topographic features.[4] Despite the countercurrent of catastrophism, uniformitarian ideas were rapidly gaining headway, even among American geologists who clung longest to catastrophic explanations.

The philosophical idealism of the German variety, closely allied with the Romantic movement, had caused many men to think in terms of a dynamic, evolutionary universe. It was the idea of nature as an organic growth, basic to the Romantic mind, that inspired a plethora of pseudo-scientific evolutionary theories as well as more respectable ones such as the nebular hypothesis, which pictured the solar system slowly evolving out of a gigantic amorphous mass by condensation and contraction. The same Romantic orientation, for instance, also influenced German historiography by stimulating an awareness that the present had its roots in the past and was a growth out of the past. As Collingwood, the philosopher of history, put it, evolution "was greatly strengthened, if not actually suggested, by the study of human history, where forms of political and social organization can be seen to have undergone an evolution of the same kind."[5] To men accustomed to thinking in terms of civil and natural history as two sides of the same coin, as had been common since the Enlightenment, evidence of evolution in the one was bound to suggest evolution in the other.

Also emerging from Romanticism was the rise of the idea of progress, which had assumed its characteristic form by the beginning of the nineteenth century. This, too, possessed evolutionary implications. The Enlightenment philosophers, bound to an essentially static view of natural processes for all their exuberant faith in their own time, never developed a conception of progress as linear development. Progress, to the eighteenth-century mind, was a reality but it was also something ephemeral and constantly subject to cyclical reverses. Persuaded that their own times were special, and profoundly contemptuous of their immediate predecessors as they were, Enlightened philosophers yet compared the eighteenth century with the height of classical antiquity and assumed that the cycle of enlightenment and decline would repeat itself. Thomas Paine, for example, assumed in his argument for American independence that there was but one time in a nation's history when it was "ripe" for a republican revolution; should that precious moment pass without action, a nation would revert to monarchical institutions as the cycle of history rolled on.* But the early nineteenth-century view, enthusiastically endorsed in the United States, proved radically different. Continual progress came to be thought inherent in the historical process; for all practical purposes, it was inevitable. It was simply in the nature of things that they should improve with the passage of time, and there was no thought of any cyclical reversion. Evolution, interpreted as the development of progressively higher forms out of lower, fit neatly with the idea of progress.

Important as were these non-scientific sources of evolutionary thought, currents in natural science itself were leading even more strikingly toward a theory of evolution. Since the geographical explorations of the fifteenth and sixteenth centuries, naturalists had mainly turned their efforts to ordering and systematizing what was found in nature. While students of the present day are inclined to dismiss the "mere classifier" as a benighted

---

*Condorcet, in France, and Benjamin Franklin, in the United States, were apparent exceptions to this generalization about the eighteenth-century conception of progress.

curiosity from the past, this work had been essential; an orderly
and systematic arrangement of life was an absolute necessity
before the investigation of evolution, or even its recognition, could
take place. Before life and its changes and transmutations could
be pursued into the past, the order of complexity of the living
world had to be thoroughly grasped and arranged in some comprehensible pattern. This had been accomplished, as we have seen,
in terms of a static view of nature that had been most useful as an
organizing concept for masses of data. But by the beginning of the
nineteenth century the utility of the older view was rapidly breaking down. Investigations since the middle of the previous century
had led to an accumulation of facts that was harder and harder
to reconcile with a static view of nature. This was particularly true
with advances in sciences such as geology, paleontology, embryology, anthropology, and philology during the late eighteenth and
early nineteenth centuries. Because so much of their subject
matter consisted of vestiges of the earth's past and the lives of
its human and animal inhabitants, it was almost inevitable that
researchers should begin to pay more and more attention to questions of origins, growth, and development. In short, these sciences
became historical in their orientations, and the scientist found
in them the inspiration and evidence necessary for constructing
a history of the earth.

The old idea that the earth had remained essentially the
same since its beginning had already become untenable with the
accumulation of geological and other data. At first, this accumulation had led only to the adoption of catastrophism and its
biological equivalent "progressionism" as ad hoc hypotheses that
did little violence to the received ideas. These compromises had
been possible because generally it was the unusual in nature that
at first brought on the awareness that something in the past had
happened to change the face of the earth and its inhabitants.
Apparent oddities of nature, strange artifacts of ancient cultures,
and newly perceived racial differences all suggested remarkable
differences between the past and the present. In addition to cyclical processes in nature real developmental and creative processes
had obviously been at work. Irreversible change, it became more
and more apparent, had indeed occurred.

Benjamin Franklin, for example, upon observing that bones
of animals natural to tropical climates had been found in temperate regions of North America, concluded that radical climatic
changes had characterized the earth's history. Conjecturing that

perhaps a shifting of the earth's poles had caused these changes, he reasoned further that any such shift had undoubtedly stimulated violent currents in the fluid center of the earth, which might in turn explain phenomena such as the catastrophic transformation of seabeds into mountains—mountains investigators sometimes found inexplicably littered with marine shells.[6] Other students approached similar conclusions. Benjamin DeWitt, upon observing in 1793 natural configurations that today are known to be results of glaciation, expressed doubt that the configurations had existed in the beginning, but must have been produced by subsequent events. Thomas Jefferson earlier had come to a similar conclusion while contemplating the passage of the Potomac River through the Blue Ridge Mountains. "The first glance of this scene," he observed in his widely read *Notes on Virginia,*

> Hurries our senses into the opinion, that this earth has been created in time, that the mountains were formed first, that the rivers began to flow afterwards, that in this place particularly they have been dammed up by the Blue Ridge of mountains and have formed an ocean which filled the whole valley; that continuing to rise they have at length broken over at this spot, and have torn the mountains down from its summit to its base.[7]

Although Jefferson himself refused to explore the fuller implications of his expressed impression, his comment is nevertheless symptomatic. That such natural scenes as these first began to break down the old, static geological theory was clearly recognized by an early critic who came upon Jefferson's passage and drew a moral from it:

> The first ambuscade of infidelity, according to custom, is among the mountains. Whenever modern philosophers talk about mountains, something impious is likely to be near at hand. Not more numerous are the streams which flow from the Alps and the Andes, than the objections which they have afforded to these sophisters against the sacred history. When mountains are mentioned in their writings, the well-meaning reader has need to guard against some wicked insinuations with as much vigilance as he would against the lurking panther, if he were passing through the forests which shade the sides of those mountains.*

---

*"Observations upon certain passages in Mr. Jefferson's Notes on Virginia which appear to have a tendency to subvert religion and establish a false philosophy"; pamphlet, dated 1804, Newberry Library Collection, Chicago.

Of course the questioning aroused by Jefferson's kind of experience involved historical concepts only in the crudest and most primitive way. Basically they posed questions of origin and not questions of historical process. It was characteristically assumed, even by the most scientifically radical, that at some moment in the past, usually by a process bordering on the miraculous, the earth had been formed with its contemporary features intact, and *then* the regular laws of nature—those now in operation—had come into existence to keep things as they were. The notion that nature's present laws could themselves account for an ongoing process of development, in which the current topography represented only a passing phase, was far too revolutionary for the essentially conservative nature of eighteenth-century thought. Those who posed questions seldom concerned themselves with the ordinary aspects of nature. These were assumed to have been fixed at their creation and were therefore not subject to explanation.[8] Nevertheless, the very act of questioning served to accustom naturalists to thinking in terms of change, however limited, and in so doing their questions clearly pointed in the direction of evolution. When men began contemplating the conventional aspects of nature in a questioning way, and when they shifted their emphasis from origin to process as Hutton and later Sir Charles Lyell did, the conclusion would be inevitable. Geology, by the 1830's, had become essentially an evolutionary science.

The application to biology of a form of evolutionism was not new either, although earlier versions had not been considered respectable by the scientific community. The Frenchman Jean-Baptiste Lamarck, a half-century earlier, and the Scotsman Robert Chambers, only fifteen years before Darwin, had published books defending evolutionary theories of the development of organic species. Indeed, Darwin's own grandfather, Erasmus Darwin had published poetic versions of biological evolution before the beginning of the nineteenth century. This means that even the most radical aspect of Darwin's theory appeared among minds to a certain extent prepared to discuss it. For years after Chambers's anonymous publication, which had touched off an extended controversy, the subject continued to come up and many of the same arguments were repeated. Most critics claimed that *Vestiges of the Natural History of Creation* was both bad science and atheistic. Invariably, they identified Chambers's ideas with those of Democritus and the Epicurean philosophers who had made the world a "fortuitous concourse of atoms." Proceeding from this point they often issued related charges: the system banished God from His own universe; it virtually annihilated religion; it pushed the Creator so far back that He was no longer needed; it was an attempt to exclude God by natural law.[9]

Underlying such allegations was the conception of God current at that time. The deistic idea of God as a first cause who, after the Creation, had refrained from "fiddling with his universe" was now in disrepute; the accepted concept was that God directly governed the universe by His day-to-day action. This new conception, as indicated in the previous chapter, had been one of man's first responses to such evidences of change that he had come to accept. William Paley's *Natural Theology*, which offered the adaptation of organisms to enviroment as evidence of devine intervention, had fastened this essentially circular mode of thinking upon the nineteenth century. Catastrophic geology, with its assumption of successive exertions of creative power as a means of accounting for the progression of fossils in the rocks, had furthered the interventionist idea by bringing the weight of science to its support. With both science and religion assuming a need for successive interventions in the natural process, it is no wonder that Chambers's and others' attempts to extend the range of natural law provoked such violent reactions. "If there is no need of a bricklayer," remarked Francis Bowen in his review of *Vestiges*, "we may discard also the brickmaker." As this background indicates, at the time Darwin published his work the problem had already been extensively discussed and large numbers of people, scientists and laymen as well, possessed well-established opinions regarding it.

As early as 1846, an American reviewer of *Vestiges* said that he had heard these ideas many times before in discussion. It was, he said, a statement of "a very common tendency of thought."[10] It is clear that many Americans had been involved in such discussions, and that some, at least, had accepted evolutionary ideas prior to this time. In the 1830's, for example, Constantine Rafinesque had been led to espouse a theory of evolution as a solution to a whole range of problems in systematic botany. S. S. Haldeman, in the same year that Chambers published, had called for a reconsideration of Lamarck's theory or at least a modification of it. Although he did not commit himself to the theory, he was convinced that numerous developments in natural history had shown "the insufficiency of the standing arguments against it, and the necessity of a thorough revision of them."[11]

What were some of the things which lured Rafinesque, Haldeman, and other Americans toward a theory of biological evolution? A point that both Rafinesque and Haldeman dwelt upon in detail was the difficulty that naturalists traditionally encoutered in arriving at and using current definitions of a species. It will be recalled that the general view of both scientists and interpreters of Genesis was that a species was a fixed and immutable unit of creation consisting of all individuals with nearly identical morphological characteristics, which had been retained with perhaps minor variations from generation to generation. In general, species were quite like those that had been known to Adam; the possibility of wide variation, thought to occur only by crossing, was severely limited by the well-known sterility of hybrids—another of God's wise contrivances, designed to keep His world the way He had made it. In the eighteenth century, when there had been an imperfect knowledge of the vast number and variety of species, both extant and fossilized, this definition possessed considerable plausibility and was applied with no great difficulty. As the number of known species multiplied, however, serious differences arose among naturalists in their attempts to define and describe species. Naturalists especially encountered difficulties in distinguishing between species and varieties. This occurred in cases where several species seemed to blend almost imperceptibly into one another and the lines of division seemed at best blurred and vague. In such instances naturalists known as "splitters" tended to multiply the number of species, while those known as "lumpers" tended to reduce the number. As these disparaging terms indicate, it was always assumed that there existed a determinate number of species in the world and that only blindness, stubbornness, or self-seeking caused men to fail to recognize a species when they saw one. But this comfortable illusion could not long persist in the face of the complexities introduced into the "chain of being" by the nineteenth century.

One problem that Americans dealt with was the Naides, a family of shells. Conchologists throughout the first half of the nineteenth century had difficulty agreeing on the exact number of genera in the family and the number of species within each genus. This was particularly true of the genus *Unio,* the same genus that had led Lamarck into his speculations. What made the problem particularly difficult was the fact that, if the various species of this genus were laid out side by side, they were found to "exhibit so gradual a change as to convince the observer of their identity as species; but if any two, near opposite extremes of the arrangement, be compared, they would be considered specifically different."[12]

Such problems as these had led many naturalists to question the real existence of species and to propose that they were only artificial groupings erected by man in order to facilitate the study of nature. Having arrived at such a radical nominalism, it was but a short step for some, as in the case of Lamarck and later Haldeman, to infer from these difficulties the possibility of a developmental theory. Those who made such inferences, however, were a minority. Most naturalists held faithfully to the concept of the reality and fixity of species. Instead of abandoning these concepts in the face of perceived difficulties, they rationalized them away by attributing them to lack of knowledge or to errors on the part of taxonomists—particularly the creation of new species where none existed.

The development theory not only offered a solution to the problem of variation in species that so greatly vexed taxonomists but it also gave meaning to a large body of important and interesting data uncovered by comparative morphologists. For some time the attention of naturalists had been drawn to the fact that all organisms seemed to be constructed according to a few basic plans. Cuvier in the animal kingdom and De Jussieu in the plant kingdom were the principal leaders in elaborating on this discovery and distinguishing the basic plans that appeared in their respective kingdoms. The study of embryology was also an important aspect of these developments, since it assisted morphologists in determining the different structural plans and the animals that belonged to them.

These facts drawn from comparative morphology were of great importance to the taxonomist. In particular, they contributed to the development of natural systems of classification—that is, systems in which those organisms were grouped together that resembled each other in large numbers of characteristics and in general structure rather than because they had a single characteristic, rather arbitrarily chosen, in common.

In such efforts the value of the information contributed by the comparative morphologists is obvious. Although Americans were relatively slow in adopting the natural system of classification, the efforts of Rafinesque, John Torrey, Asa Gray, and others caused it to gain gradual acceptance in the 1830's. By the time the *Vestiges* appeared, Torrey and Gray had already published the first parts of their *Flora of North America* and the natural system was well established.

The importance of this was great. First of all, it made the old idea of a "great chain of being" untenable. Species now had to be seen clustered about a few central types rather than forming a ladder. Naturalists also were enabled to catch their first glimpses of the phylogenetic tree with its apparent out-branching of genera and species from common sources as the limbs of a tree branch out from larger limbs and the main trunk. When this pattern became apparent it was quite natural that questions began to be asked concerning the causes that produced such relationships. Although most naturalists rejected the idea, it was by no means absent from their minds that such relationships suggested a possible descent from a common source for all animals and plants with similar morphological features. As Haldeman pointed out, acceptance of such an explanation would add new meaning to the work of the naturalist, for it would establish a material basis for the interrelationships he observed among species in nature:

> Without it every species would be isolated in creation; . . . there would be neither genus, order, nor family, no relation between the wings of a bird and the anterior limbs of a quadruped; and the seven cervical vertebrae, so constant in the mammalia, were accident. [One] might consider [Lamarck's] views as the foundation of comparative anatomy, the key to the theories of representation and types, and the basis of the classification of organized bodies.[13]

In addition to sharing this general background, Asa Gray had special interests that had brought him even closer to evolutionary views. Although he had bitterly rejected Chambers's evolutionary work as both unsound and a danger to religion in 1845 and 1846, toward the end of that decade he had become interested in the problem of the geographic distribution of species. A species, for Gray, embraced all individuals similar enough in size, color, and other physical features "so that they may be

deemed or proved to be the produce of a common parent." This
implied that those species which had wide geographic ranges
had acquired them by natural means. The problem, which had
been proposed by Alphonse De Candolle, was that of accounting
for "the resemblances between the vegetation of countries that
are apart, but between which an interchange of plants is now
impossible." In short, it was now time to investigate seriously
the problem that had been perplexing naturalists since the time
of José de Acosta.

Gray's interest in plant distribution had been heightened
in 1855 when Darwin opened a correspondence with him requesting botanical information that would enable him to determine whether certain of his ideas concerning animal variations
were true concerning plants as well. In that same year, he read
De Candolle, who gave a general summary of a whole field of
knowledge that had not been defined before. Although Humboldt
had considered plant geography a generation before in his observations on life zones and the influence of climate, until De
Candolle no connection existed between the mass of data
gathered by the great explorations and any theory about why
plants were distributed in the way they were found. De Candolle
insisted that the time for generalizing had come. In 1856, Gray
published his "Statistics of the Flora of the United States," which
Dupree has evaluated as "a landmark in the history of American
botany."* It was Darwin's enthusiasm over Gray's contributions to
man's knowledge of distribution of plants that eventually led him
to include Gray in that select group to which he revealed the
outlines of his theory on the origin of species before it was un-
veiled to the public. Darwin had been particularly impressed by
Gray's analysis of the flora of Japan and their relations to North
American flora. Of 580 plants investigated, Gray found that large
numbers were congeneric, closely related, or identical species.
Amazingly, eastern North America contained by far the greatest
number of these plants.

Gray's publication of this work placed him squarely in
opposition to America's most popular scientist, Louis Agassiz, who
insisted that even the identical species were separate creations,
bearing only an "intellectual relation" to each other. Holding to

---

*A. Hunter Dupree: *Asa Gray, 1810–1888* (Cambridge, Mass.,
1959), p. 241. Dupree emphasizes the influence of De Candolle on
Gray's work on distribution of species.

the idealistic view that species were embodied "thoughts of the Creator," Agassiz argued that each species had been created in the geographic area where it was destined to live. There could therefore be no connection between widely separated species, no matter how close the resemblance. The mere fact of physical separation was conclusive evidence of separate creation.

Gray, in reacting to Agassiz's explanation, objected because he saw that it was not really a scientific explanation at all. It superseded explanation by claiming that things are presently just as they were created in the beginning, thus leaving science with nothing to explain. His own view, which he advanced in the Japan paper, was that each species orginated in one place and spread from there as circumstances of climate and other physical conditions permitted. To explain the migration in this case, Gray appealed to geological changes in the past that might have made such migration possible, thus joining a dynamic theory of the earth with a theory of plant distribution. Relying much upon geological information supplied by James Dwight Dana, Yale geologist and Gray's co-editor of the *American Journal of Science,* Gray showed that on at least two occasions in the past conditions allowed for continuity between the flora of eastern Asia and temperate North America.

The significance of Gray's conclusion lies in the fact that a doctrine of migration undermines the doctrine of fixed species. And when this doctrine is joined to a dynamic theory of the earth, evolution is just around the corner. Once admit that plants in separate areas are descendents of common ancestors, and the next step is to account for varieties and closely related species as local variations. This was what Gray, influenced by Darwin, was ready to admit in late 1858. He introduced these views in Dccember of that ycar bcforc a discussion club of which both he and Agassiz were members. Agassiz was present, and, of course, reacted strongly. This meeting was the opening gun in the controversy over Darwinism in the United States.

By the time Gray's work on the flora of Japan was published, Darwin and Wallace had already presented their papers outlining the theory of natural selection to the Linnaean Society. Gray must have been aware that his own paper had a very important bearing on the question of evolution. By demonstrating the close similarity between plants of North America and Asia and the possibility of

an interchange between the two regions, Gray had clearly reinforced a point that was necessary for the confirmation of Darwin's
theory. The American botanist not only provided useful information in support of Darwin's theory, but Darwin's theory in turn provided a logical answer to a problem that had troubled Gray, and by
so doing it gained in persuasiveness. Although Gray had readily
accepted a common place of origin for identical species found
in widely separated areas, he had not been prepared to accept a
common place of origin for widely distributed congeneric or nearly
related species. In Darwin's theory, however, he saw a plausible
explanation of this phenomenon, and he suggested the possibility
in a footnote. It was this footnote that in 1858 introduced the new
theory of evolution to America, and it placed Gray on record as
being friendly to evolution and hostile to the rival view represented especially by Louis Agassiz. The lines of battle formed in
America even before the first copy of Darwin's book arrived.

Darwin's contributions to this long-developing line of
thought were a massive array of evidence to support the fact of
evolution and, even more importantly, a unifying hypothesis
that explained how evolution had taken place. It was the mechanism that Darwin placed greatest emphasis upon, and it was this
element that ultimately compelled belief. Natural selection, Darwin's term for the mechanism of evolution, was based on four
broad principles, all of which Darwin had borrowed and combined
in a unique pattern that he thought could explain all of organic
nature.[14]

1. In any given area a great many more life forms come
into existence than the environment can possibly support. If
all the fish eggs produced, for example, were allowed to hatch
and grow to maturity, the rivers, lakes, and oceans of the world
would very soon be unable to hold them. Even elephants, the
slowest breeding of all animals, would soon overpopulate the
earth, Darwin calculated, if they remained unchecked. This idea
was a generalization of Malthus's argument that population
always grows faster than the food supply. Where Malthus and the
classical economists had used this principle mainly to explain
why the conditions of the working classes could never rise above
the subsistence level, Darwin saw that it had broader possibilities.
Such a prodigality of life, he argued, would lead to a struggle for
existence in which only the fittest could survive. Although many
of his literal-minded followers were less cautious, Darwin used

the word "struggle" in a very broad sense—actual cases of conflict among individuals, he thought, were comparatively unimportant. The more meaningful struggle was the effort to gain
food, to protect oneself from the weather, or to procreate. The
giraffe, for example, with a slightly longer neck than his fellows,
would be able to gather leaves from higher limbs and would thus
have an advantage in the "struggle for existence."

2. From the moment of their birth, no two organisms are
exactly alike. There are minute, random variations among individuals, even of the same species, *and these variations can be
transmitted by heredity*. Darwin, of course, like everyone else of
his time, had no idea how heredity operated. The concept of the
gene was still nearly a half-century in the future. Darwin simply
postulated heredity as a fact of observation that could not, for the
moment, be explained any more than could the origin of variation.

3. In the struggle among unlike individuals, those differences that are advantageous, however minute, will enable
their possessors to survive. Those lacking the desirable characteristic will be unable to meet the exigencies of the environment,
and will die out, leaving few or no offspring.

4. The winners in the struggle for existence will transmit
their characteristics by heredity. Given enough time, an accumulation of minor modifications will finally result in the
formation of a new species and the evolutionary struggle will
begin again with some other favorable variation.

Clearly, if time is not limited, and if variation is indeed
universal, Darwin's theory could explain the entire biological
past—from the one-celled organism all the way through man. The
cumulative effect of minute variations over vast stretches of time
could be called upon to explain the development of any given
structure—say, an eye, a wing, a webbed foot, or the intricately
petaled flower of an orchid. But just as important, from the point
of view of natural science, was the ability of natural selection
to account for the current admirable adaptation of animal species
to the conditions of their existence. Naturalists had long insisted
upon the *fact* of adaptation—that is, they knew that each species
was especially suited for the inorganic environment in which it
lived, be it hot, cold, dry, or humid, and they knew that animals
were provided with appropriate mechanisms for escaping their
enemies, securing their food, and perpetuating their species.
There had even been extensive studies of internal adaptation that

had revealed a general harmony of construction, a suitability of each organ for its function, and a co-adaptation of various organs and parts. This knowledge had been extremely useful to paleontologists in reconstructing fossils, and for systematists in classification. But earlier naturalists had adopted the supernaturalistic explanation offered by natural theologians for this marvelous adaptation. Suitability of organs for function, and of the whole animal for its conditions of existence, they argued, implied conscious design, and a design implied a designer; therefore, adaptation must have been the direct work of a Creator. The question was thus effectively removed from the realm of natural science. The very fact that it was cold within the Arctic Circle, for example, was an adequate explanation for the unusually heavy body covering provided by God for those animals destined to pass their lives there. The whole work of nature was a supreme testimony to the wisdom and benevolence of God; adaptation was therefore not a matter for science to be especially concerned about.

Darwin, however, applied to these same facts a naturalistic explanation that entirely removed the need for intervention by any designer. When conditions remain constant, he argued, animals and plants naturally do not depart from their well-adapted type, for variations that make the organism less suited for the environment will place the possessor at a disadvantage in the struggle for existence. But when conditions change, some of the naturally occurring variations will prove better suited to the new conditions; such favorable variations will be selected for preservation—much the way a breeder selects from his breeding stock characteristics that please him—and the organisms will change. The change, of course, since it has no external direction, will be slow and gradual, but if continued over eons of geological time, it will produce any degree of alteration. Thus, the theory can account for slight differences between related genera and species, as well as the long-continued trends revealed by geological history.

The magnificent sweep, and yet the utter simplicity, of the theory was awe-inspiring. Bold and comprehensive as it was, it had the virtue of apparent obviousness once it was stated. "How extremely stupid not to have thought of that," was Huxley's initial reaction, and he later declared himself ready to go to the stake for the theory.

Even though in retrospect we can see that the whole trend of thought through the early nineteenth century was leading to some kind of evolutionary conclusions, and even though Darwin assembled more evidence for the hypothesis than had ever been offered before, the theory was not immediately accepted by more than a small group of scientists and philosophers. Indeed, the whole scientific and theological battery of arguments was dusted off and brought forth once more.

To present-day students, the theory of evolution by natural selection seems almost axiomatic. It is often difficult to imagine a time when evolution was not accepted, and it is difficult to attain any kind of sympathetic understanding of the point of view especially of theological opponents. As so often happens after a new theory has become established, those who had initially opposed it are now seen as obscurantists or old fogies who struck out wilfully against "the Truth." If, as Huxley said, it was "extremely stupid" not to have thought of the theory, how much more stupid must those have been who opposed it after the truth was shown to them. In short, the main portions of Darwinism have now become parts of conventional wisdom and are defended as resolutely as the earlier theories were when Darwinism first appeared.

Certainly many of the arguments directed against Darwinism were theological, and many others were dictated primarily by a misconceived ancestral pride or by self-interest. Those whose reputations had been built on the older theories—especially those who had contributed notably to elaborating the theories—had a personal stake in proving Darwin wrong, and it is impossible not to detect the malice underlying their attacks on Darwin. Clearly Agassiz felt that his reputation was at stake in combatting the theory, and his outbursts often sound more like wounded pride than anything approaching objective evaluation.

To theologians the main problem turned again on God's intervention in nature. If man was the product of natural selection, how did one account for his soul? Where had God interrupted the purely natural process to endow him with spirit? The alternative to divine intervention was to conclude that the power of thought and moral judgment, then believed to be manifestations of the soul, was inherent in matter, and no orthodox thinker could admit this. Thus to the fear that Darwin would reduce God to a being who was no God at all was added the fear that he would reduce man to a being who had no soul, and, therefore, was no man at all. Few were ready at first to accept Asa Gray's suggestion that God might effect design by a process of evolution, for they thought that this was a sure route to pantheism.

Despite the example of Agassiz, however, and despite the initial panic of theologians, the most important of the arguments against Darwinism were neither theological nor dictated primarily by self-interest or obscurantism: they were arguments advanced by scientists who were disturbed either by the incomplete agreement of the theory with familiar facts, or by Darwin's evident use of a hypothetical mode of reasoning. Opponents drew support from the history of science that clearly showed that such theories had been rejected before, and there seemed ample reason for them to think that "true science" would once again rise to the defense of the faith. Agassiz, recalling the fate of earlier efforts to establish evolutionary doctrines, declared, in what turned out to be one of the rashest predictions of the time, that "I shall outlive this mania." The *Methodist Review* summarized the history of previous development theories to make a prophecy concerning Darwin's: "The author of the 'Vestiges' repudiates Lamarck's hypothesis; Darwin rejects that of the 'Vestiges'; and his own will doubtless share the same fate."[15] Surely such thoughts must have comforted the anxious.

But there was a great deal more than pious hope for the faithful to take comfort in. Darwin's argument was not perfect— there existed many gaps of which he was painfully aware, and the paleontological, embryological, and other evidence that was to confirm the theory and secure for it the near-unanimity of support it now enjoys accumulated only gradually as the century wore on. Darwin's candor, however much it endeared him to friends like Asa Gray, was of great assistance to his opponents in their arguments against him. Critics did not have to look far for objections to the theory; they had only to read the *Origin* to discover objections that Darwin frankly anticipated and admitted he could not answer to anyone's satisfaction—not even his own.

Most evidently, as many of Darwin's critics were quick to point out, evolution by natural selection was a "mere theory" that did not "flow naturally from the facts," as the current Baconian assumptions insisted a sound theory should. Instead of merely collecting "facts" with no preconceived notion about them—the generally accepted, if simple-minded, view of the inductive process —Darwin had presented to the world one of the first examples of the long, tightly reasoned hypothetical method that has come to

be characteristic of modern science. The *Origin* is, in reality, one
long interwoven argument. Assuming that such-and-such was the
case, Darwin said time after time, we can account for such-and-
such appearances. Many who had been schooled in an earlier
version of scientific method recoiled at Darwin's violation of it.
Thus, J. Lawrence Smith, president of the American Association
for the Advancement of Science and a distinguished scientist of
the old school, regretted Darwin's "highly wrought imagination"
and considered him more of a metaphysician than a scientist.[16]
Other points of Darwin's methodology, or the apparent implications of his work for science, were also seized upon by older philosophers or specialists in the logic of science. D. R. Goodwin, for
example, the newly appointed provost of the University of Pennsylvania, expressed his concern for the doctrine of chance that
he thought was implied by Darwin's particular version of evolution.[17]

The most common type of argument used by philosophers
who were not themselves scientists was generally based on a form
of logical analysis. Mid-nineteenth-century philosophers, trained
in formal logic, were invariably on the lookout for fallacious
thinking. Since they did not possess the specialized knowledge
increasingly required for evaluating the scientists' facts, logical
consistency was perhaps the only tool the highly educated non-
scientist could use to test evidence. Critics pursuing this line, of
whom the most notable in America was Francis Bowen, professor
of logic at Harvard, could clearly show that Darwin's book did not
measure up to a logician's standards. Neither, of course, would
any other important empirical hypothesis, for one of the things we
have learned since the time of Darwin is that empirical science
and logic are two different things. To test such a hypothesis as
that of natural selection one must appeal not to logic, but to what
is, in fact, the case in nature. In the long run, this discovery may
have been the most important philosophical implication of Darwin's book.

But, methodology aside, there were some important scientific loopholes in Darwin's theory. The early acceptance of the
theory illustrates that a workable scientific theory offering advantages not possessed by older theories can be accepted even if
it is incomplete. Scientists who have accepted the theory will
simply use it where it applies and hope that the anomalies will
be removed by further research. An incomplete, although at-

tractive, theory is therefore a powerful stimulant to research, and those who criticize it have an important role in the development of science.

Some of the anomalies are, indeed, still being debated. A whole class of characteristics that had no apparent survival value—in fact, in many cases they seemed to have negative survival value—were explained by Darwin in terms of his concept of "sexual selection." These included the brighter plumage of male birds of many species, the unwieldy—although beautiful—antlers on some deer, and a whole variety of similar characteristics that seemed to function only as attractions in courtship and to be dysfunctional in every other consideration. Darwin's concept of sexual selection struck even the friendly Asa Gray as "sentimentalism," and it has bothered scientists ever since. Again, in certain populations there exist classes of sterile male workers who have what appear to be regular socially assigned functions useful only to other members of the community. When challenged to show how such traits could result from natural selection, Darwin replied that this was a case of "group selection," a process that seemed to work independently of natural selection. Scientists still debate the reality of group selection.

In the early years of the theory, one could have added a great many more apparently destructive arguments. On the theory of heredity then current, for example, one could "prove" that Darwin's version of evolution was impossible, for it was easy to show statistically that, unless new characters appeared in an inconceivably large number of individuals at the same time, they would be swamped by the old characters. This difficulty was not removed until long after the acceptance of evolution, when the work of the priest Gregor Mendel was rediscovered and his demonstration that inheritance was always in terms of unit characters that do not blend was incorporated into biological thought. Until that time, biologists had to content themselves with pointing out "the depths of our ignorance" on the subject of heredity, but this particular ignorance remained a potent source of embarrassment for them. Again, physicists had demonstrated, according to the best methods available to them, that the sun could not have existed long enough for Darwin's method to work; physicists and biologists were therefore working in separate worlds until a better method of estimating the sun's age was developed—once again, long after the acceptance of evolution.[18]

In addition to these and other gaps, which were bridged eventually by Darwin's supporters as they attempted to meet the objections of critics, there were some matters about which Darwin was simply, and demonstrably, wrong. In these cases the error was so intertwined with what was correct that many critics were moved to dismiss the theory of evolution *in toto*. We now know, for example, that he was totally wrong in his remarks upon how heredity occurs, and that he ascribed far too much active power to the environment; even more seriously, we know that the variation upon which evolution is based is not minute and continuous, as Darwin supposed, but is discrete and discontinuous. Contrary to Darwin, nature does, indeed, make leaps, and these leaps, or mutations, are the raw materials of evolution.

Given all the difficulties in Darwin's theory and the questions he left unanswered, it was only to be expected that during the early years Darwin would have some powerful opponents within the scientific camp. A great many of these scientists, to be sure, never revealed any evidence of having a clear understanding of Darwin's particular theory but contented themselves instead with general objections to "development hypotheses." Other scientists, like all of the untrained commentators, demanded too much of the new theory, insisting that it be absolutely proven before given a hearing. Still others used the familiar argument of distortion and innuendo. Yet, when all this is granted, it remains true that the best of the opponents' arguments were reasonable, and, in terms of the knowledge of the time, perhaps as good as those arguments in support of the theory. Later generations are too often disposed to believe that those who turned out to be on the "wrong" side of any scientific dispute were overly conservative. Most scientists, on the contrary, reacted as scientists should when faced with a new hypothesis carrying such profound consequences for a whole world view: cautiously and critically.

After the initial round of debates in Boston, American scientists for the most part seemed to lose interest in Darwinism. Scientists continued to do their work in terms of the traditional ideas about species and seemed comparatively unconcerned with Darwin's. Asa Gray did continue his campaign in the *American Journal of Science*, pointing out that scientists abroad were looking with increasing favor on Darwin's ideas. President George Bentham of the Linnaean Society, he reported, praised Darwin in

an address before that organization. In another number, he observed that Alphonse de Candolle considered "congeneric species"
to be derived, although he had earlier believed in the multiple
origin of species. Gaston de Saporta, the Italian botanist, was, he
reported in the same essay, abandoning his catastrophism and now
believed that life had been continuous. Gray's readers, for some
reason, did not rise to the bait.

    Although Americans were strangely silent during the Civil
War years there is evidence that they had not forgotten the
theory. The process of acceptance is difficult to trace, for naturalists seldom made formal statements of their conversion, as
some of the clergy did. Rather, they simply incorporated the new
views into their own thinking and both tested and applied Darwin's
ideas in their work. Sometimes it is only a change in terminology
to reveal that a scientist has at least been reading Darwin and has
partially accepted him, for much of a naturalist's work does not
necessarily reflect *any* theory about the origin and nature of
species.

    Yet change they did; within less than a single generation
every important scientist in America was an avowed evolutionist.
Nathaniel Southgate Shaler, a former student of Agassiz, in 1865
began explaining Darwinism to his classes at Harvard. Through
the late 60's, papers read before the major scientific societies
show a definite trend toward acceptance of evolution.[19] By 1869,
Othniel C. Marsh, the Yale paleontologist, had begun to put
together his series of fossil horses, which showed clear evidence
of development, and in 1872 and 1873 he announced the discovery of fossil birds with unmistakable teeth, thus supplying
irrefutable evidence of an early connection between birds and
reptiles. Shortly after the death of Agassiz in 1873, his own son
joined the ranks of the evolutionists and an avowed evolutionist
took over his Museum of Comparative Anatomy at Harvard.
When, in 1876, the Princeton trustees demanded a biology professor who did not believe in evolution, it was necessary to import
one from England, for none could be found in the United States.

    By 1873, President McCosh of Princeton was telling a
general conference of the Evangelical Alliance in New York that
it was useless to tell the younger naturalists that there was no
truth in the doctrine of development, "for they know that there
is truth, which is not to be set aside by denunciation." Religious
philosophers, McCosh continued, "might be more profitably em-

ployed in showing them the religious aspects of the doctrine of development; and some would be grateful to any who would help them keep their old faith in God and the Bible with their new faith in science."[20]

A major intellectual revolution had been effected; there has probably been none of comparable importance in the history of human thought which was assimilated so rapidly.

XI

Evolution and
Industrial America

A theory that drew together as many currents in nineteenth-century intellectual history as Darwin's would inevitably have implications far beyond the field of biology. Indeed, it is probable that no scientific theory ever ramified as broadly as that of evolution by natural selection, with its attendant ideas of adaptation, struggle and "fitness." Simultaneous with its acceptance, and continuing until early in the twentieth century, scholars and popularizers alike came forward to suggest the many uses to which it could be put. Charles Loring Brace, social worker and reformer, read the *Origin* thirteen times and emerged with the assurance that evolution guaranteed the final fruition of human virtue and the perfectability of man. "For if the Darwinian theory be true, the law of natural selection applies to all the moral history of mankind, as well as the physical. Evil must die ultimately, as the weaker element, in the struggle with good."[1] Evolution, it seemed, was simply one more evidence of divine purpose and the optimistic law of progress. Penologists, taking a more short-run view, began to look to the forces of heredity and environment, rather than individual free will, as the main sources of crime. They began increasingly after the 1870's to suggest that this changed emphasis demanded new custodial methods of criminal rehabilitation and increased social efforts toward the prevention of crime.

Religious writers, becoming converted to Darwinism, were sometimes able to find in it justification for traditional Christian doctrine. "What is heaven but the company of the fittest?" asked one writer.[2] After all, he observed, only a few of the best men reach that promised land. The historian John L. Motley, who had previously elevated the "law of progress" to an eternal principle, was ecstatic about the implications of Darwinism:

> To be created at once in likeness to the Omnipotent and to a fantastic brute; to be compounded thus of the bestial and the angelic; alternately dragged upward and downward by conflicting forces, presses upon us the conviction, even without divine revelation, that this world is a place of trial and of progress toward some higher sphere.[3]

More often, however, those who discussed religious implications would seize the opportunity to elaborate a totally new theology —although usually said to contain the "permanent core" of Christianity—in which God was seen as simply another name for the evolutionary process itself or perhaps the "unknowable" cosmic force standing behind the evolutionary process. Thus John Fiske, historian, philosopher, and popularizer, argued that evolution proved that religion was the final goal of social evolution, and—invoking the extremely loose test of "fitness"—that the exalted position religion occupied in the evolutionary scheme proved its importance and truth:

> He who has mastered the Darwinian theory, he who recognizes the slow and subtle process of evolution as the way in which God makes things come to pass . . . sees that in the deadly struggle for existence which has raged throughout countless aeons of time, the whole creation has been groaning and travailing together in order to bring forth that last consummate specimen of God's handiwork, the Human Soul.[4]

Heartened by his picture of the beneficence of evolution, Fiske turned to the future to envision a time when the spiritual in man would more and more prevail over his brute inheritance. Then, after many ages, yet surely and inevitably, warfare would be eliminated and there would be peace on earth for all men.

Fiske, growing up in the conventional religious atmosphere of Middletown, Connecticut, had passed through the usual religious experience in his youth, joined the church, sang in the choir, taught Sunday School and Bible Class. At an early age, he considered undertaking as his life's work the tracing out of God's Providence in history. But Fiske reached maturity at a moment when the old certainties were rapidly dissolving and men were searching for new ones. For many, the dissolution did not begin

with Darwinism; the appeal of Darwinism was simply one consequence of it. Some turned to spiritualism, some to German
philosophy, some to a misconceived kind of science. By 1858,
Fiske had turned to Comte, Buckle, and Von Humboldt for intellectual support, praising the "vast learning and comprehensive
thought" he found in Von Humboldt's *Cosmos,* the "very broad
and comprehensive generalization" on which Buckle's system was
founded. The search for a unified view of the universe that would
bring the totality of phenomena into easily comprehensible
generalizations became a recurrent theme in Fiske's intellectual
pursuits for the rest of his life.

It is thus easy to understand Fiske's enthusiasm for Herbert
Spencer, English proponent of "Social Darwinism." Spencer's
comprehensive system began with an appreciation of religion
that, with its doctrine of the Unknowable as the basic assumption
of both science and religion, seemed to Fiske to solve the problem
of Christianity's place and its relation to science by providing it
an unassailable function that seemed almost to put it at the basis
of all thought.

Of course, the religious faith that Fiske praised so highly
and found such a central place for was not the conventional Christianity of the pre-Darwinian period, but an "essential Christianity"
that Fiske characterized as consisting of two doctrines shared by
all the religions of the world. He summarized these at an 1882
banquet for Herbert Spencer in New York: (1) there is an Eternal Power which is not ourselves and (2) this Power makes for
righteousness. The idea of God, of course, would have to be
changed in this scientific religion. Fiske, as Asa Gray had suggested earlier, argued that it was no longer sensible to think of a
God outside the process, who acted at the moment of Creation and
was thereafter confined to special and unusual interventions. The
immanent concept, which pictured God continuously at work in
his universe, seemed most in accord with modern science and
the moral demands of the age.[5] "Paley's simile of the watch is no
longer applicable to such a world as this," wrote Fiske. "It must
be replaced by the simile of the flower. The universe is not a
machine, but an organism, with an indwelling principle of life. It
was not made, but it has grown."[6]

Few Americans realized the extent to which Fiske's ideas
of immanence and "essential religion" were so much in conflict
with traditional Christian notions. Such matters as the nature

of the Godhead, the reality of the atonement, the divinity of Christ—all these were but parts of the "dress of little rites and superstitions," with which men in less-enlightened days had clothed their central truths. All of this, however, was generally ignored by Fiske's appreciative audience; they noticed only that he never attacked orthodox religion, but always spoke sympathetically of it while suggesting that his ideas validated central portions of it on impeccable scientific grounds. For an age of science, there could be no better claim than this. Fiske was widely hailed as a scientist and philosopher who was lending his support to religion. His religious lectures in the 1880's were the beginning of his great popularity, and they indicate how much Americans were in need of the kind of assurances he could bring them in dealing with their perplexing world.

Although the earliest and perhaps the most striking of the effects of Darwinism were on religion and ethics, it is safe to say that no area of social thought escaped the pervasive influence of the theory. Each of the social sciences, for example, eagerly adopted the approach of the social evolutionist, who looked upon human institutions as organic things adapting themselves to an ever-changing environment. In his influential seminar at Johns Hopkins University, Herbert Baxter Adams stressed the "germ theory of politics," which looked upon American institutions as having evolved from primitive German origins. "Government," wrote the young political scientist Woodrow Wilson, "is accountable to Darwin, not to Newton. It is modified by its environment, necessitated by its tasks, shaped to its functions by the sheer pressure of life. . . . Living political constitutions must be Darwinian in structure and in practice."[7]

From the organic analogy it followed that institutions, like biological organisms, evolve upward by stages in a purposeful direction; although some social scientists disliked the supernatural flavor in the idea of purpose, they remained in a minority. The symmetry of pattern suggested by Darwinism both stimulated the social sciences to unprecedented development and rearranged them around a common theme. "Mental philosophy," for example, had within less than a generation become "scientific psychology": a laboratory science that studied the mind as an evolving organism or habit-system. The "psyche" was no longer the "soul" that Aristotle and scholastics had talked about, but a biological process evolving with the animal form and adapting itself to the environment. And by its side, said David Starr Jordan happily, "ethics and pedagogics are ranging themselves—the scientific study of children, and the study of the laws of right, by the same methods as those we use to test the laws of chemical affinity."[8]

In the New Pedagogy of John Dewey, the child was adapted to democratic living through a flexible curriculum and cooperative school institutions intended to encourage both self-expression and a socialized sense of responsibility. The biological concepts of continuity and inheritance had made the child distinctive—worth studying for his own sake—in marked contrast to the older associationist psychology, which held that the child was merely an adult in miniature. The child was thought to be closer to the origins of the human mind, for the mind of the adult is too much overlaid by experience to reflect the basic forces of biological inheritance. It was primarily this belief that the child was more attuned to the primeval mind, rather than a concern for improvements in education, that provided the critical push toward the systematic study of children that was developing concurrently with Dewey's New Pedagogy.[9]

The two halves of geography—natural and human—were for a time neatly brought together by the infusion of evolutionary ideas of adaptation, struggle, and environmental pressure. In philosophy, the pragmatist sought to discover truth by seeing how it survived the test of conflict with other truths, leaving the fittest to emerge victor. This idea was clearly expressed by the phrase of Oliver Wendell Holmes, Jr., the "free trade in ideas," and later in the rise of sociological jurisprudence, which treated law as an evolving set of practices adapting themselves to a changing world—not fixed axioms from which principles were deduced. It followed for Holmes, whose many dissenting opinions from the bench of the Supreme Court indicate that he generally practiced what he preached, that the function of a judge was to take into consideration the social needs of the developing community and determine the meaning of a law in terms of its social consequences as of today.

Even though the core of this new outlook—social evolution—was an old idea, it was not until the time of Herbert Spencer, whose first book appeared in 1850, that we find the idea of social evolution linked to the idea of organic evolution. Comte and earlier writers had tended to keep the two processes separate in their minds. Spencer was a staunch adherent of the free enterprise school of political economy founded by Adam Smith, Thomas Malthus, and David Ricardo, and it is clear that his ideas were predicated more on classical economics than on biology. In essence, he gave the doctrines of the classical school an evolutionary and sociological twist, applying their concepts to the analysis of social evolution. According to Spencer, competition between individuals and races had provided from earliest times the impetus to social progress. Primitive man had been wild and savage, as his barbarous condition required. The government of primitive man had been based on force and fear. In the course of time, however, population pressure precipitated a struggle for existence, placing a premium on ingenuity and a capacity for voluntary cooperation at the tribal level. Since some tribes and races responded to this challenge more effectively than others, they survived and carried forward the banner of progress. Eventually, said Spencer, as voluntary cooperation supplanted force and fear as the basis of social order, human nature would be so transformed by competitive elimination that government would become unnecessary. Each individual would respect the rights of others as they respected his. All this would happen, of course, only if government gave a fair field to all and favors to none, so that competition could produce its beneficent effects. Thus, nine years before Darwin's *Origin of Species* appeared, Herbert Spencer had made population pressure, struggle for existence, and survival of the fittest the key concepts in an emerging theory of social evolution.[10]

Darwin's book impelled Spencer to elaborate. He was soon able to proclaim a grand synthesis of biological and social theory in terms of universal competition and survival of the fittest—a synthesis that he immediately identified with his idea of laissez-faire. Adopting Darwin's explanation of biological progress, Spencer proceeded to draw the analogy to social progress:

> As with organic evolution, so with super-organic evolution. Though, taking the entire assemblage of societies, evolution may be held inevitable as an ultimate effect of the cooperating factors, intrinsic and extrinsic, acting on them all through indefinite periods of time; yet it cannot be held inevitable in each particular society, or even probable. A social organism, like an individual organism, undergoes modifications until it comes into equilibrium with environing conditions; and thereupon continues without further change of structure. When the conditions are changed meteorologically, or geologi-

cally, or by alterations in the Flora and Fauna, or by migration consequent on pressure of population, or by flight before usurping races, some change of social structure is entailed. But this change does not necessarily imply advance. Often it is towards neither a higher nor a lower structure. Where the habitat entails modes of life that are inferior, some degradation results. Only occasionally is the new com- bination of factors such as to cause a change constituting a step in social evolution, and initiating a social type which spreads and supplants inferior social types. For with these superorganic aggregates, as with the organic aggregates, progression in some produces retrogression in others: the more-evolved societies drive the less-evolved societies into unfavourable habitats; and so entail on them decrease of size, or decay of structure.[11]

Endless variations on this theme emerged from American pens. Thus Lewis Henry Morgan, in his study of *Ancient Society* (1878) postulating the psychic unity of all men, taught that the human race had advanced through three stages, which he named: *savagery,* ending with the invention of pottery; *barbarism,* ending with the invention of the alphabet; and *civilization,* the present happy stage of Western man, which presumably would never end and toward which all "lower" societies were striving. Brooks Adams directly adopted Morgan's pattern, adding only the built-in certainty of cyclic development. Civilization, he thought, would inevitably regress to barbarism as it crumbled under the pressures of economic centralization.[12] Whatever the differences among social theorists on the mechanics of social evolution, common to them all was the conception of determinate stages, analogous to stages of organic growth, through which all societies must pass. Except in the case of inveterate pessimists like Brooks Adams —who were extremely rare—this was a comforting doctrine that had the double virtue of assuring Western man he was the highest product of the evolutionary process and would never be surpassed. More "primitive" races could progress only by imitating his model and would in all likelihood never catch up.

Philosophers ransacked the writings of ethnologists and anthropologists for evidence that the known types of human society could be placed in an evolutionary series with late-nine- teenth-century Western European and American society as its climax. Usually, it was not only the Western European–American who was found to be at the top of the evolutionary scale, but the Anglo-Saxon—a "race" so defined as to include all the English-speaking peoples and frequently also the German. This highest "race" was sometimes referred to as the Teutonic, sometimes the Nordic, depending upon the writer's notions about its origin. In any case, however, roughly the same groups were being designated. Thus Josiah Strong, secretary of the Evangelical Society of the United States, found support in Darwin for his belief that the United States was now the bastion of Anglo-Saxonism that would save the world from "inferior races." As the increasingly most populous race, as well as the carriers of the most advanced ideas of civil liberty and of a "pure spiritual Christianity," Anglo-Saxons would burst forth from America to proselytize—or annihilate—the races of the world. "Does it not look," Strong asked in a book that sold 130,000 copies in its first five years, "as if God were not only preparing in our Anglo-Saxon civilization the die with which to stamp the peoples of the earth, but as if He were also massing behind that die the mighty power with which to press it?" Races of manifested inferiority were merely precursors of the superior; they were voices crying in the wilderness: "Prepare ye the way of the Lord!" With the world population beginning to press upon the subsistence level, the day of the great race conflict was imminent. The outcome of this, according to Strong, was clear:

> Then will the world enter upon a new stage of its history— *the final competition of races for which the Anglo-Saxon is being schooled.* If I read not amiss, this powerful race will move down upon Mexico, down upon Central and South America, out upon the islands of the sea, over upon Africa and beyond. And can any one doubt that the result of this competition of races will be the "survival of the fittest"?[13]

The seeds of colonialism, discrimination, and immigration restriction were all contained in this view of Strong, and they all came to fruition within the next few years. Prompted by growing urban problems at home, many persons with knowledge of immigrant poverty became persuaded by spokesmen for labor that the Oriental immigrant on the West Coast and the Eastern and Southern Europeans in the East would so depress the wages of labor that industrial chaos and widespread suffering would result. At the same time, driven by an expanding industrial capacity to seek access to markets overseas, Americans readily adopted the light which "science" now seemed to cast on the race question: if the Anglo-Saxon race had been the agent of progress in the past, so the argument went, it followed that further progress depended upon maintaining its dominance and keeping it "pure." "You have your 'white man's burden' to bear in India," wrote William Z. Ripley to the English people; "we have ours to bear with the American Negro and the Filipinos."[14] Dissimilar races, like the new immigration from Eastern and Southern Europe, could only weaken the blood and militate against the divinely ordained progress. "They are beaten men from beaten races," declared John Fiske after his ascension to the presidency of the Immigration Restriction League in 1894.[15]

Sociologist Edward A. Ross, disturbed by the new immigrants being attracted by thousands to work in the factories and populate the urban areas that had been springing up and expanding at a dizzy pace since the Civil War, colorfully described a gathering of Eastern Europeans:

> To the practiced eye, the physiognomy of certain groups un-mistakeably proclaims inferiority of type. I have seen gatherings of the foreign-born in which narrow and sloping foreheads were the rule. The shortness and smallness of the crania were very noticeable. There was much facial asymmetry. Among the women, beauty, aside from the fleeting, epidermal bloom of girlhood, was quite lacking. In every face there was some-thing wrong—lips thick, mouth coarse, upper lip too long, cheek-bones to high, chin poorly formed, the bridge of the nose hollowed, the base of the nose tilted, or else the whole face prognathous. There were so many sugar-loaf heads, moon faces, slit mouths, lantern-jaws, and goose-bill noses that one might imagine a malicious jinn had amused himself by cast-ing human beings in a set of skew-molds discarded by the Creator.[16]

Such "degraded" types could, of course, not be safely left to control their own institutions. "Democratic ideals among a homo-population of Nordic blood, as in England and America, is one thing," wrote Madison Grant in his popular book The Passing of the Great Race (1916), "but it is quite another for the white man to share his blood with, or intrust his ideals to brown, yellow, black, or red men. This is suicide pure and simple." The nation-state was itself uniquely the creation of this Anglo-Saxon race, and as John W. Burgess, professor of political science at Columbia University, said, this race therefore had the right to as-sume leadership in the establishment and administration of states, and to determine the limits of political participation to be allowed

other races of the world. Senator Albert J. Beveridge from Indiana
stated the same point somewhat more eloquently: "God has not
been preparing the English-speaking and Teutonic peoples for a
thousand years for nothing but vain and idle self-contemplation.
. . . He has made us the master organizers of the world to establish
system where chaos reigns. . . . He has made us adepts in government that we may administer government among savages and
senile peoples."[17]

Although he did not adopt the blatant racism that often
accompanied it, it was Thorstein Veblen who, more than any other
American, made a conscientious transfer to social theory of
Darwinian concepts of struggle, natural selection, and adaptation
to environment. In accordance with the pervasive naturalistic
point of view, Veblen conceived of the social process as an
environment, partly human and partly non-human, acting upon
human beings and shaping their institutions, that he defined
simply as prevalent habits of thought about particular relations
and functions of the individual with the community. To Veblen,
the Darwinian method revealed the impersonal sequence of
mechanical cause and effect and therefore dispensed with a search
for universal purposes and belief in a "natural order" under the
direction of Providence. Social evolution was simply a process
of mental adaptation on the part of individuals under pressure
of environmental circumstances. Thought was epiphenomenal in
nature—an adapting mechanism that responded to the environment. And, as Veblen finally stated in *The Instinct of Workman-
ship and the State of the Industrial Arts* (1914), it was primarily
technology to which the human mind had to adapt. Since the
technology of the moment shaped the institutions of the morrow
by selection, the "cultural lag"—that is, a disequilibrium between
culture and material environment—became a law of nature.

In focusing upon technology as the primary motive force,
Veblen was reasserting a common American opinion. Americans
in the late nineteenth century, seeing evidences of technological
progress all about them—this was the era of the telephone, the
typewriter, the high-speed press, the steam turbine, and the
ubiquitous electric gadget—could not help but be impressed by
it. Earlier, Friedrich Engels had begun the process of Darwinizing
Marxism by emphasizing technology as the human animal's
mode of adjusting to the environment. But American theorists
were more inclined to emphasize the active role of technology

rather than the instrumental one. Postulating a closely integrated society, as the organic analogy demanded, technological innovations were seen as useful variations which effected prompt adjustments and adaptations throughout the whole body of society. Generally it was believed that progress was to be achieved by closing the gap between changing material structure and lagging social institutions. Louis Henry Morgan held that society evolved through "mechanical invention." Brooks Adams stated it as occurring through the discovery and dissemination of knowledge, most of which turned out to be technological. The American socialist-historian Algie M. Simons, postulating the organic unity of society, asserted that technological innovation was the basic cause of social change. Social structure consisted of a technological base and a series of social classes whose mutual relationship was determined by their respective roles in the productive process. A struggle for power among these classes was the raw material of politics. The outcome of the struggle consisted of institutional realignments that we customarily characterize as turning points of history. The social classes that arose out of the productive process struggled for economic power and, consequently, for political power to confirm their economic advantages. These struggles resulted in a pattern of social evolutionary development that formed a succession of well-marked historical stages, reminiscent of the anthropological stages from savagery and barbarism to civilization.[18]

The growing tendency to focus upon technology as the active agent of social change both made it possible to bring together classical economics and organic evolution and also changed the old doctrine of moral improvement of man into a theory of material progress. Despite the evidences of philistinism in early-nineteenth-century America, prophets of the nineteenth century clung, in general, to that notion of a beneficent natural order which had so intoxicated the Age of Reason. The concept of natural laws was still present in the late nineteenth century, but drained entirely of ethical content; and the millennium now envisioned as the capstone of progress was different from any that the Enlightenment philosophers or the prophets of the Romantic era would have recognized. Civilization was now equated with industrialization and progress was defined as the accumulation of capital and the proliferation of industrial inventions.[19]

This was a view appropriate to that period in American history. The United States in the first few decades after the Civil

War was transformed from a developing country into the foremost industrial nation of the world. A complete transportation net, the beginnings of the generation of electrical power and its transmission, the creation of new industries, the modernization of the farm plant: all these were accomplished within the span of a single generation. During the Civil War, Congress had opened the vast Western public domain to free or inexpensive settlement and had permitted the quick exploitation of its natural resources of timber, stone, coal, and other minerals; had laid out a plan and supported the building of trunk railroads to the Pacific across lands still uninhabited; had set up high tariff walls behind which infant industries could thrive and already-developed ones could expand; had encouraged the importation of a cheap working force by passing a contract labor law; and had established national control over banking in order to expand and regularize the money supply.

Under such conditions, and with such inducements before individuals, it is not surprising that an acquisitive, materialist philosophy came to dominate the country. Individuals, relatively unhampered by governmental restrictions and only moderately so by taxation, eagerly took advantage of the opportunity offered. Huge fortunes were accumulated, giant industries were begun, and, of course, gross inequities resulted. The mores of the period —in the articulations of its spokesmen, whether economists, spiritual leaders, or academics; in the work of its legislatures and the rulings of its courts—gave approval to acquisition, unequal wealth, and the competitiveness and ruthlessness of the period's entrepreneurs.

The antisocial behavior of businessmen in the Gilded Age is often said to have been based on the philosophy of Social Darwinism. The "Robber Barons" of that generation are portrayed as having made a literal application of the ruthless struggle in the animal world to human society. Their rugged individualism and their apparent disregard for common decency in their business dealings was the result of this direct application: a transference of the law of the jungle to the business world. Certainly if one looks to the scholarly expositions of rugged individualism, one can find a great deal to sustain this thesis. "The millionaires are a product of natural selection," wrote William Graham Sumner, Yale sociologist who has been the archetype of the Social Darwinist. "They get high wages and live in luxury," Sumner continued, "but the bargain is a good one for society."

It is idle folly to meet these phenomena with wailings about
the danger of the accumulation of great wealth in a few hands.
The phenomena themselves prove that we have tasks to perform which require large aggregations of capital. Moreover, the
capital, to be effective, must be in a few hands, for the simple
reason that there are very few men who are able to handle
great aggregations of capital. . . . The men who are competent
to organize great enterprises and to handle great amounts of
capital must be found by natural selection, not by political
election. . . . The aggregation of large amounts of capital
in a few hands is the first condition of the fulfillment of the
most important tasks of civilization which now confront us.

In good Darwinian fashion Sumner was scornful of any
and all proposals "whose aim is to save individuals from any
of the difficulties or hardships of the struggle for existence and
the competition for life. . . . It is not at all the function of the
State to make men happy. They must make themselves happy
in their own way, and at their own risk." Sumner's argument
was simply that if the social and economic order is left to run
itself, the fittest will survive, and among them those who are
most capable will gain the greatest advantage. "Fitness" for
Sumner, of course, signified merely the "ability to contribute
to the material welfare of society." "Let it be understood that we
cannot go outside of this alternative," Sumner warned: " . . . liberty, inequality, survival of the fittest; not liberty, equality,
survival of the unfittest. The former carries society forward and
favors all its best members; the latter carries society downwards
and favors all its worst members."[20]
In large part, the findings of Darwin seemed to complement and confirm the hypotheses of the classical economists—
since Darwin's inspiration was in part from them, it is not at
all surprising that this should have been so. Political economy
had long taught that maximum utility for society as a whole
would be achieved if economic forces were allowed to work
themselves out without restriction and without guidance save
by the enlightened self interest of the individual. Social Darwinism added enormously to the prestige of non-interference by
making it the *sine qua non* of all human progress. An unfettered
industrial order would insure not only an optimum product in
the world of today, but a perfect race and a perfect social order
in the world of tomorrow. It was an engaging, even sometimes
an inspiring, conception. Certainly those who looked upon them-

selves as belonging to an élite were flattered to think that their success stemmed from the survival of the fittest. And it was good to be able to look upon oneself as an agent of progress.

Even better, it was a conception made to order for an industrial age. Its character, its terminology, its symbols were completely secular, purporting to rest on empirical truth, on concrete, scientific findings. No appeal need be taken to an abstract moral law for verification of the rules that governed a just society. One needed only look to nature herself to trace the inexorable workings of those rules in the geologic record. Facts were what Americans in the late nineteenth century understood best, and facts were what the Social Darwinists promised to give them. Abstract ethics, having lost its religious basis, had lost much of its charm; here, on the other hand, was the basis for an ethics empirically derived. "This was a vast stride," said Henry Adams. "Unbroken evolution under uniform conditions pleased everyone—except curates and bishops; it was the very best substitute for religion; a safe, conservative, practical, thoroughly Common-Law deity."[21] Besides fitting so neatly with the inclinations of industrial America, the major tenets of Social Darwinism were acceptable almost without question because, as Robert Wiebe has suggested, for all the theory's harsh qualities, it drew upon a rich tradition of village values. Equal opportunity for each man; a test of individual merit; wealth as a reward for virtue; credit for hard work, frugality, and dedication; a premium upon efficiency; a government that minded its own business; a belief in society's progressive improvement—all these read like "a catalogue of mid nineteenth century virtues."[22] And although the philosophers who worked with these new materials —men like Spencer in England and Fiske in America—often envisioned a flowering of individual personality at the rainbow's end—a "divine event," in Spencer's terms—what most men understood of Social Darwinism was its promise of constantly increasing material well-being. "I remember that light came in as a flood and all was clear," said Andrew Carnegie. "Not only had I got rid of theology and the supernatural, but I had found the truth of evolution. All is well since all grows better."[23]

As the conservatives employed it, the Darwinian revelation supported all their traditional premises. In nature, the fittest rose to positions of dominance, the less fit were eliminated. Thus the species slowly improved through natural selection, so long as no extraneous influence interfered. As in nature so it was in the business world. "The growth of a large business is merely a survival of the fittest," said Rockefeller, drawing upon an apt analogy to explain the dominance of the Standard Oil Company. "The American beauty rose can be produced in the splendor and fragrance which bring cheer to its beholder only by sacrificing the early buds which grow up around it. This is *not* an evil tendency in business. It is merely the working-out of a law of nature and a law of God."[24]

At a blow, then, the timeworn presumptions of American conservatism were given new confirmation. The doctrine of steady progress served nobly as a defense of the established order. Since the order was itself the promise of continued progress, all suggestions for reform could be branded as reactionary meddling which would plunge the race to doom. "Whatever is, is right" because "what is" is a product of nature and a guarantee of a better future. "Fitness" was defined in terms of material success, because nature is incapable of recognizing any other standard. Here again was the tendency to equate civilization with industrialization, and progress with "the accumulation of capital and the proliferation of industrial inventions." The élite, the saints of the new religion, therefore, were those who proved their native superiority by their survival value. This will be recognized as the Puritan doctrine of "election" in modern dress; the supporting rationale was different, but the implications were almost indistinguishable. Inequality was no longer a dismal necessity as the economists had argued; it was a disguised blessing that helped move society onward and upward. The claim of the great body of the people to control the social order they lived in was manifestly unwarranted; the inferiority of the masses was attested by their economic position, and the great social decisions must be left to those who had won the right to make them. The masses, however, would not simply be abandoned to their fate, for, as Andrew Carnegie explained in a book of 1900, the wealthy man had a duty to use his accumulated wealth in the public interest; ideally he should dispose of it all before his death, for having proved his superior virtue by accumulating his wealth, it followed that he would best know how to use it in the interest of society.

Ministers of the Gospel were expected to make the unsuccessful contented with their lot, to assure the successful that, in Bishop Lawrence's words, "Godliness is in league with riches," and to instruct the young in industry, frugality, and honesty. The concluding passage from William Makepeace Thayer's *Tact, Push and Principle* (1880) makes this duty very clear:

> It is quite evident . . . that religion requires the following very reasonable things of every young man, namely: that he should make the most of himself possible; that he should watch and improve his opportunities; that he should be industrious, upright, faithful, and prompt; that he should task his talents, whether one or ten, to the utmost; that he should waste neither time nor money; that *duty*, and not pleasure or ease—should be his watchword. . . . Religion uses all the just motives of worldly wisdom, and adds thereto those higher motives that immortality creates. Indeed, we might say that religion demands success.

On similar grounds, the property right earned nature's sanction. Those most qualified to control property were those who had demonstrated their capacity in the competitive struggle. Movements to deprive them of control were ill-advised in a double sense: first, because such action might disturb the cosmic plan and inhibit progress; but, second, because in some way not always clear the acquisition of property somehow invested the owner with a moral right to hold his prize. This curious blending of an empirically derived moralism with deductions from the facts themselves was characteristic of the new commercial apologia. Pretending to reject abstract moral concepts, its exponents introduced one by the back door.

There was, they argued, no injustice in the distribution of this world's goods. "They are not equally distributed but it does not follow they are unjustly distributed," explained Charles Elliott Perkins, taking the idea of "social justice" to task:

> Is the rainfall unjustly distributed when an honest farmer loses his crop by drouth? Is the law of gravitation unjust when a child accidentally falls out of a second story window and is injured for life? . . . If a man by hard work and intelligence, honestly acquires property and takes care of it, while his neighbor, equally honest and intelligent, acquires property and fails to take care of it, are the products of the industry of both of them unjustly distributed?[25]

Those oppressed by monopoly were to be convinced that their hardships were divine visitations, sent by the Lord who first made the laws, not by the men who obeyed them. Thus, those

who were crushed by competition, regimented, or underpaid
were to find solace in the fact that their defeat came from the
nature of things and that their sacrifices were essential to
national welfare. Without a struggle for existence there could
be no progress, and the struggle required that some be victims
and that many fall by the wayside.

Businessmen were happy to accept the philosophy for
whatever support it offered them. Like every ruling class, they
felt the need of a philosophy to justify, for the long run, activities
that appeared at the moment costly or corrupt; they naturally
wished to make the lesser seem the greater good, the private
seem the national profit. It was both useful and comforting
to believe that the inequities of the moment were a part of the
ordained nature of things, and that the individual businessman
could neither be blamed nor controlled. "Social forces cannot be
created by enactment," said E. L. Youmans, "and when dealing
with the production, distribution, and commercial activities of
the community, legislation can do little more than interfere
with their natural course."[26]

Businessmen were naturally flattered by being referred to
as the outcome of an evolutionary process which had been working inexorably for the benefit of the race. They were often willing
to refer to themselves as evolutionists, and many contributed
handsomely to Herbert Spencer, prophet of the new order.
Such terms as "survival of the fittest" appeared frequently in
their writings; and, of course, in business *practice*—especially
in the newer industries—a kind of "struggle for existence" was
a daily occurrence. But all this need not imply an acceptance
by businessmen or even a real acquaintance with this general
philosophy. It is unlikely that very many captains of industry in
the 1870's and 80's knew enough of either Darwin or Spencer
to turn biology to self-justification. In the depths of one late-
nineteenth-century depression, the *Commercial and Financial
Chronicle* (1874) expressly repudiated Darwinism as an explanation of the current business failure. It did grant that the
philosophy would reinforce such adages as "experience keeps
a dear school, but she teaches well"; that is to say, Darwinism
probably did no more for the business community than to furnish
a new terminology for old ideas. The sentiments of William
Makepeace Thayer may all be found in the writings of Benjamin
Franklin, who formulated them without benefit of evolutionary

thought and no formal concept of a "struggle for existence." The well-known hostility of businessmen to government was simply a part of their optimistic belief that natural laws worked for the best interests of all—a view that fits John Fiske's version of Darwinism and may have drawn some support from it, but which fits a number of other things as well, such as classical economics, Protestant theology, and Jeffersonian political theory.

The Robber Baron was, in part, a real role, although it seems to have been only vaguely and indirectly related to Darwinism. This role formation was guided and sanctioned by three cultural themes, which long antedated Darwinism and which were shared by most of the larger society. There was, in the first place, the concept of the autonomous economy that was self-adjusting—a theory that had a long history of development in classical economic thought. Secondly, there was the belief that profit or material gain was the only reliable incentive for action—a view that Americans had lived by for most of their history. And finally, there was the idea that progress came through competition and survival of the fittest.[27] The only thing new here was the term "survival of the fittest," and it had been coined by Herbert Spencer nine years before the publication of Darwin's book. All else had been commonplace in Western European and American society for nearly a century before the rise of the Robber Baron.

Darwinism was primarily useful as an academic explanation for the industrial situation in the late nineteenth century, although it seemed to have little part in shaping either the situation itself or the responses of men to it. To a generation engrossed in the competitive pursuit of industrial wealth, it gave cosmic sanction to free competition. In an age of science, it "scientifically" justified ceaseless exploitation. The most that can be said for Darwinism is that it *seemed* to contemporaries to bolster these pre-existing commitments by stating them in "scientific" terms.

And one can make the same point about the other "implications" of Darwinism. There was no inherent necessity in Darwin's theory that it be applied to social theory. To be sure, Darwin did believe that evolution promised progress and he fully believed that man should be incorporated into nature. But he had always been wary of direct applications of his ideas to social or political matters. Furthermore, not one of the American reviewers of the first edition of the *Origin* made any attempt whatever to discover social ideas in the theory. Americans and Western Europeans in general had always considered themselves superior to other peoples. Experience with a new class of immigrants whose cultural mores seemed to threaten American values would have assured some kind of unfriendly reaction to them, whether Darwin had published or not. Imperialism arose out of national rivalry, Christian arrogance, and presumed commercial necessity, not from Darwinism. International strife, to many, must have seemed the quickest way out of the commercial and territorial tangles of half a century. Neither Darwinism nor any concept from natural science seems to have been important in shaping these attitudes or activities.

What *is* important, however, is that when a justification for such activities was sought, it was *science* that was invoked in the late nineteenth century, as an earlier generation would have invoked religion, tradition, or prescriptive right. Science was coming more and more to supply American social thought with its vocabulary and its supply of images. It served as a major source of metaphor and, like figures borrowed from any area, the analogies drawn from science variously suggested, explained, and justified social categories and values. This reflects the changing position of science in the hierarchy of American values; for since that time it has had a growing role as an absolute, able to justify and expected to motivate individual behavior.[28]

XII

~~~~~~

The Implications of
Professionalism

❦ American scientists emerged from the evolutionary controversy with a new position of importance in their society. This
was matched by a new position of importance in government, by
an essentially new orientation toward their own work, and,
consequently, by a new set of problems—or, rather, by problems
that had been secondary to their predecessors.

By the end of the 1870's the position and the prospects of
the American scientist were probably at a high point. Possibilities for employment had increased enormously in the past few
decades. To cite only a few of many possible indices, by 1880
there were approximately 400 colleges and universities employing
at least one scientist each, normal schools in the Northeast were
emphasizing science heavily in their instruction, and by 1882
there were 144 observatories in the country, all presumably
providing facilities, if not always paid employment. By the
1880's science had become of such relevance that the possible
establishment of a federal Department of Science, at Cabinet
level, was being seriously discussed, and both state and federal
governments were spending increasingly large sums on scientific
research.

In many ways, the Civil War itself helped to bring about
this happy state of affairs, for the war was unique in its technological problems and their impact on the government's ability
to use science. The first American use of aerial observation in
wartime, the first important use of the telegraph, problems of
disease in the mass army, efforts by both sides to use new
explosives, and problems of logistics and supply all forced

science upon government attention. But even more important, developments during the war left three remarkable scientists each at the head of an important and integrated scientific organization within or quite close to the administrative framework in Washington. The men, Joseph Henry, Alexander Dallas Bache, and Charles H. Davis, were all members of an élite group that called itself the Lazzaroni, an informal group of Washington- and Cambridge-based professional scientists who since the 1850's had sought to control the scientific organizations of the country and to create new institutions more adequate to the position they thought scientists should occupy in society.*

Alexander Dallas Bache, initially faced with the prospect of having his U.S. Coast Survey completely ruined by the war, managed in the end to turn it yet to advantage by cooperating with the "great movements of the day." He made special surveys, distributed maps, and sent out parties of his assistants to serve directly with forces in the field, extending his activities to inland rivers as well as the coast. On his initiative the Navy and the Army both had ready access to the Survey's information and, in a special commission that he founded, had an organization capable of translating these data into effective military decisions. The reports of the commission covered the whole coast, and one of the fruits of its work was DuPont's successful attack on Port Royal.[1] In consequence of his services to the government, Bache emerged from the war with his Coast Survey in a stronger position than ever before.

The abrupt departure of Matthew Fontaine Maury, who joined the Confederacy, also provided a chance for some regrouping to aid science. Charles Henry Davis, a warm friend of the civilian leaders of science, became head of the new Bureau of Navigation, which included the Naval Observatory, the hydrographic functions that under Maury had been attached to it, and the *Nautical Almanac*. Astronomy, after twenty years of neglect while Maury was mapping the seas, once more became the major interest of the observatory. With Davis as a leader the surveying-oriented science in the Navy gained an

---

* Mark Beach, in a paper read at Northwestern University in April 1970, argued that the Lazzaroni was merely a social group of no political importance. His argument was both ingenious and interesting, but it failed to convince me, His article, entitled "Was there a Scientific Lazzaroni," is soon to be published in a volume edited by me for the Northwestern University Press, and tentatively entitled *Science in Nineteenth-Century America: A Reappraisal*.

administrative position much more favorable than it had known in the days of its surreptitious development. Not only did he have higher rank in the Navy than Maury—and the trust of civilian scientists, which Maury had always lacked—but Davis's influence was spread by his incidental duties; for example, his membership on the Ironclad Commission and the Coast Survey's commission.

Joseph Henry, Secretary of the Smithsonian, was also called into additional government duties by the needs of war. His reports did much to convince the Army that ballooning was practical and militarily useful. A personal acquaintance with Lincoln, which ripened during the war years, brought many requests from the White House, and Henry continued active in his research for the Lighthouse Board, especially on navigation in Confederate waters. By far the largest amount of work done by Henry for the government came from his membership in the Navy's Permanent Commission, "to which all subjects of a scientific character on which the Government may require information may be referred." Made up of Davis, Henry, and Bache, the commissioners possessed explicit authority "to call in associates to aid in their investigations and inquiries." Neither members nor associates were to receive any compensation. The commission immediately went to work, and for the next two months, according to Henry, it "occupied nearly all my time not devoted to the Institution and more than I could well spare." The commissioners examined inventions, made tests if required, and in all wrote 257 reports; in effect the commission served as the nearest thing to a central wartime scientific agency achieved during the Civil War.[2]

These three government scientists—Henry, Bache, and Davis—exemplified the rise of the professional scientist in government service. And, as A. Hunter Dupree has observed, at the same time they represented an effective lowering of the level at which decisions respecting science were made. In the time of Thomas Jefferson, the President had a real understanding of scientific matters and it was usually he who set policy. In Van Buren's time, Joel Poinsett, Secretary of War, was making the real decision at Cabinet level. But by the Civil War, neither the President nor any Cabinet member was able to give systematic attention to science and such decisions fell, for the first time, into the hands of professionals who had emerged at the level of bureau heads. Thus from their position at the intersection of two basic drifts, Henry, Bache, and Davis were able to undertake what had always failed before: the coordination of the government's scientific policy. The Civil War thus emerges as a genuine "watershed" in the institutional history of American science.[3]

Although all scientists profited by the new situation, chemists, especially, by the 1880's had won recognition as scientific specialists in government service, and this was in large part due to wartime innovations. The Bureau of Agriculture, finally established in 1862 after the core of opposition to it had seceded, provided an early home for chemists, entomologists, botanists, and other scientific specialists, and was on the verge of an expansion that would make it by 1913 a twenty-four-million-dollar business with 14,478 employees, of whom 1,812 were engaged in scientific investigation and research. This is larger than the whole number of American scientists known to be active in the first five decades of the nineteenth century. Already by the late 1860's, the Bureau of Agriculture botanist C. C. Parry was creating a national herbarium within the agency. Although Parry was eventually fired by an unsympathetic Commissioner of Agriculture who believed that the "routine operations of a mere herbarium botanist are practically unimportant," Parry's successor, George Vasey, continued the old policy for another twenty years, when the herbarium was finally transferred to the Smithsonian.[4] A newly consolidated Geological Survey, under the leadership of John Wesley Powell, by 1884 had an annual budget of $500,000 and was embarking on an ambitious scheme to prepare a geological map of the entire United States and its territories. Besides employment for geologists and other naturalists, the Survey in 1880 had opened in Denver its first chemistry laboratory.[5]

The land-grant colleges, also begun during the Civil War with Southern congressional opposition removed, had in effect legitimized the teaching of science and provided homes for many scientists. The culmination of a long-standing pressure that had made itself felt even before the 1850's, the land-grant college idea received its final form from Representative Morrill of Vermont, who had been its leading advocate for some years. The act, passed in 1862, called for the donation of public lands "to the several States and Territories which may provide colleges for the benefit of agriculture and the mechanic arts." Although at first Morrill's idea seems to have emphasized direct instruction to the agricultural and industrial laboring classes, by 1867 he was quoted as saying that the institutions envisioned in the act were not agricultural schools but "*colleges*, in which science and not the classics should be the leading idea." Despite the limited growth of the colleges before the end of the century, their very existence was a symbolic success for science. As educator David Starr Jordan observed, science had been given definite rights in the curriculum, "where before it seemed to exist by sufferance."[6]

Civil War legislation—in particular, the Morrill Act and the act establishing an agricultural bureau—had been significant also as marking a genuine turning point for science in the government. Prior to this time, scientific institutions had had a questionable constitutional status because they were tied to the much-disputed question of "internal improvements." All the accomplishments of the prewar years—the Smithsonian, the Coast Survey, the Naval Observatory, the agricultural research program within the Patent Office, the Western surveys—had evaded the constitutional issue by being tied to something else. All except the Smithsonian had been conceived as temporary in nature, regardless of the well-known tendency for "temporary" agencies to become permanent. But in establishing the Bureau of Agriculture and in granting public lands for colleges, the Congress had for the first time put itself on record unequivocally as sponsoring scientific research and as establishing permanent agencies for the conduct of such research. From this time on, Congress proved itself at least occasionally willing to establish permanent agencies with ample grants of power explicitly stated in organic acts. With the Constitution no longer a stumbling block, the era of bureau building had begun.[7]

Building on the base provided by Civil War developments, a number of organizations dedicated to the advancement of science and the interests of scientists had proved to be viable concerns by the 1870's. The AAAS, which since its organization in 1847 had emphasized the *advancement* rather than the *diffusion* of science, was, thanks to the revolution in transportation, annually drawing hundreds to its meetings; with this base of support, it acted as an effective pressure group to promote federally funded research. Earlier its political power had been limited by the great expanse of the country, for a large part of its membership could not be relied upon to attend from year to year. The principle of peripatetic meetings meant that there

could be no real continuity in attendance. In the latter part of
the century, however, annual meetings of the AAAS, and the
multitude of specialist societies that grew out of it, not only
provided fellowship, a sense of participating in a common
enterprise, and knowledge of the latest work of colleagues; they
also gave scientists the chance to make a united attack on the
technical problems of their science, such as standardization of
nomenclature and the development of specialized bibliographies.
As the specialist societies toward the end of the century began
to take on more of the professional work, the AAAS, although
continuing as a powerful pressure group, became somewhat more
popular in its orientation, seeking to provide a "channel of communication between the purely abstract scientific work of the
very limited number . . . and the great intelligent public upon
whom such men must, after all, depend for their support and
final appreciation."[8]

On another level, the National Academy of Sciences, apparently a permanent fixture on the American scene after its
reorganization under the leadership of Joseph Henry between
1867 and 1872, was evolving into an honorary organization
for recognizing and furthering "abstract science."[9] At the in-
stigation of Bache, Agassiz, Benjamin Peirce, Davis, and Benjamin Apthorp Gould, an act incorporating an academy of
fifty members, to act as advisers to the government, had been
smuggled through during the closing moments of Congress in
1863. In its original conception, it was to emphasize service to
the government. Henry's subsequent reorganization aimed at
shifting its emphasis from governmental service, a function that
it had never filled to any extent, to "original research," with no
one "elected into it who had not earned the distinction by actual
discoveries enlarging the field of human knowledge." Henry
also enlarged its membership from fifty to seventy-five, opening
the way for the admission of several who had been passed over
earlier because of the essentially conspiratorial nature of its
founding. By the 1870's, it was well on its way toward fulfilling
the long-standing dream of the Lazzaroni, as stated by Gould
in 1869, by providing a "recognized tribunal, whose judgment
might be provisionally accepted upon matters requiring scientific
knowledge for their decision, which might command public
confidence by the character and attainments of its members, and
which could represent, advocate and maintain the interests of
science with the public and with the government."[10] In terms

of jobs, recognition, and the existence of organizations to speak for their interests, American scientists had never been so well situated. They were therefore in a position to make demands that would have been considered utopian twenty years before.

The situation of scientists was made even better by the fact that in the seventh decade of the century, there was no longer any necessity of publicizing the utility of science. Science had long since passed so completely beyond the common understanding that little chance existed for meaningful intellectual contact with the public. Although Louis Agassiz and others were still proclaiming the "democratic" nature of American science— and Agassiz continued to hope that fishermen and farmers would use his books quite as extensively as students and fellow-professionals—the increasingly esoteric nature of their own published works indicated that this minority was deluding itself. In the absence of contact on this level, professionals of the preceding generation had devoted a great amount of time and effort to persuading an uncomprehending public to support the work of the scientific community because its results would be useful. Living in a society with built-in pressures upon the individual to do useful work, scientists had had to greatly exaggerate the utility of their work simply because much of its immediate utility was questionable. It is natural, given the democratic assumptions, that those who contributed little should have claimed much, and that they should have been especially vocal in their insistence that they, too, conformed to the values of their society. One looks in vain for actual applications of theoretical science, as opposed to products of mechanical ingenuity, before the middle of the nineteenth century. By the last quarter of the century such applications were so obvious that it was no longer necessary to make a point of them. The contributions of geology to mining technology were by then common knowledge; the chemical, chemical-process, and electrical industries depended directly upon nineteenth-century scientific discoveries; the application of scientific principles had made its impact on engineering, resulting in increased safety, economy, and assurance; scientists seemed to be firmly entrenched in the government and by common consent were recognized as useful, if sometimes cantankerous, public servants.

By 1883, a physicist like Henry A. Rowland could safely refuse to dignify "telegraphs, electric lights, and such conveniences, by the name of science"[11] because the public well understood that these mysterious conveniences were scientific products; and their morning newspapers, with news from the far corners of the world, served as daily reminders of the power of science. The admission by Andrew Carnegie, master of production and thus an embodiment of the ideal of useful work, that he would prefer to lose his buildings and his plants than the services of his scientists, symbolizes the new public role of the scientist. The popularizers, engineers, and applied scientists— whom Henry Rowland despised—had already done their work so well that Rowland and others like him could be spared the necessity of doing it. In short, the claim of utility had to be insistently made in the earlier part of the century simply because it was not even approximately true; the coming of the fact, in a number of areas, in the latter part of the century made formal enunciation unnecessary, and thus for the first time made it possible for a new ideal to be developed.

As one of the major earlier forms of justification had become unnecessary, the other had become virtually impossible. After the flurry caused by the evolution controversy, the old conflict between science and religion had, at least on the higher intellectual level, been settled. From the 1870's on, the things that were science's were generally rendered unto science, and the same was true of religion. John Fiske, Paul Carus, and a few others remained to preach a religion of science, and they frequently had the power to stir the emotions of lay audiences. But very few clergymen and even fewer professional scientists were impressed. Theologians had generally found it advisable to surrender all of nature to science, reserving for themselves a domain of purposes and values outside of nature. A statement by Henry Drummond, prominent British clergyman who adhered to both evolution and Christianity, is illustrative of what came to replace the older view of natural theology that science and religion, if each were "true," must agree on all points:

> Nature in Genesis has no link with geology, seeks none and needs none: man has no link with biology, and misses none. What he really needs and really misses—for he can get it nowhere else—Genesis gives him; it links Nature and man with their Maker. And this is the one high sense in which Genesis can be said to be scientific.[12]

Speaking for the scientists, J. Lawrence Smith told his fellow-members of the American Association for the Advancement of Science that the old task of "reconciling science and religion, which had been a major preoccupation of the preceding generation," was a "mischievous work" and that in his opinion there was "less connection between science and religion than there is between jurisprudence and astronomy." The sooner this was understood, the better Smith thought it would be for both.[13] Although Smith's statement does seem to be a declaration of independence on the part of science, such statements of the period should always be taken in context. It is clear, in the remainder of Smith's address, that he was fully as concerned for religion, if the old association were continued, as he was for science. Thinking no doubt of the recent revolution in biological thought, he observed that change was in the very nature of science: what was accepted as scientific truth by one generation was often rejected by another. But religion should not change; it should be concerned with eternal verities like faith, hope, and love. The best way to preserve them, as many were coming to recognize, was not to tie them to *any* scientific doctrine.

If the Copernican hypothesis, Newtonian mechanics, and uniformitarian geology had not proved conclusive enough, the coming of evolution had finally demonstrated that no amount of "reinterpretation" could any longer hide the fact that the Bible could not be read as a scientific document, for as a scientific document it was seriously lacking, bound as it was by the knowledge of the ancient Hebrews. The "higher criticism," having its origin in Germany prior to the coming of Darwinism but drawing powerful support from it, had demonstrated to all but the most obdurate that the Hebrews had simply written their inherited cosmology into the sacred literature. One important result of the evolution controversy was to free religion from a dependence on science—a dependence that since the seventeenth century had resulted in one retreat after another—and, therefore, to free scientists *qua* scientists of any religious obligations.

Although a great many exceptions remained, and some still do, the older natural theology was virtually dead by the mid-1870's—John Fiske could preach his religion of science only by discarding most of conventional Christianity—and it died by mutual agreement on the part of scientists and theologians. A source of frequent conflict had thus disappeared, but so had the most common justification for abstract science. If "reconciliation" was no longer desirable, then neither was a "scientific demonstration" of the power and character of God. There was therefore no longer any serious possibility of an appeal to the religious element in American culture.

Out of this background there developed a new ideal of the scientist and of his role in society. Previously, science had been successfully "sold" to the public in terms of its contribution to important American values—utilitarian, equalitarian, religious, or even as a means of social control, depending upon the speaker's estimate of his audience. But in the 1870's, for the first time great numbers of scientific spokesmen began to vocally resent this dependence upon values extraneous to science. The decade, in a word, witnessed the development, as a generally shared ideology, of the notion of science for science's sake. The core of the new value theory of pure science consisted of two related assertions: (1) new knowledge should be evaluated according to its significance for existing theory and in no other terms; and (2) scientists should be evaluated solely on the basis of their contributions of new knowledge. The revolutionary character of these assertions was unmistakable. Science was no longer to be pursued as a means of solving some material problem or of illustrating some biblical text; it was to be pursued simply because the truth—which was what science was thought to be uniquely about—was lovely in itself, and because it was praiseworthy to add what one could to the always developing cathedral of knowledge. Applied science, it followed, was a distinctly lower enterprise, for the extraneous political or social criteria introduced in the selection of problems might misdirect the scientist's work by having him spend time on researches of minor theoretical value.

Chemists, leading contributors to practicality in late-nineteenth-century science, were also among those who spoke out most vociferously in favor of the new ideal. In all their recommendations for new research projects, for example, they made no effort to guide American chemistry toward investigations of an immediate practical nature; they displayed no impulse to define their task as contributing to any practical goals. Rather, they recommended work on the spatial arangement of atoms, the determination of chemical constants, and reaction speeds—all in the realm of "pure science."[14]

Even a prominent government scientist such as Simon Newcomb, a man who headed a governmental observatory dedicated to practical science, could firmly adhere to the ideal. "All modern experience shows that nature does not make known her secrets to those who court her for lucre," he wrote. "The work of him who enters the field of science with any other immediate motive than the advance of knowledge will lack that thoroughness and comprehensiveness which is necessary to success. Disinterestedness is the first condition of the highest forms of research."

Understandably, given the democratic context, the representatives of those sciences that had already "proved" themselves in practical terms were able to make the most unqualified statements of the new ideal. Representatives of the emerging social sciences, for example, still clung to the ideal of practicality, although even they tended to be somewhat defensive about it. "Our Section I (Economic Science) may be only the bread and butter section," argued a leading economist in 1885, "we may deal only with the prosaic subject of providing food, shelter and clothing; we may be only trying to find out how to save a few dollars' worth more or less to the community, but what would become of all the rest of you without us?"[15]

In one sense, this new orientation could be regarded as simply the scientific analog of the general fragmentation of life and thought that was occurring in the 1870's—this was also the decade of the rise of "art for art's sake," "profit for profit's sake," and similar viewpoints in other areas. But more meaningfully, the ideal can be related to a new professional consciousness emerging among American scientists, as it was among other groups during the same period. Indeed, as one political historian has pointed out, the strongest political ambitions of the time concerned occupational autonomy. For such groups as doctors, lawyers, and teachers, this entailed legal sanctions for their own standards of entry and proficiency. Accredited members of the group would administer the laws, passing upon the qualifications of applicants and judging any violations within the profession. Scientists, like other academic professionals, by controlling degrees and jobs could enjoy similar privileges without a need for legislation.[16] For almost the first time, scientists had begun to think of themselves as a *disciplinary* group, and they began in the 1870's to care more for the approval and esteem of their own colleagues than they did for the approbation of the society that surrounded them. Success, for them, came to mean ac-

ceptance as a creative scholar by one's disciplinary peers. Con-
cretely, this demanded the publication of books and articles
as well as the research support that alone could make publication
possible. These new demands revealed in a particularly stark
fashion what has come to be the basic dilemma of American
science—namely, that scientists, dependent upon public support
but also striving for intellectual independence from the source
of that support, consequently suffer a degree of isolation and
alienation from the larger society.[17]

The avowed model of the new generation of scientists
was the German university system with its research emphasis,
its laboratories, and its seminars. Certainly the contrast between
the situation in America and the life American scholars found
in Germany was a dramatic one, and certainly their experiences
abroad must have fired the ambitions of returning students to
remake the American system. Their German teachers, said one
agricultural chemist, gave them "a contempt for that superficial
smattering of everything without even an idea of what thoroughness is in anything which is too characteristic of our American
system of education." Once exposed to the laboratories of Liebig
and Wohler, these young scholars could no longer applaud the
vague groupings of American science. It is hardly surprising
that, fired with ambitions, they began as soon as they returned
home to write memorials, to cultivate politicians, and to seize
every possible opportunity in which to spread the gospel of
research. Nor is it surprising that they began to instill in their
students a reverence for original investigation and a contempt
for the restricting American conditions. To accept those conditions would have been to deny the values that now ruled the
world of international science. "I can never adjust myself to my
surroundings here," wrote a Göttingen graduate from the rural
Kentucky academy where he taught, "to do so would be to
proclaim my stay in Europe a failure."[18]

Although Germany supplied the most obvious example,
and many directly tried to model their discipline after the German
style, it was not merely a matter of "copying." Professionalism
had an indigenous development in America and it would no
doubt have arisen without the German example. It is most likely
the case that so many American scientists went to Germany to
study in the 1870's because they wished to obtain the kind of
professional education they knew was obtainable only there. At

any rate, products of the Sheffield Scientific School at Yale and of Agassiz's teaching at Harvard were cooperating with their German-trained colleagues to spread the new orientation across the country.

The proliferation of colleges, technical schools, and other institutions during the middle of the nineteenth century had at first been hailed with joy by the scientific community. Scientists had been, in fact, in the forefront of such expansion, and for obvious reasons. Virtually all of the colleges provided employment for at least one professor of science, and these appointees were generally successful in expanding the scientific content of the curriculum. The multiplication of such schools had been one of the major factors in the rising status of the scientific community. But, with the spread of the new ideal, these schools more and more came to be regarded as unsatisfactory. The collegiate professor of science was regarded primarily as a *teacher*, and whatever scientific investigation he undertook remained in the tradition of the amateur—an avocation that one pursued after working hours, that was in no way thought to be related to his main business. As Stanford University president David Starr Jordan testified, he had begun about 1870 to prepare himself for *two* essentially unrelated professions: naturalist and college professor.

The new ideal can be expressed in the words of Charles S. Minot, president of the ultra-professional American Society of Naturalists. The qualifications for a university professor, he observed, were two in number: (1) the ability to carry on original researches himself; and (2) the ability to train others to carry out original work.[20] Another scientist stated the ideal in an even more extreme form in the very title of his address before the Texas Academy of Science: "Original Research and Creative Authorship the Essence of University Teaching." Wherever one turns in the scientific literature of the last quarter of the century, one finds such statements. Gone, apparently forever, was the balance of teaching, research, and application that had been the highest ideal of early professors like Joseph Henry. It had been replaced now by purely professional considerations, such as contribution to science and preparation for the reproduction of a self-sustaining scientific community.

But the college system of the period was totally unable to accommodate such an ideal. When scientists of Jordan's generation began to press for the new conception of the professor, they found that the very success of the educational movement had now become the greatest stumbling block. In 1876, to take one example, there were over 500 institutions claiming the name of college or university, most of which had sprung up overnight, responding to no particular need other than the divisiveness of an expansive democracy and a proliferation of religious sects. They were, therefore, for the most part small, impoverished, and certainly unable—even had they been so inclined—to provide adequate research facilities for their respective professors of sciences. Neither were the curricula at all adequate in terms of the new generation's needs. Of nearly 400 such institutions that T. C. Mendenhall surveyed in 1882, almost all offered some instruction in physics. Of these, only 6 met his standards of adequate preparation for graduate study, and less than 30 met what he termed *minimal* standards, meaning only that some laboratory instruction was offered in addition to lectures.

It had seemed natural to applaud the growth in numbers of the colleges, astronomical observatories, and other institutions in the day when the lone researcher, working with simple equipment and few expenses, was able to contribute importantly to scientific knowledge. But even by the 1870's that day was plainly on its way out. Frank W. Clarke, later a distinguished government chemist, in an address to the AAAS in 1878 spoke of the urgent general need for the large, endowed research laboratory: a place where the fundamental data of chemistry and physics could be accurately established "without more than casual reference to particular industrial questions or to theories." The necessary apparatus for determining physical constants was too expensive for individuals to own, he said, and the work could be successfully done only by groups of trained specialists together with assistants and other staff members.

Five years later, Rowland was deploring the "folly" of filling the country with telescopes and calling them observatories—a few "first class institutions" would be preferable to the multitude of inferior ones then existing. The same was true of the colleges and universities: the total wealth of the hundreds of such institutions in the country, he calculated, would be sufficient for one great university, four smaller ones, and perhaps twenty-six colleges. Some work could still be done on a shoestring, he concluded, "but not the highest kind."[21] The situation was so bad, Clarke had concluded by 1876, that the college system was itself as great a drawback to American scientific growth as any other factor. He could only suggest that perhaps the best way to deal with the colleges was to tax most of them out of existence.[22]

Pure science did not fare much better in terms of college curricula. Although scientists earlier in the century had been in the forefront of the drive for "practical education"—including a de-emphasis of the classics, a reshaping of the colleges to better prepare students for "life," and establishment of the practically oriented technical schools—they now found that the fruit of their labor was actually destructive of the new professional ideal. Nicholas Murray Butler saw that the rapid growth of technical schools was the "main obstacle to the full establishment in America of the pursuit of science for its own sake."[23] For the same argument that threatened to cast the classics out of education could be used to attack the equally "useless" abstract science. As C. Hart Merriam bluntly put the matter, pure science, which serves only the specialist, was not suited to the college curriculum. The tendency of the times, he said, "is to render under-graduate courses more practical, so that the knowledge acquired may be useful later in life." Not 10 per cent of the biological instruction favored by the professionals could possibly be of any value in later life to anyone not destined to become a specialist, so Merriam said, and certainly no more than 1 per cent of those taught biology became specialists. This implied, of course, that 90 per cent of the really professional training was useless to 99 per cent of the people subjected to it. Merriam, a naturalist who found "delight in contemplating the aspects of nature" and in "studying the forms, habits, and relationships of animals and plants," was particularly distressed by the new emphasis, by "self-styled biologists," on histology and embryology. The "self-styled biologists" were clearly on the wave of the future, but older naturalists understandably felt that narrow specialization would make them "blind to the principal facts and harmonies of nature," as Merriam put it. The compound microscope and other mysterious paraphernalia, the intricacies of new laboratory techniques, the preoccupation with the lower life forms—all this was alien both to the scientific generalist and to ordinary students who did not intend a career as professional biologists.[24]

A story told by Ira Remsen of his early days at Johns
Hopkins aptly illustrates the new orientation of professionals
toward their subject. A young man had come to consult with
Remsen about the possibility of studying with him:

> He regarded me with some curiosity, and after a time he ven-
> tured to say: "Professor, I should like to enter the Johns Hop-
> kins University, but your work doesn't seem to be practical and
> others are saying the same thing." I acknowledged the truth of
> the observation, and added that I feared it was an incurable
> case, that there was, in fact, no prospect of my work ever be-
> coming practical in the sense in which I supposed he used that
> much abused word.[25]

B. A. Gould had warned his colleagues as early as 1869
that their late enthusiasm for curriculum changes had been
misplaced and had now become a positive danger. The current crusade against classical culture "bode no good to science,"
he observed. "The Champions in this crusade occupy simply the
utilitarian ground, and their alleged advocacy of science is in
fact scarcely more than an advocacy of the useful arts. . . .
The crusade is not in behalf of this or that form of intellectual
progress; it is against such intellectual culture as has not some
tangible end, capable of being represented in dollars, or finding
expression in some form of physical well-being."[26]

In view of the perceived inadequacy of the technical
curriculum, with its emphasis upon the "practical," it comes as
no surprise occasionally to find scientists lined up with classi-
cists at the more practically oriented schools in defense of a
traditional curriculum. The dissension became so great in at
least one case—at the Kansas State Agricultural College—that
Benjamin F. Mudge, a nationally known authority on Kansas
paleontology, was fired for opposing the introduction of what
the regents termed "science" into the curriculum. The regents,
who were also moving against the teaching of Latin and Greek,
accused Mudge of teaching "the abstract sciences in a loose
and unprofitable way" and insisted that scientific instruction in
Kansas should concentrate on "practical agriculture and the
mechanic arts."[27]

The younger generation of professionals in the post-Civil
War decades had moved completely away from both the concept
of practical education and the old liberal-arts notion of giving
the masses the smattering of scientific knowledge necessary for
them to "appreciate" nature. Those who discussed teaching

problems in public quite generally chose to consider their subject "more especially from the standpoint of the preparation
for professional occupation." That is to say, the main function
of the college was now seen to be that of providing facilities to
enable the profession to reproduce itself. The ideal could be
realized only at Johns Hopkins and a very few other places, but,
nevertheless, it was a generally accepted standard by which
one's success as a teacher was to be measured. Only a few
natural scientists, usually those who still had one foot in the
social sciences, continued to think in terms of the older ideal
of science in a liberal education. The astronomer Simon Newcomb, who wrote an economics text on the side, was one such
exception, and his point of view—"a wide and liberal training
in the scientific spirit and the scientific method"—was that
generally held by social scientists at about the beginning of
this century.

In the same way that professional aspirations foundered
on the democratic educational assumptions of the colleges,
they also encountered difficulties with the other great patron
of research, the Federal Government. Once again the problem
was compounded by the fact that an earlier argument had been,
by any ordinary standards, astonishingly successful. By the end
of the nineteenth century, the United States government had
become possibly the world's greatest supporter of scientific research. But in keeping with the egalitarian democratic context,
every single advance in federal support had been justified by
some presumed public purpose of a practical character that
the research would serve. That a democratic government should
become "patron" of a privileged group of pure scientists was
unthinkable. This orientation, however, came increasingly under
fire with the rise of the pure science ideal. Alexander Winchell,
in 1886, could see no reason for praising government science,
for he could not think of a single case where any public provision had been made for pure research, and surely no credit
was due for supporting the practical. He pointed with alarm,
furthermore, to the fact that popular and legislative prejudice
against pure research, and the corresponding effort to limit the
activities of governmental support, were apparently increasing.[28]

There is no doubt that Winchell was correct in his assessment of the situation. The establishment of the Allison Commission, a congressional committee appointed to study the administration and organization of scientific agencies of the gov-

ernment; the cuts in appropriations for scientific bureaus in the 1880's; the several cases of dismissal of research-oriented scientists—all testify to a growing public suspicion of government science. As late as 1903, a Committee on the Organization of Government Scientific Work adopted as a basic assumption in its report (never published) that "research in pure science on broad and general grounds is more properly within the scope of private institutions, and that in general the work of scientific research on the part of the Government should be limited nearly to utilitarian purposes evidently for the general welfare."[29]

But although Winchell was correct in his description of the situation, it is not so clear that he was correct in ascribing it to a growing anti-intellectualism on the part of the American people—or, for that matter, to any change on the part of the public at large. Scientists had successfully gained support for a great many enterprises on the grounds that they would provide material returns for the money expended. Geological research is a good case in point. Despite Winchell's disparaging remarks, the Coast Survey *had* originally been defended by scientists in terms of its benefits to "harbors, commerce and national defense," and the Geological Survey, like the numerous state surveys, *had* been advanced as a means of discovering the material wealth of the land. What had happened had been entirely predictable. State legislators, for example, who had been prevailed upon to finance geological surveys in expectation of immediate returns in discovered mineral wealth, lost interest when such dramatic results were not forthcoming—as was usually the case in the early days. Time after time, legislators refused to finance publication of the complete results of a survey, insisting upon publishing only those parts dealing with "economic geology." If government work was being done in the public interest, as the scientists had claimed it was, then that work should be restricted to enterprises clearly in the public interest. So went the not entirely unreasonable argument. Geologists found, increasingly, that the work that really needed to be done, from their point of view, did not meet the requirements of being clearly in the public interest. The paleontological work of the Geological Survey and the Survey's efforts to reconstruct the geological history of the continent came under continuous attack. As one angry reader of *Science* magazine explained, "Among those who have given it any attention, with whom I converse or correspond, not one expresses satisfaction [with the Survey] and generally they have only words of severe condemnation." The writer himself could not even discover a genuine attempt to make a geological survey in the published volumes; instead, he said, these contained "theoretical discussions about the glacial period, that have no economical value, and which period, I think, is fiction, and they contain a vast amount of extremely localized and temporary matter of no general utility."[30]

Other government scientists encountered similar difficulties. The case was the same with all but the most obviously practical astronomical work of the Naval Observatory. Agricultural experiment stations were expected to confine their efforts to reaching "an empirical solution of one problem after another" to the neglect of a rational, scientific approach to agricultural science.[31] As one critic put it in a letter to the editor of *Science* magazine, the scientific bureaus were established by the United States in order to do practical work, and the government's only authorized scientific investigation was in connection with making such practical work possible. "It is a step in a radically new direction to introduce the prosecution of investigations *per se,*" warned the reader of *Science.*[32]

Although historians do not customarily recognize the fact, it was indeed a change in direction for which scientists of that period argued, a change necessitated by the near-exhaustion of the *scientific* possibilities of the old-fashioned survey, the prototype of government science in the early nineteenth century. Once again, in order to understand the discontent of the scientists, one must refer to the changed orientation on their part, and observe that the new orientation imposed requirements in utter conflict with prevailing democratic assumptions about the political process. It was the pure-science ideal that now made the formerly satisfactory situation seem unsatisfactory, and it was the new demands upon government introduced by the ideal that led to frequent clashes with legislators, administrators, and the general public.

The pure-science ideal demands that science be as thoroughly separated from the political as it is from the religious or the utilitarian. Democratic politics demands that *no* expenditure of public funds be separated from political control, or, to state it another way, that no power be granted without responsibility—which always includes public accountability. With such diametrically opposed assumptions, conflict is inevitable.

As even so staunch a supporter of pure research as the editor of *Science* admitted, government scientific work, unfortunately, was "far removed from that public criticism which is so conducive to efficiency in other branches of the service."[33] On the other hand, there was broad agreement among most representatives of the scientific community that their work *must be* so removed; the scientific establishment, it was often said, should be kept "safe from political spoilsmen." The preceding words, quoted from an article in the *New York Evening Post,* introduced a vigorous defense of T. C. Mendenhall's qualifications as superintendent of the Coast Survey in response to a rumor that a scheme was afoot, backed by the chairman of the Democratic National Committee, to replace him with one of his associates.

In part, the effort to avoid political involvement was a matter of self-defense. For example, in 1888 when John Wesley Powell threatened to encroach upon the political question of distribution of public lands, the Congress, spurred on by Western interests, not only turned against Powell and his irrigation survey but turned upon scientific agencies apparently unrelated to Powell's activities. Powell had envisioned the development, through topographic and hydrologic studies, of largely independent agrarian communities, each controlling the water of its own drainage basin through systems of reservoirs and canals. Western legislators, wishing to get on with the rapid exploitation of the public domain, were particularly incensed by Powell's suggestion that any further disposal await the conclusion of the studies; neither did they like his reference to the Great Basin as an "arid region" with special problems, for this might discourage the flow of immigrants. As a result of the outcry against Powell's excursion into political affairs, the irrigation survey was allowed to languish in the Bureau of Agriculture, and drastic cuts were made in the budgets of the Coast Survey, the Lighthouse Commission, the Smithsonian Institution, and the Naval Observatory.[34] A few such examples of congressional wrath would surely inculcate caution.

But there were other factors inherent in the notion of pure science that also made conflict inevitable. Two things were held to be absolutely necessary for pure scientific research: long-range planning and flexibility. The first required that scientists be free of the limitations on tenure and the annual appropriations customary for other government operations. Control of scientific work by the military was especially criticized because it seemed to make impossible the kinds of investigation that require long-range planning. Flexibility, the second requirement, meant that overly detailed instructions should not come with an appropriation and that a rigid accounting should not be demanded. To state the problem in the manner that most often occurred to critics, the needs of pure science demanded an undemocratic suspension of the rules on behalf of a select group.

According to the ideology of the scientists, the ideal appropriation for a scientific bureau would be granted as a lump sum, marked simply "for research," and its expenditure would be entrusted to a professional scientist in charge of the bureau. If cuts in the appropriation must be made, as the editors of *Science* remarked in connection with the Geological Survey, this should be an over-all cut, not a paring of some specific operation. In such a way, maximum freedom could be retained for moving in any direction that seemed advisable to the director.[35]

That one politically adept bureau chief, John Wesley Powell, had—until his downfall in the late 80's—actually achieved this much freedom made it seem all the more compelling that the practice be made general. Such an administrative head as that desired by the professional scientists would naturally have to be a trained professional himself, not a layman whose main job it was to watch out for the public interest. The latter, of course, is the traditional view of the department head in a democratic framework. The notion, for example, demands that the Secretary of the Navy be a civilian, not too closely identified with naval interests; it still calls forth outraged protests when a trade union leader is named Secretary of Labor; and it was a major ingredient of public protests against the developing federal scientific establishment. Thus Simon Newcomb spoke in a letter to a fellow-scientist of "the unavailing efforts of the president [Cleveland] to get somebody absolutely ignorant of all scientific matters to act as fish commissioner." He was thankful for Cleveland's "final surrender to the force of circumstances," which Newcomb regarded as a very healthy sign, giving him reason to entertain some hope that before the end of the President's second term he might yet reach the conclusion "that honest administration is not wholly inconsistent with scientific knowledge."[36] The prejudice, however, was very strong. As one lay correspondent of *Science* put it in commending the appointment of a non-scientist as Commissioner of Agriculture, "technical experts . . . are, as a rule, those gentlemen who have bees in their bonnets." Unlike the scientific administrator, the reader said, the new Commissioner of Agriculture was "without a pet hobby." The editors, naturally, issued a sharp rejoinder. The worst thing about a non-scientist as administrator, they said, was that "any broad, well-planned policy is practically out of the question." All the non-specialist could do was either irritate and hinder by "ill-judged interference" or else leave matters to take their own course, in which case he was useless.[37]

Ideally, from the point of view of the scientists, both the personnel and the work of scientific agencies should be kept completely "out of politics," for the close connection with politics undoubtedly was a hindrance to conducting research for the government. The chances of politically motivated dismissals were thought to be too great to make federal jobs desirable, and frequent personnel changes, it was pointed out, were absolutely "fatal to good, long-continued work." Bureau heads should be named either by the National Academy of Sciences or an appropriate specialist society. In this way, the business of any bureau "would be more likely to run smoothly, like the work of the Smithsonian Institution."[38]

To many, the notion of public accountability seemed both unseemly and ridiculous when applied to the highly specialized, technical work being done by federal bureaus. How, Alexander Agassiz asked rhetorically, could "a clerk in the auditor's department" be expected to pass upon the work of heads of bureaus? More than the dignity of bureau chiefs, however, was thought to be at stake. It was generally held that "science cannot be carried forward by prescribing too definitely the task of scientific men." If there were to be real progress, the methods of free inquiry would have to be employed, not those of "petty regulation." This principle, above all, ought to be held in mind whatever measures might be adopted with respect to reorganization of the government bureaus. As B. A. Gould explained in answering the charge that he had violated the instructions of his superiors on the Sanitary Commission, he had found it impossible to predict in advance the expense of any investigation. "The best computers make mistakes," he said. "One research frequently entails others, & quite as often it conducts to purely negative results, which have nothing tangible to represent them."[39] Powell, in his interrogation before the Allison Commission, spoke to the same point: "Now, the work of the Survey is of such a character that it cannot be fully specified and planned in advance, from the fact that to a large extent it is research for facts and principles not yet discovered. . . . Only the general object in view and the general line of investigation to be pursued can be designated . . ."[40]

It goes without saying that Powell was technically correct, but it could only appear to critics, such as the editors of the *Popular Science Monthly*, whose Social Darwinian principles usually reflected the public's point of view, that an effort was being made to establish "scientific pontiffs" in Washington and that scientists were merely trying to gain some private advantage. The editors were willing to tolerate whatever government science activity was indispensable, but, in general, the less there was the better. For them, even the formation of the National Academy of Sciences had been too decisive a step toward centralization. For the most part, however, their anti-centralization, anti-governmental feeling was focused on the scientific bureaucrat in Washington. The same rules should be applied to the scientific bureau as to any other, the editors reasoned:

> An unsupervised and irresponsible scientific department at Washington would be run in the interests of its sharpest managers, would be filled with sinecures, give the least results at the grandest expense, while the results would be aggravated by the sense of exemption from criticism.[41]

The claim that scientific work should be exempt from public criticism—an exemption considered a prime necessity by the scientific community even before the development of the pure-science ideal—was an outrage to the sense of justice of an uncomprehending public that, through both cultural heritage and training, opposed the concept of élites of any kind. The novelist, so one critic said, does not demand that he be read only by novelists. Painters, sculptors, and historians make no such demand. Why, he asked, should only the scientist insist that no one except "his own brethren" be allowed to form an opinion of him and of his work?[42] This was a particularly unacceptable position, so a correspondent of *Science* urged, when public funds committed to allegedly public purposes were involved. Any public matter, including the Geological Survey, "is a proper subject of criticism, by any citizen."[43]

XIII

〜〜〜〜〜〜〜〜〜〜〜〜〜〜〜〜〜〜〜〜〜〜

Science, Scientism, and Planned Progress: The Uses of Science in Progressive Ideology

"It is our confident claim," wrote Theodore Roosevelt, the darling of Progressive reformers, "that applied science, if carried out according to our program, will succeed in achieving for humanity, above all for the city industrial worker, results even surpassing in value those today in effect on the farm."[1]

These words, written in 1908 to William Howard Taft, illustrate a commonplace of Progressive thought and, at the same time, suggest the profound distance that had developed between most professional scientists and the rest of society. They also illustrate once again the magic that had come to be wrought by the word "science." Most scientists, as a consequence of their drive for professionalism, had by the end of the century become quite wary of trying to solve social problems by "applied science." It was not that they doubted the power of science; on the contrary, they had a faith in it unsurpassed by any preceding generation, and they were sure that, if problems ever were to be solved, science would play a commanding role in their solution. But such results, they held, should be incidental, not the object of the scientist's work. The scientist was a seeker after the truth; any extraneous considerations might bias the researcher and impede the progress of science. If social problems influenced the scientist in framing his research, then the purity of the developing body of knowledge would be seriously undermined. The term "applied science" itself had taken on invidious connotations—something not quite worthy of the highest intellect.

Professionals had, in fact, as a byproduct of their drive for autonomy, become one of the most socially conservative groups in the country. In the early days, when the line between professional and amateur remained blurred, natural historians and natural philosophers had often been attracted by atypical beliefs and reformist public activities. David Rittenhouse during the previous century had been a leader in Jacobean societies. William MacLure, Thomas Say, Gerard Troost, and others had participated in the New Harmony experiment; MacLure himself had been active in virtually every available kind of reform during his long career at the beginning of the century. Even Benjamin Silliman, conservative in most matters, had dabbled in educational reform. The word "philosopher" had, in fact, become virtually synonymous with "revolutionary" to the conservative American in the opening decade of the nineteenth century. But in contrast to the earlier pattern, most of the professional scientists of the late nineteenth century confined their innovating within the boundaries of their discipline. The true heirs of Rittenhouse, MacLure, and their kind were the reform-minded social scientists, who were not usually held in high esteem by their colleagues in the natural and physical sciences.[2] While sociologist Lester Frank Ward had visions of the smoothly functioning, progressive order that could be achieved by the rational application of the methods and insights of science, physicist Henry A. Rowland concerned himself solely with the limitations imposed by a democratic social order on the ideal development of his discipline. John Dewey, philosopher and psychologist, sought to spread the beneficent influence of the public school, including the community college, to every hamlet in the nation, while Frank W. Clarke, chemist, wished to tax most of the colleges out of existence because of their inadequacy in scientific instruction. The division between the two points of view was fundamental, and the position they took would appear to place most professional-minded scientists outside the mainstream of American developments during the Progressive Era. Yet, the paradox is that nothing was more important to that era than "science." It was a word to conjure with; a word to sweep away all opposition by labeling it "benighted," "Romantic," or "obscurantist"; a word to legitimize any program no matter what fundamental reorientations it might entail or what sacrifices it might call upon particular groups to make. In the name of science, one might reorganize a city government, fundamentally alter the relations between labor and management, revolutionize a school curriculum, or consign whole races of men to genetic inferiority. If the "science" involved in Progressivism seldom possessed the rigor of a Newtonian law, few would notice; the name had a magic of its own which made questioning irrelevant.

A part of the appeal of science to reform-minded citizens is explained simply by the strong Progressive sense of power, for science seemed to promise the power to control both man and nature. Technology, especially, fired Americans' imaginations, giving them unlimited hopes for the future. Americans were increasingly impressed by the visible manifestations of the power of technology that had appeared at an accelerating rate since the third quarter of the nineteenth century—the same manifestations that had cleared the way for the rise of a pure science ideal among American scientists. The telephones, telegraphs, and electric lights of which Rowland had spoken so deprecatingly were more powerful arguments for science from the layman's viewpoint than was the most elegant theorem of mathematical physics. It was not the laws of thermodynamics but the steam engines and railways, not Maxwell's equations but the harnessing of electricity for everyday purposes and the use of the internal combustion engine, not the scientific aerodynamics but the Wright brothers' fabulous machine that inspired visions of an idyllic future. Science and technology by the beginning of the twentieth century had begun to permeate all of life, and this wrought both obvious and subtle psychological effects. Where once the hero of adolescent fiction had been Horatio Alger's youthful business entrepreneur, by the 1910's, Tom Swift, boy scientist and inventor, had emerged as the ideal for American youngsters in a series of best-selling books.

The "lesson" of the nineteenth century and its achievements, so author after author concluded in the plethora of century-end reviews, was that science was the mainspring of inevitable progress. To a large group of intellectual leaders, including even those scientists who themselves scorned practicality, the key to progress was the increased application of the tools, knowledge, and methods of science to all spheres of human activity. The general public readily believed such assertions, for to it science was universally confused with technology and upheld as a wonder-worker that promised more of the startling innovations they had experienced during their own time. Subtle distinctions, they thought, only confused the issue. In science, as in everything else, there was only the "good" and the "bad"; the good was the true and the bad the false; the good promoted human life and comfort, the bad did not.

Motivated by the prospects of what science promised, Americans applied it to problems on a scale unthinkable a generation before. By the beginning of the twentieth century, science was firmly entrenched in dozens of government agencies and legislation based on science was becoming more common. Public health was perhaps the most obvious example. Those who had known the harsher side of life in the nineteenth century now saw children rescued from certain death by the miracle of diphtheria inoculation. The public rightly celebrated the conquest of yellow fever and hookworm and confidently expected that other diseases would also soon be vanquished by the irresistible power of science. Not only were the construction of water and sewage systems accelerated and public medical programs expanded, but the technique of chemical analysis had led to pure food and drug legislation in the states, beginning in 1892, and on the federal level in 1906. The germ theory of disease had led to a reorientation in thinking about public health problems.[3]

To the average American, even more impressive demonstrations of the power of science were provided by World War I, which had the further effect of tending to soften the long-standing public fear of experts. "The present national crisis brings home to us the crying needs of the nation in availing itself of the knowledge and ability at its command," said one scientist late in 1917. He ventured the opinion that, as soon as the problems were formulated and given to the engineers, "few will remain unsolved long."[4] The scope and importance of the scientists' contributions to warfare had never been greater, and thanks to the popular press and the deliberate public relations activities of scientific societies, the achievements of scientists had never been better publicized. One of America's first wartime scientific projects was the assembling of leading physicists to devise a more effective means of submarine detection. Leaving their university posts to work in cooperation with English and French physicists who had been struggling with the problem for some time, scientists from the United States devised a much more effective method of detecting and locating submarines than had previously been developed. Even though it involved no new science, the importance of the device that they developed—and its public relations value—is obvious, for the greatest threat to the Allied cause was German submarine warfare. American physicists also worked on new instruments to improve the accuracy of aircraft bombing, to improve aerial photography and mapping, and to improve aerial navigation. The prewar leadership of Germany in the production of fine optical glass created a shortage of that essential material after 1914, which in turn led to great efforts to develop this industry in the United States. Physicists cooperated with the manufacturers with such success that by the end of the war the United States was actually exporting some optical glass in addition to providing for all domestic needs.

Undoubtedly the most publicized contributions to the war were those made by chemists. The conflict was, in fact, sometimes referred to with pride as "the chemist's war." Besides improving explosives, chemists were called into service to devise means of protection against poison gas and to develop gases that could be used against the enemy. Chemists were widely acclaimed for their work at the time; it was only later that Americans began to have second thoughts about the gas research and, consequently, about chemists.

It was not only the physical scientists who were called from their familiar routines and placed at tasks with special bearing upon the war; botanists, entomologists, zoologists, and nutritional experts made important contributions toward improving wartime food production, preservation, and distribution. Bacteriologists helped with problems of sanitary engineering and of disease prevention and control in the Army as well as among undernourished civilian populations. Physicians devised new and more effective methods for treatment of conventional battle wounds along with the new and exotic gas poisoning and "shell shock." As weather prediction assumed greater importance because of gas and aerial warfare, meteorologists were called upon to contribute their services.[5]

It is true that wartime mobilization of science, especially by Germany, inspired many attacks on science as such, but these were generally expressed before United States entry

into the conflict. At any rate, most Americans were easily persuaded that the evil was not the fault of science but of *misdirected* science in the hands of evil men. For the most part
they observed what was happening, marveled at what they
saw, and refused to be disillusioned.

Such concrete evidences of the potency of science increased the authority—both within the scientific community and
without—of anyone who spoke in the name of science. A
Cornell University scientist caught the mood perfectly when
he observed that one of the most surprising outcomes of the
war had been "the sudden and I believe permanent enthrone-
ment of science in the activities of humanity."[6] The chairman
of the National Research Council, in drawing up a balance
sheet of the effects of the war on science, mentioned as gains
the better public appreciation of the usefulness of science, increased governmental appropriations for research in several countries, the increased appreciation of the value of science in industry
which had led to a multiplication of industrial laboratories, and
the recognition by educational institutions that science should
have a larger role in the curriculum.[7] Many Americans, believing that science dealt only with objective truth, were prepared
to listen with credulity to the great physiologist Jacques Loeb,
who believed he had found the mechanism of the life process
itself and who proposed to draw conclusions from the discovery
that are now largely repudiated by biologists. Adherents of the
eugenics movement, which drew for its support upon a sizable
fraction of the national scientific community, believed that
new knowledge of genetics would permit a change in the physical
inheritance of man himself, thus carrying reform to extremes
that not even the most sanguine educators had envisaged. The
immigration restriction law of 1924 was one result of their
efforts; eugenic legislation in more than a dozen states was
another.[8]

The popularity of such movements was possible simply
because the American public in the twentieth century has been
marked by a childlike faith in science. This has caused it to
adopt enthusiastically any panacea that can be associated,
however loosely, with science. Naturally, charlatans as well
as honestly misled scientists abound to take advantage of this
receptivity. Every marked scientific advance that touches upon
an area of vital interest to the public attracts its charlatan or

marginal practitioner who hopes to exploit that interest. A comment in a popular magazine of 1922 catches exactly this feature of the American mind:

> How swiftly the spotlight of popular interest shifts from one part of the stage to another! The eyes of distressed humanity turn eagerly toward any quarter that appears to promise health and happiness . . . Those who recently were reading Freud and Jung have now taken up with Berman and Harrow [pioneer endocrinologists]. Those who formerly were rushing to have complexes extracted are now anxious to have glands implanted.[9]

As the author indicated, the discovery that gland dysfunctions could cause certain diseases and severe personality changes had led to the inference of a whole psychology based on glands. Sensational stories about treatment with gland extract and gland transplantation overshadowed the general theory. Even serious intellectuals often took the exaggerated claims of endocrine enthusiasts surprisingly seriously, as they did any other claim which carried the magic stamp of "science."

This aspect of the popular enthusiasm for science has often been overlooked by those historians who celebrate science as a series of "advances." In fact, every major advance has carried with it unintentional consequences that have often been unfortunate. The germ theory of disease, for example, one result of which has been to increase vastly the curative ability of physicians, had an even earlier effect in stimulating patent-medicine quackery. Even before most American physicians had accepted the theory, William Radam's "Microbe Killer," was on the market as a sovereign remedy for every known kind of disease, all of which were said to stem from the action of "germs." Similarly, the gas research of the First World War led not only to better-equipped police departments, but also to heavy sales for "chlorine chambers," which were guaranteed to cure the common cold.

The theory of the unconscious was another that powerfully stimulated quackery, of which Coueism was the best-known form. An importation from England of the 1920's, Coueism was based on the standard psychology of suggestion, which in turn had been inferred largely from the phenomenon of hypnosis, which is suggestion in an extreme form. In addition to the conscious mind, so the theory went, there is a subconscious mind which influences behavior. A repeated suggestion aimed at the subconscious will eventually influence it. A well-known theory for many years before Émile Coue, it had been employed in a type of psychotherapy based on suggestions made by physicians. Coue's innovation was a kind of do-it-yourself plan which would lead to self-improvement. Each morning, Coue taught, one should repeat to himself, "Day by day, in every way, I'm getting better and better." Presumably, this auto-suggestion cured both physical and psychological difficulties and had the added benefit of improving the character. The enthusiasm and exaggeration to which this idea was subject was suggested by a rhymster for *Life* Magazine:

> Would you be freed from every kink
> of woe and make your forces double?
> Bamboozle dark Subconscious Mind,
> That ever-present source of trouble . . .
>
> No matter what your goal or aim,
> You must not doubt yourself a minute,
> But say, "Of course I'll win the game."
> Subconscious Mind will make you win it.[10]

But one does not have to search on the fringes of respectability for such evidences of enthusiasm. On the contrary, the most respected thinkers of the day, and the most important movements of the period, were wholly captured by the magic of "science."

In the nineteenth century, despite the frequent obeisance made to science and technology, social forces in America tended to shape, or at least restrict, the development of science. In the early years of the twentieth century, the relationship became more reciprocal, and in important ways: science had a substantial influence upon American social policies. As early as 1898, W J McGee, according to Theodore Roosevelt the "scientific brains" of the conservation movement, realized what had happened. "America has become a nation of science," he said. "There is no industry, from agriculture to architecture, that is not shaped by research and its results; there is not one of our fifteen millions of families that does not enjoy the benefits of scientific advancement; there is no law on our statues, no motive in our conduct, that has not been made juster by the straightforward and unselfish habit of thought fostered by scientific methods."[11]

As a tool of social change, science appeared in two aspects to reformers of the period: as accumulated and systematized factual material to be used as a basis for decision, whether regarding the construction of a bridge or a railroad tariff schedule; and as a method to be applied to difficult unsolved problems, technical or social. The influential philosopher of science Karl Pearson had argued that science consisted solely of a method, not of its material; and furthermore, *all* materials were potentially part of the domain of science.* Significantly, Progressive leaders believed that the successes of science were based upon a known technique termed "the scientific method," and that this method could be applied to all human problems. By the beginning of the Progressive Era, American scientific popularizers were generally agreed that the scientific method, which they presented as a simplistic adding up of "facts" relevant to a problem, was the only gateway to determining truth. Science, which had been distinguished from other forms of knowledge in the early nineteenth century, had now come once again to be identified with *all* knowledge.

Johns Hopkins chemist Ira Remsen, for example, maintained that the great generalizations which had been reached through the persistent efforts of scientific investigators were "the foundations of all profitable thought." And this was a man who refused to enter into patent litigation because he "would not sully his hands with business," and who carefully explained to prospective students at Johns Hopkins that he was not interested in practicality.[12] It was not necessary to be directly interested; practical benefits would automatically result from the advance of science. Confidence in the scientific method was well expressed by Thomas C. Chamberlin, one of the nation's leading geologists at the turn of the century and president of the University of Wisconsin from 1887 to 1892:

> There has grown up with the sciences a distinctive mode of work, known as the scientific method. It is distinguished from previous methods in its constant appeal to observation and experiment, and to rigorous induction from these. Now, this scientific method is confessedly superior to all previous methods . . . It is being gradually introduced into other departments of

---

* Karl Pearson's book, *The Grammar of Science* (London, 1892), was widely read in America. In 1896, the *Popular Science Monthly* ran a series of his articles.

> thought and will at length beyond question come to be practi-
> cally the one universal method of research. To the acquirement
> of this method the study of the natural sciences is the obvious
> avenue.[13]

Holding the presidency of the University of Wisconsin at the
very time when the curriculum was being expanded and graduate study developed, Chamberlin was able to impress his ideas
upon the development of the university. It is not surprising that
a few years later the university became virtually a laboratory
for Progressivism in Wisconsin, and that the faculty, many of
whom had been selected by Chamberlin, became the most active in the nation in advising Progressive legislators and governors.

The belief that science contributed to the intellectual
progress of mankind by encouraging habits of thinking which
tended to make men more rational, 'to make them base their
thinking on facts, observations, and experiments, was generally
accompanied by the assumption that the scientific method could
be applied to mental and social areas of study. As Simon Newcomb put it, in calling for an emphasis on science in liberal
education:

> All such political questions as those of the tariff and the cur-
> rency are, in their nature, scientific questions. They are not
> matters of sentiment or feeling, which can be decided by popu-
> lar vote, but questions of fact, as effected by the mutual inter-
> action of a complicated series of causes. The only way to get at
> the truth is to analyze these causes into their component ele-
> ments, and see in what manner each acts by itself; and how
> that action is modified by the presence of the others; in other
> words, we must do what Galileo and Newton did to arrive at the
> truths of nature.[14]

From this, it followed that the leaders in society had to be trained
in the methods of science. The result would be a more efficient
society in which basic technological decisions were made through
a single, central authority in a rational and scientific method.
"Is it too much to hope," asked P. G. Nutting, that the day when
all the great problems would be automatically placed in the hands
of trained specialists? "The problems should not be entrusted,"
he continued, to "self seeking politicians, nor yet men with mere
theories, but [to] engineers with a real command of fundamental
principles, men with an unbroken record of big achievements
and no failures, men ever ready to stake their all on their ability

to handle problems in their specialty."[15] This was the crux of what one historian has labeled the "gospel of efficiency."

W J McGee, a self-trained geologist and ethnologist, had early emerged as the leader of those concerned with making science more "consciously pragmatic," as one writer put it. More than any other single individual, McGee became the prophet of the new world which conscious purpose, science, and human reason could create out of the chaos of a laissez-faire economy where short-run individual interest provided no thought for the morrow. McGee, the key figure in disseminating the expanding concepts of the conservation movement, exhibited much of the passion for direct social planning held by his contemporary, the sociologist Lester Frank Ward. Again and again at resource conclaves McGee emphasized his faith in applied knowledge: ". . . the course of nature has come to be investigated in order that it may be redirected along lines contributing to human welfare . . ." The conservation movement he considered a momentous step in human progress because it involved "a conscious and purposeful entering into control over nature through the natural resources, for the direct benefit of mankind." The Inland Waterways Commission, with its vision of massive river development, he maintained, involved the "highest application of applied anthropology thus far attempted." In his speeches and writings McGee formalized the spirit of efficient planning implicit in the policies of Roosevelt resource administrators.[16]

Becoming acting chief of the Bureau of Ethnology in 1892, McGee hoped to use the bureau as a kind of social science laboratory for the development of techniques to implement social planning and control. The organic analogy, which he adopted from evolutionary theory, implied for this reform-minded scientist that, under the relatively simple conditions prevailing in primitive societies, one could study many of the problems that in more complex forms baffled modern civilizations. Causes could be more readily isolated under simple conditions, solutions could be worked out, and, it was never doubted, these solutions could be easily translated into the more complex terms appropriate for modern civilization. The Indian cultures of the Southwest, which an earlier generation had studied mostly as curiosities on the verge of extinction, would play a noble role in solving the problems of American society.

As the example of McGee illustrates, the earlier concern

of social science for "finding out how to make men live according to nature" was changed in the late nineteenth century to efforts to discover how to direct natural and social processes rationally for the benefit of members of society. The anthropologist E. B. Tylor had foreshadowed this change when, in his *Primitive Culture* (1874), he studied prehistoric civilizations in order to ferret out those elements of his own culture which were mere survivals of a more backward and less civilized age. Having been exposed as anachronisms, so the reasoning went, the survivals would speedily disappear and the society would become correspondingly more rational, more "modern." "The science of culture," he said, "is essentially a reformer's science."

The changing view of social science paralleled the development of sociology as an independent discipline in the decades after 1885, a discipline oriented primarily to the study of social problems. The syllabus of early courses in sociology at the University of Wisconsin and the University of Chicago, which became the center of the new school, read like catalogues of the major problems of industrial society; and the students of Albion Small, E. A. Ross, and C. H. Cooley were imbued with the ideas of meliorism, social control, and planning. Ross especially, who invented the term "social control," taught that out of a knowledge of the scope and nature of the informal controls operating in society came the possibility of exercising management through deliberate manipulation. The task of the sociologist is to discover how societal control is exercised and to give this information to moral and opinion leaders, who would use it to accelerate the "progress" of society.[17] Meanwhile, James Harvey Robinson and Charles A. Beard were bringing the same reformist approach to the study of history, which they saw as a pragmatic weapon for explaining the present and controlling the future. J. M. Clark, Richard T. Ely, and others were founding a new economics dedicated to the same idea—that human institutions were subject to human control by scientific methods, and that man's future progress could be directed by the application of human reason. As Ely somewhat cryptically explained, the purpose of the American Economic Association was to "bring science to the aid of Christianity."[18]

It was this trend that came to a height with John B. Watson's behavioristic psychology developed in the first two decades of the twentieth century. "The interest of the behaviorist in man's

doings," said Watson, "is more than the interest of the spectator—
he wants to control man's reactions as physical scientists want
to control and manipulate other natural phenomena." To do this,
Watson explained, one must gather scientific data by experimental
methods.[19] "The belief that we are at last on the track of psychological laws for controlling the minds of our fellow men," wrote
Abram Lipsky in 1925, "has brought about a revolution in the
popular attitude towards the science that teaches how to do it."
Out of this change, he explained, had developed a universal interest in psychoanalysis, psychotherapy, hypnotism, character
analysis, mob psychology, and salesmanship—"all connoting a
technique with which one man controls the minds of others."[20]

Other bureaus of government, under the impetus of Roosevelt's appointees, were undergoing the same kind of changes that
McGee wished for his Bureau of Ethnology and which the leaders
of the emerging social sciences wished for their disciplines. The
Bureau of Labor, explained Roosevelt in 1908, "for the past twenty
years of necessity largely a statistical bureau, is practically a Department of Sociology, aiming not only to secure exact information about industrial conditions, but to discover remedies for
industrial evils."[21] The Bureau of the Census had, since 1880,
begun to add volumes on resources to the ordinary decennial
activity, and by 1903 had been given a permanent life in the
Department of Commerce and Labor, where it rapidly developed
into an agency capable of providing the continuous flow of information necessary in the Progressive Era. Other bureaus, such as
the Bureau of Mines and the Biological Survey, were soon at work
gathering more specific information relevant to particular problems.

It was McGee who, through his contact with Gifford Pinchot, was instrumental in interesting Theodore Roosevelt in
conservation, the widespread movement on many fronts which
became almost synonymous with Progressivism. In 1907, he
drafted a plan for the creation of the Inland Waterways Commission, which was the first national agency to be charged with the
consideration of resources utilization as a unified problem.
Because of the ad hoc nature of the growth of government programs in the late nineteenth century, coordination of efforts, or
even a full exchange of information, had been extremely rare.
One agency concerned itself with water power, another with
irrigation, and still another with flood control and the improve-

ment of navigation. The new idea, evolved by hydrographers of the Geological Survey, was simply that water was a single resource of many potential uses; and they had begun to develop the concept of multiple-purpose river basin development, which only came to fruition decades later with the Tennesee Valley Authority.

McGee, in planning the mandate of the Inland Waterways Commission, represented perfectly the Progressive drive toward rationality, consolidation, and efficiency, as did Pinchot when he began to insist that Western land problems, including grazing, forestry, and immigration, would have to be thought of in terms of the general problem of "the best use of every part of the public domain."[22] McGee also organized the Conference of Governors held at the White House in 1908, out of which came the National Conservation Commission, an even more ambitious effort at consolidation.

This commission, under the leadership of Gifford Pinchot, made the exhaustive inventory of American natural resources that scientists had been urging for decades. With experts as effective heads of sections, it reported on the supply of resources, their rate of use, and the probable date of their exhaustion. Efficient management, the experts argued, required an exact knowledge and careful classification of resources.

Although scientists had characteristically been early and ardent conservationists, they had lacked the organized base of power to secure legislation before the twentieth century. As early as the 1840's, Joseph Henry had made a report on forest trees and their economic uses one of the early projects of the Smithsonian. At the AAAS meeting in 1873, a member had spoken on "The Duty of Governments in the Preservation of Forests," and at the same meeting a committee had been appointed to memorialize Congress and state legislatures on the preservation of forests.[23] The effective proposal for a national inventory had finally come, however, from the engineering societies which had maintained contact with the Roosevelt administration for a number of years through a National Advisory Board on Fuels and Structural Materials established to supervise the quality of materials used by the government. It was the four major engineering societies (American Society of Civil Engineers, American Society of Mechanical Engineers, American Institute of Electrical Engineers, and the American Institute of Mining Engineers), in fact, which

spearheaded the entire drive for efficiency. They took a keen interest in resource problems, publicized conservation affairs in their
journals, and defended the Roosevelt administration from attack.
Looking upon conservation as essentially an engineering problem
which would represent a major step in the progress of civilization,
they argued that it would bring conscious foresight and intelligence into the direction of human affairs.[24]

Not primarily interested in either aesthetics or in preservation, the engineers tended to stress efficiency and rational use for
maximum sustained yield. During a conservation meeting in
1908, the president of the ASCE expressed the prevailing attitude
of the members of his profession on the question of aesthetics vs.
productivity. Lord Kelvin, he declared, was once asked how water
power development at Niagara Falls would effect its natural
beauty: "His reply was that of a true engineer: 'What has that
got to do with it? I consider it almost an international crime that
so much energy has been allowed to go to waste.'" Archeologists,
the society's president complained, had argued that irrigation
works on the Nile would inundate ancient ruins, but "engineers,"
he assured his audience, "will naturally consign all such archaic
questions to the oblivion of the past, and concern themselves
with that which confers the greatest good upon the greatest
number."[25]

In the same vein, Gifford Pinchot argued that forestry did
not concern planting roadside trees, or making parks and gardens,
but involved scientific, sustained-yield timber management. This
is why Pinchot and the engineers, unlike the campaigners for
preservation, were able to enlist representatives of big industry
in their campaign, for it was clearly they who could best afford
to experiment with long-term methods and who therefore would
be able to profit most from the efficient management stressed by
Pinchot.

The Progressives who came to see conservation as a unified
scientific management of the national wealth also began to look
upon health as a resource and a part of the general welfare. in
which scientific research could produce dramatic results. A foundation of basic data had been building up since 1880, when John
Shaw Billings, an Army surgeon and head of the Army medical
library, organized a vital statistics program in cooperation with the
Tenth Census. The germ theory of disease, becoming established
and taking on something of the nature of a fad in the two decades

after 1880, lent new credence to the notion that scientific research could solve the age-old problem of sickness and premature death. Consequently, some of the social legislation of the Progressive Era that produced the bitterest controversy and the most far-reaching court decisions was enacted in the interest of health. In 1906, Professor J. P. Norton read a paper to the AAAS suggesting a national department of health. Norton, an economist rather than a medical man, stressed losses in efficiency from perventable ill-ness in his appeal. The resulting Committee of One Hundred on National Health of the AAAS began a crusade for the conservation of human resources which paralleled and cooperated with those whose main concern was environment. Irving Fisher, another Yale economist, became the chairman of the committee and the author of its manifesto, *A Report on National Vitality: Its Wastes and Conservation*. Calling upon the Federal Government to "re-move the reproach that more pains are now taken to protect the health of farm animals than of human beings," he urged the building of "more and greater laboratories for research in pre-ventive medicine and public hygiene."[26]

The widespread interest in conservation was only one mani-festation of a much broader phenomenon of the period: the reverence for expertise and the corresponding belief in broad executive discretion, which became central features of the Progressive movement. In their search for ways to relieve the weakness, inefficiency, and corruption that seemed to plague democratic government in the face of mounting concentrations of economic power during an industrial age, Progressives fixed upon the trained expert who would be swayed only by scientific facts. Presumably, training in the scientific method not only in-creased the critical faculties of an individual, but made him immune to the corruptions of power that plagued ordinary poli-ticians. "Exact thinking," wrote an English scientist, "leads in the main to correct conduct."[27] Since the fundamental character-istic of the scientific method is honesty, explained Ira Remsen, "the constant use of the scientific method must in the end leave its impress upon him who uses it." If one spent his life in accord-ance with scientific teachings, "it would practically conform to the teachings of the highest types of religion."[28]

Research, another magic word of the period, would allow the expert to solve problems—in government, in industry, in social enterprises of every sort. The ideal leader was no longer the

back-slapping "people's choice" of an earlier generation. He was
a highly trained expert who conducted research on a problem
and cranked out decisions like some well-oiled machine. The
accepted model of the expert was, of course, the scientist in
his capacity of specialist. A decision based on science would of
itself be in the public interest. "Laws may be made by the Congress, interpreted by the courts and executed by the president,"
said a writer in the *Popular Science Monthly* in 1906:

> But they should be based on scientific investigation and carried
> out by scientific experts. . . . There are . . . sound reasons for
> keeping the government of a state or nation free from politics
> and conducting its affairs with such skill and efficiency as are
> attainable. . . . It would be well if we could separate those
> questions which must for the present be settled by party govern-
> ment from those which should be decided by expert knowl-
> edge, and if the latter could be settled by men having the
> necessary special training. And of course nearly all the execu-
> tive work of the government should be done by experts, and a
> large part by those who are technically men of science.[29]

The assumption was that those questions which must "for the
present" be settled by ordinary political means would sooner or
later wither away under the clear light of science and the only
business of elected representatives would be to see that the experts
remained truly disinterested in the application of their expertise.

This attitude, a direct forerunner of the various "tech-
nocracy" movements that sprang up during the next quarter-
century implied a hostility for the professional politician, who
was thought always to be a tool of special interests and incapable
of working simply for the common good. The editors of engineering magazines, for example, had been arguing for years that an
independent commission of experts should be appointed to plan
and execute river and harbor developments. When Representative
Francis G. Newlands proposed that the Federal Government
finance irrigation through a reclamation fund composed of the
proceeds of the sale of Western public lands, his bill stipulated
that the Secretary of the Interior should have complete discretion
in selecting projects. Congressional control of annual appropriations, he thought, would produce the same inefficiency, confusion,
and delay prevalent in river and harbor work. By removing irrigation from public control, he hoped that expert knowledge and
planning, rather than the customary logrolling would prevail.[30]
The Newlands Act, passed in 1902, was one of the first major
triumphs of Progressivism.

Under American conditions, the reverence for expertise also led to a revival of interest in direct democracy as a way to circumvent the politician and his devious ways. The disinterested expert would do his work by bringing the people to share a common understanding of their own best interests, and under direct democracy the people would then act wisely on the basis of the value-free advice they had received. This was the paradox of the movement: essentially élitist, and carrying with it a profound suspicion of popular government as practiced in the United States, it yet called for a degree of democracy greater than that ever practiced in the United States or any other country. The difference between this and earlier calls for direct democracy, however, was that the Progressive version was not thought of as a substitute for the government in Washington; it was, rather, a way of strengthening that government.

Later students have demonstrated that the devices of direct democracy—initiative, referendum, and recall—actually permit a well-informed and organized minority to impose its will upon an unorganized majority, and there were some, like the historian Charles A. Beard, who saw this clearly at the time and rejoiced at the possibility. Others called for a redefinition of democracy to fit the new circumstances. Henry Laurence Gantt, head of an organization of engineers and sympathetic reformers with the announced intention of acquiring political power, made no effort to conceal his hostility to conventional political democracy. He scorned the "debating society theory of government" on the grounds that "true democracy" was to be ruled by "facts" and not "opinion." True democracy, he said, "is attained only when men are endowed with authority in proportion to their ability to use it efficiently and their willingness to promote the public good. Such men are natural leaders whom all will follow."[31] But probably the majority of those calling for both more expertise and more democracy had no such thoughts. Convinced for the most part that there was a "one best way" to accomplish anything, as Frank Gilbreth, exponent of scientific management, put it, they were sincere in their belief that the people would act wisely once they knew the facts. To "act wisely" was, by definition, to accept the advice of the expert for only he had all the facts necessary for a rational decision.

Because of their commitment to both democracy and expertise, reform-minded scientists were of necessity involved in an intensive cultivation of public opinion. In the interests of river basin control, McGee was active in promoting the organization of pressure groups, such as the Rivers and Harbors Congress and the Lakes to Gulf Deep Waterway Association. He also wrote popular articles for magazines, for unlike the purely professional scientists of the period, reformers believed that popularization was both a valuable and a necessary part of their job.

Concern for the elimination of natural-resources waste combined with the reverence for expertise to produce a wider gospel of efficiency in every phase of human life. As Joseph N. Teal, a prominent Oregon conservationist, declared, "I hope that the time will come when efficiency in all directions will be given consideration. When that time comes we will begin to get our money's worth for what we spend." The ultimate goal, argued McGee, was absolute efficiency. "The perfect machine," he wrote, "is the fruit of the ages . . ." The editor of an engineering journal that heartily supported resource conservation proclaimed: "The Millenium will have been reached when humanity shall have learned how to eliminate all useless waste . . . When humanity shall have learned to apply the commonsense and scientific rules of efficiency to the care of body and mind and the labors of body and mind, then indeed will we be nearing the condition of perfect." And Theodore Roosevelt declared at the Conference of Governors in Washington in May 1908: "Finally, let us remember that the conservation of natural resources . . . is yet but part of another and greater problem . . . the problem of national efficiency, the patriotic duty of insuring the safety and continuance of the Nation."[32]

Like most early Progressives, who were inculcated with the laissez-faire doctrines of the late nineteenth century, and with a belief in the efficacy of "the facts" typical of the Progressive Era, scientists in the government bureaus had at first opposed legislation, insisting that "research" and "publication" would suffice to bring the benefits of science to the people. If scientists of sufficient stature within their professions were appointed, they would be able to exert all the influence necessary without being "burdened" with executive power, said the editor of *Science* magazine, writing in 1885 of a proposed national board of health. "The fact that the members of this association would be also members of powerful state organizations, would secure the co-operation of the various states, and would legitimately control, in a high degree, congressional action," he explained. And as a board of consultation, he continued, it would "speak with an influence that no department at Washington could afford to neglect."[33]

Both S. W. Johnson, pioneer analytical chemist, and Harvey Wiley, head of the Department of Agriculture's Bureau of Chemistry, had at first thought within this framework. They wished simply to expose wrongdoers to the public eye, publishing analyses of fertilizers, foods, or drugs and trusting to an enlightened public to choose the good and eschew the bad. As Johnson said, it was only necessary to "publish the character of their goods . . . and public opinion inflicts due punishment." In the beginning, he even opposed licensing and inspection of manufacturers' samples as undue interferences with individual liberty.

But chemical analysis, especially when buried in a government document of limited circulation, did not make obvious to the Connecticut farmer the superiority of one brand of fertilizer over another; neither could the urban consumer know with certainty the level of benzoate of soda considered safe for his catsup. Legislators, more finely attuned to public demands than scientists, began to give to important segments of public opinion the legislation they wanted. Johnson, during the 1880's, as head of the Connecticut Experiment Station found himself inevitably drawn into politics. He was asked for aid in devising an equitable tax on fertilizer manufacturers, asked for an opinion on how to relieve small dealers, and, finally in 1894 drafted the Connecticut Pure Food Law, which was passed the following year. Soon he was recommending that all states adopt a uniform law using that of Connecticut as a model; later, he naturally supported federal legislation.[34]

Wiley experienced the same difficulties in remaining clear of politics. The ethics of pure food, he had explained in 1899, required honest labeling, but forbade any restrictive or prohibitory legislation. "It is not for me to tell my neighbor what he shall eat, what he shall drink, what his religion shall be, or what his politics," he protested. "These are matters which I think every man should be left to settle for himself."[35]

But Wiley, as chief chemist of the Department of Agriculture, found his agency more and more tied to positive government legislation. Neither the frequent bulletins—the most important running to 1,417 pages of facts concerning chemical composition of a wide range of products—nor Wiley's popular articles, nor his testimony before government committees could adequately inform the public. By the turn of the century, petitions for pure food legislation had been received from the National Dairy Association, the National Association of Retail Druggists, and the National Grocers' Association, none of whom found the informative bulletins adequate to protect them. The states, too, were troubled, because, in spite of their own pure food laws, the transition to a national economy made it impossible for them to control food coming in from other states. From a research and information agency, the Bureau of Chemistry gradually evolved into a regulatory agency. The Pure Food and Drug Law of 1906, which prohibited interstate commerce in adulterated and misbranded foods and drugs, provided for criminal penalties and authorized seizure of offending products. The determination of whether a food was adulterated or misbranded was put expressly in the hands of the Bureau of Chemistry. The growth of the Bureau of Chemistry in the years after 1900 provides a measure of the importance of the regulatory function for the growth of government science. At the outset of Secretary Wilson's regime in 1897, the bureau had a staff of 20; by 1902 it had increased to 50. In 1906, it had 110 men and an appropriation of $130,000. Four years after passage of the federal law, the figures had jumped to 425 men operating with an annual appropriation of $697,920. The employees included Wiley's celebrated "poison squad"—a group of specially selected young men who tested the physiological effects of questionable food additives on their own bodies.

Meanwhile, under the pressure of government regulation, food manufacturers had begun to hire their own chemists to study the composition of various fruits and foods and to provide expert testimony counter to that of the government chemists. If the spectacle of rival "experts" disagreeing about the "facts" embarrassed the proponents of expertise and rationality, it did not visibly dampen their ardor for still more expertise. The business scientists could always be written off as tools of the "interests" and not the disinterested expert, whose opinions always, by definition, served the public.

As this chapter hopes to make clear, nothing marked the Progressive movement more than its interest in "efficiency" and its identification of science with efficiency. The scientific method applied to social problems, to business, to politics, to education— in short, to every aspect of life—would result in increased efficiency and, presumably, a more rational social order. Efficiency Societies sprang up in major cities all over the country, journals were published, and earnest plans were made. In April 1914, in the midst of the popular furor, an efficiency exposition was held at the Grand Central Palace in New York, with Frederick Winslow Taylor as the main speaker. Before it closed, 69,000 people had attended and its success prompted a similar exposition in the Midwest. Lecture courses in efficiency became familiar offerings in YMCA's all over the country. Few thought it rash when Louis Brandeis, in a famous railroad rate case, promised to show the railroads how to save $1,000,000 a day by efficient management and make unnecessary the raising of rates.[36]

In politics there came the call for "scientific city government," which usually turned out to mean government by experts who utilized social science techniques in making the everyday decisions of government. The City Manager form, conceived as a way to take municipal administration "out of politics" and substitute the value-free judgment of a trained expert, came into great vogue and reformers dreamed that the city boss and the urban machine were soon to be relics of a barely remembered history. After 1916, hundreds of cities followed the lead of New York in passing zoning legislation, based on a systematic study of the height, size, and use of buildings throughout the city. Out of this was to come the "City efficient," which had replaced the city beautiful as the ideal in the minds of reformers.[37]

In business the mood called for "scientific management," which in the conception of Frederick Winslow Taylor meant breaking a job down into its component parts, discovering the "one best way" to do each step, and providing incentives for workers to do it that way. Taylor, proposing to make "each workman's interest the same as that of his employers," conceived of the factory as one big machine, with all tasks organized and distributed properly, and with men of special training placed to see that the gears meshed. Taylor believed that he had fashioned his methods after the exact sciences—experiment, measurement, generalization, checking on conclusions—in the hope of discovering laws of management which, like laws of nature, would be impartial and above class prejudice. The result would be a neat, orderly world of maximum productivity. It was only necessary that the workers "do what they are told promptly and without asking questions or making suggestions . . . It is absolutely necessary for every man in an organization to become one of a train of gearwheels."[38]

The same drive permeated educational thought of the time, where there were continuing efforts to construct a curriculum on "scientific" grounds. True to their emphasis upon expertise, leaders in the Progressive Era tended to believe that education, by bringing information to bear on both personal situation and social issues, could change the course of events. They held the comfortable belief that the world was essentially rational, and that problems all had solutions which "the cold light of facts" would surely reveal. Institutions of higher learning were often characterized by the Progressives as "the University in the service of society." At least one state university—Wisconsin—consciously tried to become a laboratory for Progressive ideas.

But it was elementary education that interested Progressives the most. If young minds could be trained in the ways of right thinking before they had a chance to be corrupted by their environment, what problems might not be solved by the next generation? It was a heady thought, which reformers seized with characteristic enthusiasm. But what information should individuals have when they finished school and were ready to take up the tasks of civilization? Science and "scientific method" had already provided the answer. "Whatever exists at all exists in some amount," wrote the psychologist E. L. Thorndike in 1918:

> To know it thoroughly involves knowing its quantity as well as its quality. Education is concerned with changes in human beings; a change is a difference between two conditions; each of these conditions is known to us only by the products produced by it—things made, words spoken, acts performed, and the like. To measure any of these products means to define its amount in some way so that competent persons will know how large it is, better than they would without measurement . . . and that this knowledge may be recorded and used. This is the general Credo of those who, in the last decade, have been busy trying to extend and improve measurements of educational products.[39]

It had been an exciting decade indeed for those who shared Thorndike's dream of a genuine science of pedagogy, and the key to the ferment had been the rapid development of intelligence and aptitude tests for use in the schools. Tests themselves, of course, had long been known to European and American psychologists: Francis Galton had experimented with them, and the American psychologist James McKeen Cattell had apparently used them in his laboratory at the University of Pennsylvania as early as 1885. But the real excitement began after 1905, when the French psychologists Alfred Binet and Théodore Simon conceived the idea of an intelligence scale: a series of problems of graded difficulty, each one corresponding to the norms of a different mental level. Once Binet's work had been translated, American educators quickly recognized that the scale idea could be applied not only to intelligence but to achievement as well, and they accordingly began developing instruments for appraising virtually every aspect of educational practice.

There were numerous adaptations and refinements of the Binet scale, the most important of which was the Stanford Revision, described by Lewis Terman, who popularized the idea of the I(ntelligence) Q(uotient) in 1916. Meanwhile, Thorndike and his students developed scales for measuring achievement in arithmetic, handwriting, spelling, drawing, reading, and language ability. The work was quickly taken up at other universities and the development of tests spread like a craze, so much so that by 1918, the National Society for the Study of Education could describe over a hundred standardized tests designed to measure achievement in the principal elementary and secondary school subjects, and these represented a culling of the best.

All this activity would undoubtedly have remained a professional phenomenon had it not been for the historical intervention of World War I. Following the American declaration in 1917, the American Psychological Association offered its services to the Army to construct group intelligence tests for Army recruits. Under the direction of Robert Yerkes, president of the Association, a number of instruments were developed, the most famous of which was the Army Scale Alpha test. These tests were administered to hundreds of thousands of recruits, and while they were initially designed to weed out the grossly incompetent and to cream off the exceptional for officers' training, they were ultimately used for a much greater variety of classificatory purposes, especially after the psychologists undertook to assign mental ages to the several letter ratings on the Scale Alpha. The results of this procedure, which assumed that scales derived from school children who took the Stanford-Binet could be used to judge the mental ages of adults long out of school, were disastrous, indicating that the average mental age of Americans was fourteen. Among other things, this led some to argue that most Americans were uneducable beyond a certain level, and it led others to question the viability of democracy.[40] But most did not reach such pessimistic conclusions. They used the results, rather, to fortify conclusions that they had already reached regarding the need for "education for life."

The height of the drive began in 1911, when the National

Education Association appointed a Committee on the Economy of Time in Education. The fundamental assumption of the Committee, common in educational circles at the time, was that it was uneconomical to teach a child something that he did not need to know; economy would result from the selection of only that knowledge which was directly serviceable. For most, this meant discarding the traditional academic orientation and making the life of society the curriculum of the schools. Adopting a familiar Progressive tactic, they therefore embarked on a vast Baconian attempt to describe what people actually do in the course of "life in society" and to derive from this analysis the content of the school curriculum. Equipped with standardized checklists and questionnaires, an army of investigators marched out in the scientific spirit to survey the community's ways of making a living and its habits of daily life. Naturally, the best of the recently developed achievement scales were applied whenever possible, and the results reported in an impressive array of charts, graphs, and tables that lent an unmistakable aura of precision to the undertaking. Responses were recorded, correlations tabulated, and results translated into lists of skills, habits, attitudes, and appreciations which were to be taught in particular school districts. Courses of study thus prepared appeared throughout the nation.

A 1918 report developed a mathematics course by tabulating data from four typical sources: a standard cookbook, payrolls from a number of artificial flower and feather factories, markeddown sales advertisements, and a general hardware catalogue. The author then proceeded to draw inferences for the design of an arithmetic syllabus, assuming that his data were the "concrete stuff out of which arise the arithmetical problems of housewives, wage earners, consumers, and retail hardware dealers." It was such reasoning, he argued, that had to form the basis for the ultimate determination of the elementary-school arithmetic program. What the report did for mathematics, others did for English, geography, history, civics, literature, and physical education. Brought together in the committee's reports of 1917 and 1918, these analyses added up to a formidable criticism of conventional syllabi and textbooks, and the inevitable conclusion that experts should devise a wholly new curriculum in terms of these new findings of science.[41]

Even though the immediate thrust of these educational efforts was overwhelmingly reformist, they still illustrate one of the central difficulties of all the Progressive efforts to make science the basis of social reform. For in the particular method it had chosen to determine "minimal essentials" of education, however precise and scientific it may have appeared, the committee had ended by defining the goals of education in terms of life as it was, and hence by proposing a curriculum that would accommodate youngsters to the existing system, with no emphasis on improving it. In the Progressive understanding, at least, most efforts to use "science" resulted in a similarly conservative reform, for the empirical data that science had to use concerned life as it was.

The achievement and intelligence tests, formulated in terms of the dominant Anglo-Saxon culture, seriously misrepresented the potential of all other cultural groups, and were used to nourish racist ideology for decades. Scientific city government, designed to keep administration out of "politics," only made the administration more sensitive to the "politics" of a particular group and less sensitive than before to other groups. Frederick Winslow Taylor, who thought of himself as both reformer and scientist, actually not only reduced the working man to "one of a train of gear wheels" but militated against the fullest exploitation of science. Taylor was, in fact, suspicious of new and radical innovations in machinery and instead laid emphasis upon perfecting the tools at hand to a point of high efficiency.[42] New tools introduced uncertainty—and the man dedicated to rationality does not like uncertainty.

But even though the "science" of the Progressives proved time and time again to be a poor guide for suggesting specific social policies, by the beginning of the twentieth century a government that could not *use* science and deal with the problems posed by an increasingly complex technology had become unthinkable. In providing for this need, the Progressives were more successful than they were in guiding the Republic speedily toward Utopia. The scientific bureaus which they founded, and the institutions outside of government which they either created or strengthened, may have been their most important lasting achievements. As one historian has pointed out, the government's scientific establishment was virtually complete by 1916, and the foundations of every important scientific institution that existed at the beginning of the Second World War were already laid in the last days of peace before the United States entered the First.[43]

Scientists have profited from this relationship they inherited from the Progressives, but in order to enjoy its benefits they have had to compromise as well. For in the government agencies, scientists soon found that they had to compete with all other interest groups to whom the agency is relevant; "value free" decisions cannot be made by a democratic government, and probably by no other form of government. The scientists' hopes for autonomy, which they regard as being necessary for the ideal development of their science, are curbed not only by the non-scientist officials in the executive hierarchy and by a host of congressional committees, but also by interest-group associations in the science bureau's own constituency. Thus, the Bureau of Mines must listen as attentively to the American Mining Congress and the United Mine Workers as it does to the American Institute of Mining Engineers; the Bureau of Standards is as responsible to a great many industry associations as it is to the American Physical Society or the American Chemical Society; the Weather Bureau has to serve the needs of the Air Transport Association and the Farm Bureau Federation as well as those of the American Meteorological Society.

Science is able to remain "pure" only so long as it is without power. Its purity diminishes in direct proportion to the growth in its relevancy to the life of the society and the vital interests of groups in the society. The gaining of power by any science agency has therefore been accompanied by increasing demands that it serve the needs of those over whom it has power. This means that with the acquisition of power, there is an inevitable shift of emphasis in research from basic to applied. Thus, when geologists and hydrologists of the Geological Survey persuaded the government of the need for controlling irrigation in the West, the result was a new Reclamation Service which grew up within the framework of the Geological Survey to administer the program. Research of any kind played but a minimal role in the Reclamation Service, and the parent organization itself began to devote more and more of its energies to the "classification of public lands," which although mentioned in the organic act of 1879 that established the Survey, had only been practiced during one short interval in the past. In 1906, however, the Survey began the major applied task of segregating potential coal beds. Later it added water power sites and lands bearing oil, phosphate, potash, and various other minerals. On the basis of this applied research— really prospecting—a reservation policy similar to that for the forests developed.[44] The chemical branch of the Survey likewise yielded to pressures toward utility. Organized in 1879 as a pure research branch, largely devoted to providing chemical analyses for identification of mineral substances, the branch continued to furnish minutely detailed analyses of rocks, waters, etc., until 1904, when the "technologic branch" began to be developed. Studies were made of coals to determine their availability for briquetting or for the manufacture of producer gas; one laboratory was devoted to studies of mine explosions, another passed upon the quality of structural materials used in public works; another studied petroleum fields to determine the physical properties of the oils and their distillation properties. The same shift in emphasis occurred within such agencies as the Forest Service, which evolved from a purely research agency under Bernhard Fernow to an administrative agency under the more politically powerful Gifford Pinchot; and within the Biological Survey, which began as a basic science program and, with the gaining of regulatory power after 1913, shifted the bulk of its activities markedly in the direction of applications.

The partnership that developed between science and government during the Progressive Era has therefore been an uneasy one, and its legacy is still with us. One might say that the subsequent history of government science agencies has been an effort to work out the tension between basic and applied that first emerged during that period. By and large, the applied has been victorious, and basic science has found its home in the universities and, to a smaller extent, in the research laboratories of great industries. A new role for government has been found, especially since World War II, as it became more and more a patron of basic research conducted in the laboratories of universities and private research centers. This has created problems of its own, as the following chapter will demonstrate.

The other legacy of the Progressive Era—the image of science as wonder-worker—is also still with us, and it too has continued to create problems for the scientist in his relationship with the public. This is another theme of the chapter to follow. Viewed from the perspective of science in the early 70's, the Progressive Era does, indeed, appear to be a watershed. This is so because, however naïve the Progressives were in their enthusiasm, they were right in their basic assumption. America in the twentieth century has become, as W J McGee observed two years before this century began, "a nation of science."

XIV

〜〜〜〜〜〜〜〜〜〜〜〜〜〜〜〜〜〜〜〜〜〜〜〜〜〜〜〜〜〜〜〜

Epilogue: Science in a Corporate Age

❡ In 1881, a young American physicist then studying in Germany received a grant of $200 from Alexander Graham Bell to conduct an experiment on one of the most fascinating questions of nineteenth-century physics: the reality of the ether. The ether was a mysterious, jellylike, invisible entity which was thought to fill all of space; it was even present in solid matter. The vibrations set up in this ether made it possible to explain how the wavelike radiations of light could be carried through millions of miles without weakening or diluting their initial energy. Although the behavior of light seemed to demand some such medium, Albert A. Michelson doubted its existence, and he designed a relatively simple experiment which he thought might resolve the question unconditionally.

With his $200 provided by Bell, Michelson had a machine of his own design, called the interferometer, constructed by a Berlin manufacturer, and he took it to the observatory at Potsdam for the crucial experiment. His conclusion, published in the August 1881 issue of the *American Journal of Science*, was that "the hypothesis of a stationary ether is erroneous." Although Michelson later repeated the experiment, with more sophisticated apparatus, in collaboration with Edward Williams Morley it was the first experiment which, as Albert Einstein remarked, "showed that a profound change of the basic concepts of physics was inevitable" and led eventually to Michelson's becoming the first American recipient of a Nobel prize.[1]

If one contrasts this picture of the young physicist, working alone on a two-hundred-dollar grant provided by a private patron to produce results that shook the world of physics, with that of the modern American research laboratory, the enormity of the changes that have come about in American science in the twentieth century will be evident. To cite one example, the discovery of the "omega minus," which added only minutely to knowledge about atomic particles, was announced in a joint publication by 33 researchers who worked with one hundred thousand photographs taken with the aid of a thirty-million-dollar atom smasher. And, since the ratio of research physicists to total laboratory staff is now about 1 to 7, a piece of research with 33 authors involves on the average something like 231 people in the laboratory.[2]

As this example illustrates, American science in the twentieth century has become big, complex, and expensive. It has also become relevant to the ordinary lives of men to an unprecedented extent. One could say, without too much exaggeration, that the course of American history since the last quarter of the nineteenth century has been a story of increasing acceptance and incorporation of scientific knowledge—and scientists—into the practical institutions of society, including both business and government. Science has become involved in our domestic politics, in international relations, in virtually every institution which vitally affects men in the mid-twentieth century. The result of these changes is that science has become largely a corporate enterprise, dependent upon government and industry for its continued support and subject to the same pressures toward bureaucratization, centralized control, and politicization as other institutions in modern American life. The two sources of change—increasing support from outside and increasing growth inside—are operating to open wide the older closed-system scientific community and to alter its entire structure. All of these characteristics were evident by 1940; developments since World War II, impressive though they are, have been changes in magnitude, not in kind.

The first impetus toward a reorganization of science came from the giant industries that emerged in the late nineteenth century. As the wildcat days of the post-Civil War industrial boom gave way to a period of consolidation and steady growth, it became possible for the rising captains of industry to think in terms of the deliberate manipulation of men and materials in order to maximize profits. The industrial research laboratory was one important innovation which came from such motivations. It can be understood as simply a part of the general rationalization of economic life that accompanied the rise of Carnegie, Rockefeller, and other giants of the period.

The industrial research laboratory had first appeared in the German chemical industry in the final third of the nineteenth century, particularly in firms producing coal-tar dyes. Stiff competition, rapidly changing technology, a good supply of trained chemists, and the traditions of German graduate scientific instruction apparently encouraged this development. In this country, Edison's laboratory, established at Menlo Park in 1876, provided a striking example of an organized research establishment; its head casually informed a friend one day that he proposed to turn out "a minor invention every ten days and a big thing every six months or so." But Edison was almost *sui generis* at that time; few others were able to finance such invention factories out of profits from previous inventions. It was left to large corporations developing everywhere on the American scene at the end of the century to introduce organized industrial research.

During the last quarter of the nineteenth century, American industry, having little understanding of the new powers of science, only rarely turned for assistance to university professors and commercial research chemists. Science was of great importance in the development of certain American industries at that time, but it was the lone scientist and the independent inventor who made the discoveries which became the basis for new industries. These scattered, unorganized researchers, often empirical "boil and stir" men, were financed by industrial capitalists only after their discoveries had been made and tested. No one sought to give direction to the work, and it was accordingly characterized by inefficiency, misdirection, duplication—and freedom, although perhaps not much remuneration, for the individual investigator.

Andrew Carnegie had early seen the value of scientists, and had employed several, but he, like Edison, was practically alone among late-nineteenth-century industrialists. The new pattern did not really begin to develop until early in the twentieth century, when American industrial firms began, on a considerable scale, to organize their own research departments and to employ university-trained scientists; often, in fact, enticing them from academic posts with the offer of larger salaries and better equip-

ment. The first industries to do so were those which had themselves been born in the laboratory, like the electrical industry
and the chemical industry. In 1900, General Electric founded its
Research Laboratory under the direction of the German-trained
chemist Willis R. Whitney, who joined GE from the Massachusetts
Institute of Technology. DuPont in 1902 established its Eastern
Laboratory under Charles L. Reese, a Heidelberg graduate. American Telephone and Telegraph began research work in this decade
also, and in 1912 Eastman Kodak brought in C. E. Kenneth Mees
from an English subsidiary to begin its organized research.

It is significant that most of the early laboratories appeared
in industries such as electricity or chemicals. Not only were
they relatively new industries, with no traditional inertia from
the past, but in these industries there were large firms with
sufficient financial resources and stability to support the laboratories; and a rapidly changing, competitive technology made
successful research imperative for the sponsoring firm. Older
industries, such as iron and steel, and those where no giants
existed were relatively slow to bring science into their activities,
and a great differential still exists.[3]

Although a solid foundation had been laid before the
First World War, the new pattern for the use of science spread
slowly. But after that conflict, when the actual and potential
uses of science for industry were demonstrated beyond any
doubt, the movement of scientists into industry became an irresistible tide. With a tangible record of achievement in military
technology and in some civilian areas as well—in optics and
industrial chemicals, for example—the physical sciences stood
higher in American esteem than ever before. As Robert A.
Millikin, a major organizer of the government science effort
during the war, noted, it had been the first time that the whole
scientific brains of any nation was systematically mobilized for
the express purpose of finding immediate new ways of applying the accumulated scientific knowledge of the world to the
ends of war.[4]

Although little of theoretical value came from the effort,
the extraordinary practical success of the mobilization was a
lesson not lost either on industry or on the scientific community.
Research was infused so thoroughly into the economy that industrialists became accustomed to calling routinely upon science for aid, especially in production. As for the scientists, they

became used to working together for the quick solution of an immediate problem, often cooperating across the lines of the accustomed disciplines. The idea that some problems could only be solved by massed and cooordinated scientific resources became firmly implanted upon the American mind, and its imprint has been felt ever since. From that time on, with only a minor setback during the Depression, the number of organized research departments in industrial organizations increased very rapidly, from about 300 in 1920 to 3,480 in 1940. Over the same period, the personnel employed in scientific research in industry increased from approximately 9,300 to 70,000. By 1967, reflecting the added boost provided by the Second World War, the Cold War, and the space program, there were nearly 400,000 scientists and engineers working in industrial research laboratories on a budget of over $16,000,000,000.[5]

More than half of all professional scientists in the United States are now employed by private industry, including 40 per cent of all physics Ph.D.s and 60 per cent of all chemistry Ph.D.s. Even mathematicians in industry have been multiplying by a factor of 12 every twenty-five years for the past three-quarters of a century. Contrary to the classical pattern—and the pattern upon which the scientific ethos is still based—the industrial research laboratory, not the university, is now the principal habitat of the scientifically trained American. This means that the scientist today is typically an "organization man," who in real life bears little resemblance to the frock-coated, absent-minded professor still enshrined in the mythology. And, as Walter Hirsh pointed out in 1965, it is "simply not realistic" to speak of a scientific ethos which does not take account of the specific organizational setting in which it operates. One cannot expect the chemist who works for DuPont or the physicist who works at IBM to have the same motivations as a Darwin or a Newton.[6] The chances are that the industrial scientist will be subtly, yet measurably, affected by the business orientation of his employer; and the business virtues are sharply at odds with the virtues of the scientific ethos.

The most serious discrepancy between the ideal and the current reality is that by far the majority of those scientists in industrial research laboratories—perhaps three-fourths of them—are not involved in the research for which they were trained in college and which they were led to believe was the only worthwhile pursuit for a scientist. Instead of becoming immersed in basic research—contributing to the fund of knowledge as their professors urged them to do—they become technical troubleshooters, do various kinds of applied work, or even go into management or sales. "Undirected basic research," the dream of the pure scientist, is possible at only a few industrial laboratories and, even there, for only a few scientists. Only a large company with a fairly stable business and good profit margins, or a regulated monopoly like AT&T, which can charge cost of research to its customers, can afford to invest large sums of money in undirected basic research. And in a day of Big Science, the investment of small sums is pointless, for good scientists will seldom work where they will have no chance to talk with other people doing basic research in the same or in closely related areas, and where they cannot have access to the increasingly expensive equipment they need. Even a big company cannot reasonably expect to profit directly by investing in basic research unless, like RCA, DuPont, or GE, it markets a wide variety of products—or a few products, such as computers, embodying a good deal of advanced technology—that stand to be improved (or made obsolete) by advances in scientific knowledge.[7]

There are, to be sure, examples of a large payoff from the support of basic science. The synthesis and production of the first truly synthetic fiber, nylon, for example, is one of the most frequently cited cases of successful industrial research in the twentieth century. Although it is often used as a demonstration of the "triumph" of basic research, the story of nylon is really a triumph of large-scale finance, and it demonstrates as well as anything else the gigantic investment required in order to profit directly from basic science. Nylon had its origin in DuPont's decision, in 1927, to begin basic as well as applied research. One of its new chemists, Wallace H. Carothers, selected as his problem the synthesis of polymers or long-chain molecules, a subject of interest to the DuPont Company, which was already involved with semi-synthetic fibers, viscose, and acetate. Carothers, over a period of years, produced a large number of both polyesters and polyamides, but only in 1935 did he come up with a series of polyamides, one of which, Nylon "66" showed real possibilities. Even then it took another four years, the work of some 230 technicians, and some $20,000,000 to learn how to produce the raw materials on a commercial scale, solve other production problems, devise new methods of converting Nylon into commercially acceptable fabrics, and get it through the pilot-plant stage. Since then, Nylon has amply repaid DuPont's patience, but a smaller company could not have afforded to wait.

Since there are few companies that can afford the patience of DuPont, there are few scientists in industry who are privileged to follow their own interests, and they remain, as a recent student has concluded, "one of the most disgruntled groups on industry's payroll."[8] Their discontent is intensified by the insecurity felt by even those scientists who are privileged to follow their own interests, for they never know how long they will keep this privilege. Because research has such a speculative quality in the short run, there is a tendency in industrial research for boards of directors to pour in money in good times and to cut it off in bad times. During the "dark days" of 1931, for example, the order came to the research departments of many industries to "cut out work on all projects not producing profit."[9] Today, when the continuation of a project frequently depends upon the company's success in getting a government contract, this insecurity is intensified.

The same pattern of development that occurred in American industry in the twentieth century is evident in government. Even before the turn of the century, the federal and state governments employed scientists in a variety of bureaus and agencies, which included laboratories to carry on independent work. There were so many at the federal level, in fact, that the trend toward consolidation and bureaucratization had already begun. As early as 1879, for example, four rival geological organizations had been abolished and a U.S. Geological Survey had been created to deal with "questions relating to the geological structure and natural resources of the public domain." By 1900, the Department of Agriculture had spawned so many separate divisions, often operating on parallel lines, that some reorganization was carried out. For example, the Divisions of Botany, Pomology, Vegetable Physiology and Pathology, Agrostology, Gardens and Grounds, and Seeds—together having appropriations of $268,400—carried out parallel functions. The Secretary of Agriculture used his power to group these into a Bureau of Plant Industry, analogous to the already-existing Bureau of Animal Industry; and Congress gave formal recognition to the regrouping in the next appropriation act. These two bureaus, along with the Bureau of Entomology, then became the core of the department's research establishment.[10]

Even broader schemes of consolidation were being discussed by the turn of the century, none of which came to fruition. There is, in fact, a long history of abortive efforts both by government and private organizations to further centralize government science. The Allison Commission, a joint commission of Congress established in 1884 "to consider the present organization of the Signal Service, Geological Survey, Coast and Geodetic Survey, and the Hydrographic Office of the Navy Department, with a view to secure greater efficiency and economy of administration of the public service," adjourned two years later with mountains of testimony but no practical recommendations. Another commission, appointed in 1903 under the leadership of Gifford Pinchot, studied the organization of the government science establishment and made recommendations for reorganization, but its report remained unpublished. In other quarters, a department of science at the Cabinet level was being proposed and the issue was debated through several years in *Science* magazine. This solution was recommended by a commission of the National Academy of Science as early as 1885, and by Donald F. Hornig, science adviser to President Lyndon Johnson, as late as 1968.[11]

Despite the continuing evidences of consolidation, none of the over-all schemes for the centralization of government science was initiated, and it has continued to grow chaotically and without apparent direction down to the present. Individual Bureaus have become larger, and each bureau is increasingly centralized and formally organized, but thus far a complete reorganization has been resisted. A recent reorganization of the scientific activities of the Department of Commerce, which involved the creation of the Environmental Science Services Administration, was met with suspicion, if not outright hostility, by influential members of the scientific community; and the effort to include related agencies from other departments was defeated. For the most part, scientists prefer a multiplicity of scientific agencies, on the very practical grounds that worthwhile work which might not be supported by one agency may find a patron in another.[12]

Although the roots go much deeper, the present situation had its origin in the Progressive Era. Progressives, with their concern for the problems emerging from the complex society of the twentieth century, had created or reinvigorated a great many scientific bureaus, which were shaped in large measure by the relations of government and large-scale industry. The National Bureau of Standards, the Bureau of the Census, the Bureau of Mines, and the National Advisory Committee for Aeronautics are only among the better known of many such problem-oriented agencies that began the deliberate use of science shortly after 1900. Significantly, all were closely tied to the needs of industry, and both in terms of organization and aims, they reflected their industrial ties.

Over-all government aid for science, decreasing in the 1920's, increased again in the 1930's during the depths of Depression, and a phenomenal increase of scientific personnel began. Chemists, geologists, and mineralogists filled the ranks of the expanding Bureau of Mines and the various bureaus of the Department of Agriculture that dealt with mineral resources and conservation; mathematicians, economists, and statisticians entered the government through the Bureau of the Budget and the Census Bureau; physiologists and biologists were drawn into the expanded Public Health Service, through the National Institutes of Health. Scientists also benefitted by the establishment of the Naval Research Laboratory in 1923, the National Research Fund in 1926, the Science Advisory Board for the TVA in 1933, the Agricultural Research Center in 1934, and the massive study financed by Congress in 1935 under the National Resources Committee. Finally, the government in a report on the results of a long-term study recognized the need for basic research with the publication in 1938 of *Research— A National Resource*.[13]

The massive investment in science under the stimulus of World War II and the Cold War has carried the trend toward increasing investment in science to a point where it has become obvious that it cannot continue at the same exponential pace. The trend in government expenditures for research and development, beginning from a base of only about $3,000,000 in 1900, climbed steadily upward until 1935 when it had reached $50,-000,000. In that year the curve turned sharply upward and began to level off only in the mid-1960's. By 1965, the U.S. budget for science had risen to 2.5 per cent of the Gross National Product; in 1969, while still somewhat larger in absolute terms, it had fallen to 2 per cent. The full repercussions of this leveling off have not yet been felt, but since scientists have been both psychologically and institutionally conditioned for an ever expanding market, the effects are bound to be severe.

Thus far, the squeeze is being felt more by young researchers who have not yet established their reputations. Going through graduate school during the times of plenty, taught to rely upon the expensive equipment that only increasing govermental investment in science can maintain, they discover in many cases that there is simply no place for them on the research front. In 1968, for the first time in decades, graduates in the natural sciences encountered difficulty finding jobs. According to a spokesman for the American Institute of Physics, 30 per cent of 1968 physics Ph.D.s were without jobs upon completion of their doctoral work and about one out of ten of these unemployed physicists were still unemployed a year later.[14] Universities, too, having assumed the burden of maintaining expensive equipment and having become accustomed to a continuing expansion of funds available for research, are beginning to feel a squeeze and, in some cases, have found it impossible to keep past commitments. Many are beginning to severely restrict graduate enrollment, and, with government withdrawal of much of its aid to graduate schools in 1970, enrollments will go down still further. As numerous scientists have observed, a crisis situation seems to be fast approaching.

One of the most marked consequences of the burgeoning of scientific activity in the twentieth century and the consequent involvement of science with other institutions of society has been an increasing tendency toward centralization and bureaucratic control. This is an increasingly important pattern in American scientific research, and it reflects important changes that have occurred in science itself and in its relations to the rest of society. Beginning with the 1840's, American science had been transformed by specialization and professionalization. At the beginning of the twentieth century, yet another generation of scientists was riding the wave of a new organizational revolution that was to be even more serious in its effects than the first. The revolution was part of a larger movement in modern society—the change toward

an increasing "bureaucratization of the world"—but it is of special
significance for science. These developments are attributable to a
number of factors, some emanating from within science and some
from without. For one thing, a certain amount of centralization
seems to be inherent in bigness. Bigness implies specialization and
division of labor, and everywhere in social life the division of
labor creates the need for coordination and control. During World
War II, for example, one study of the Office of Scientific Research
and Development reported that "the administration of OSRD
resolved itself into the triumvirate of Bush, Conant and Compton."
Indeed, Vannevar Bush himself pointed out that although there
were approximately 30,000 scientists and engineers working on
new weapons and medicine during the war, there were "roughly
thirty-five men in the senior positions" of control.[15] However
absolute the anti-authoritarian values of scientists may be, this is
only what is required for so large an enterprise if it is to be
successful.

Such an inherent need for organization and control was
seen clearly by even such an alleged "rugged individualist" as
Andrew Carnegie, who founded the first great centralized research
institution in the United States. The first policy statement of the
Carnegie Institute's board of trustees, dictated by Secretary Charles
D. Walcott in 1903 and thoroughly approved by Carnegie, made
clear the orientation of the institute: "Hitherto, with few exceptions research has been a matter of individual enterprise, each
man taking up that problem which chance or taste led him to."
But in modern times, the stakes of research were too high to risk
chance success, Walcott explained. Modern science, like modern
business, required centralized planning and direction. "The most
effective way to find and develop the exceptional man is to put
promising men at work . . . under proper . . . supervision. The
men who cannot fulfill their promise will soon drop out, and by
the survival of the fittest the capable, exceptional man will
appear."

According to Walcott, there were two basic approaches to
scientific discovery: individualism, "the old view that one man
can develop and carry forward any line of research"; and col-
lectivism, "the modern idea of cooperation and community of
effort." Walcott thought, apparently correctly, that most of the
younger American scientists, thirty-five to fifty years, were
thoroughgoing collectivists, while their elders remained loyal
to the individualist traditions of an earlier day.

This approach ran counter to one of the scientific community's most cherished traditions: that of the unfettered scientific genius seeking knowledge according to his own dictates. Some scientists protested that a central disbursing agency like the Carnegie Institute might dry up local sources of financial support; a few took the theoretical position that centralized support was inappropriate for a democracy, and that a "machine-like organization" would be fatal to scientific freedom.

The Carnegie trustees were correct: machine-like or not, organization in research and manpower procurement was as necessary in modern society as was organization in business enterprise and social welfare. In accordance with the guidelines set down by Walcott, the foundation decided to give priority to sustaining a few large research projects rather than making individual grants. In fact, like the Rockefeller Institute for Medical Research, the Carnegie became more an operating than a grant-making institution, and it operated along clear "collectivist" lines. In its research departments, teams of specialists tackled major scientific problems. The non-magnetic brigantine *Carnegie* criss-crossed the oceans charting the earth's magnetic field. At Cold Spring Harbor, New York, Charles B. Davenport supervised far-reaching studies of experimental evolution in plants, animals, and man; and he and his co-workers gathered the data that was to make him America's most outspoken eugenist within a few years. Atop Mount Wilson in California, George Ellery Hale directed a team of workers at a great solar observatory.[16]

Developments during the 1920's emphasized the new team approach to science. When the Rockefeller Foundation, for example, asked a group of scientists for advice concerning the best way to aid American science, they were told that it was necessary to build up excellence in the research departments of a few American universities. Thus originated the Rockefeller Foundation National Research Fellowships, of which fifty were given annually for postgraduate work. The fellow would spend two or three years in research at the American university of his choice. Students, the scientists reasoned, would go where they believed they could find the best facilities and the most stimulating contacts, and in this way excellence would be built up in six to twelve existing universities.[17]

Although the scientists concerned clearly saw the need to work in a team situation, it is ironic that they thought their plan to be decentralizing, for they rejected the idea of establishing a centralized research institute for physics and chemistry. In a certain sense, perhaps, their plan was decentralizing, but the outcome was to build not one, but several centralized research institutions. Looking back on the results of the National Research Fellowships from the vantage point of 1950, Millikin, who had chaired the committee of scientists consulted by the Rockefeller Foundation, pronounced that they had been "the most effective agency in the scientific development of American life and civilization that has appeared on the American scene in my lifetime."[18] Millikin's judgment was probably overenthusiastic, but only slightly so. For the same universities that had developed an advanced research capacity with the aid of the Rockefeller Fellowships—Harvard, MIT, Cal Tech, Berkeley, Chicago, and a few others—were in a position to profit by the increased government funds during World War II, and they still absorb the bulk of governmental research funds.

But whether it is one university, one research institution or several, especially in application to specific practical problems it is clear that a formal organization is necessary. In such areas as the chemical and electrical industries, in medicine, and in agriculture, for example, fundamental science has been applicable for quite some time. Such areas of application are increasing in number, for the growth of modern science has meant that in an ever larger number of practical situations, theories of science are useful.

When a social crisis endows the end of research with unusual urgency, the need for formal organization in achieving that end is especially clear. During World War II, for example, large organizations were speedily set up in the fields of electronic and atomic research in order to achieve such ends as radar, proximity fuses, and an atomic bomb. And it is this type of social situation in which formal social organization is most efficient. The original plan for war research had been to leave each scientist at his own university, for that had appeared to be the happiest situation for the scientists themselves. But as the volume of work increased, as the urgency was felt with even greater force, the need for consultation, for mutual help, and for frequent contact among scientists working toward the same end required that they all be brought together in the same large organization. "The benefits to be derived from teamwork of sizeable groups," said the historian of OSRD, "were too great to be neglected." Large formal organizations for research were accordingly set up at the University of Illinois, Chicago, Northwestern, MIT, and Harvard. On the staff of the Radiation Laboratory at MIT alone there were men from sixty-nine different academic institutions.[19]

What was true of wartime research is true of any applied research deemed to be important. "In many fields," remarked Frank B. Jewett, former Director of the Bell Telephone Laboratories and president of the AAAS, "the products will be such as to involve a wide range of physical, chemical, and biological problems so interwoven as to call for scientific attack from many angles, and so we will have large research organizations with specialists and specialized facilities in many fields, all organized to function as a coordinated unit." Experience, he explained, had shown that this was the most "powerful, effective, and economical method of handling complex problems. It is greatly superior to any scheme of farming out portions of the problem to individual laboratories. This results from the fact that at all stages of the work the several elements react on one another and that what can be done in one field determines what can or cannot be done in another."[20]

But if bureaucratic organization is efficient, it is not without its problems. Nowhere are the problems of bureaucracy more evident than in the multitude of agencies which were either created or reinvigorated in the post-Sputnik era. The entire space program, in fact, has had administrative problems unprecedented in magnitude, if not in kind. The National Aeronautics and Space Administration (NASA), created in 1958 to have responsibility for civilian space activities, functions as an allocating, coordinating, evaluating, and planning administrative superstructure, which places manufacturing or hardware largely in the hands of private-enterprise contractors. In its Project Mercury—the effort between 1958 and 1963 to place a man in orbit—NASA mobilized a dozen prime contractors, about 75 major subcontractors, and some 7,200 third-tier subcontractors and venders, under whose employ some 2,000,000 persons, at one time or another, had a direct hand in the project.[21] NASA administrators have had to innovate in solving such problems as the coordination of production schedules of the thousands of industrial contractors and the allocation of enormous public resources within NASA and among the competing contractors. Other problems have included the creation of an institutional framework within which engineers and

scientists would flourish, the maintenance of constructive relations with the Department of Defense, which has charge of military space activities, and inspiring the confidence of Congress
and the public.

These problems have only been made manageable by adopting a different form of organization than that usually used by such
agencies. The scientific community, with its commitment to keep
science "out of politics," had long preferred a multi-member,
executive unit composed of private citizens with professional
scientific backgrounds. This had been the pattern in NACA, the
AEC, and NSF. But NASA was created in terms which the
Bureau of the Budget had long held as a basic principle of
public administration: it was created as a single-executive structure, which placed reliance on generalized administrative competence and limited the plural-member, specialized body to a
purely advisory role. The administrative structure provided in the
bill would be most amenable to clear executive coordination and
control, thereby avoiding problems with vested interests, potential
military domination, and powerful congressional alliances.

But despite its enormous fund-raising success, and despite
the glamour of its projects, NASA has not always been entirely
successful in solving its problems. For one thing, the space program does not have the unequivocal support of the scientific
community. Most scientists had been lukewarm at best to Project
Mercury; the only scientific advice on the project had come from
governmental scientists or special-interest groups—rocketry associations, for example. The moon program was worked out over
a hectic weekend in May 1961 at the Pentagon following Alan
Shepard's successful suborbital flight. In effect, it was a political
response to the Gagarin flight and the Cuban disaster; a diplomatic venture rather than a scientific one. The Department of
Defense and NASA representatives agreed that "national security
or national prestige" in a broad sense was involved, and that the
United States had to do something that appeared dramatically
significant on a world-wide basis. Secretary McNamara wondered
if the human landing on the moon was a big-enough jump, and
he suggested the possibility of a manned planetary exploration as
the American goal. More cautious counsel prevailed, however,
and the moon was settled on as the target.

The plan that was presented to President Kennedy, and
which ultimately led to the manned landing in 1969—an event

that did call worldwide attention to the United States—was not a scientific program at all. "The scientists cringe when you call it science," admitted a NASA administrator. Life scientists, not yet rolling in opulence, are understandably grieved by the attention lavished on physics, but it is not only they who complain. Scientists argue that radio astronomy and observational astronomy have in recent years contributed more to man's knowledge of the universe than the space venture is ever likely to add; and they further point out that most of the money for NASA is spent on development anyway, rather than on basic research; and that questions of much more pressing scientific interest lie in other areas which are neglected because of the preoccupation with space.[22]

Yet at the same time, scientists know that Congress does not regard the expenditures for space science in the same category as those, for example, by the National Science Foundation—an agency more to their liking because of its commitment to "undirected basic research." As Frederick Seitz, president of the prestigious National Academy of Sciences, pointed out in 1966, the most that those scientists who attack the space program can hope to achieve is to have the budget for space science cut by perhaps 20 per cent. Such a cut would simply guarantee that the scientifically most interesting aspects of the work would be eliminated, for the basic science portion of the budget is the most vulnerable. Clearly, such cuts would not produce comparable add-on's to other budgets significant for science.[23] The tendency, therefore, is for scientists to accept the space program as the best they can get.

Given the nature of the decision-making process in American democracy, scientists who are dependent upon government support face the same kind of uncertainty as those in industry. Congressional interest in a particular scientific program is always a matter of reaction; it therefore tends to take on a spasmodic character. The American temperament is more at home with the "crash project" than with sustained efforts over long periods of time. Sometimes, to be sure, the "crash project"—round-the-clock efforts on parallel approaches to an objective, no matter what the cost—can be dramatically effective. The Manhattan project, for example, showed the genuine possibilities in a creative relationship between American government, industry, and the academic community; and the success of that project perhaps explains why men living in its afterglow make the mistake of believing that large amounts of money and effort will automatically overcome even the most stubborn technical obstacles and make up for years of inactivity. There is, therefore, no real drive to support science that is divorced from some presumed crisis. Even with the space program, for example, Congress tends to become lethargic in its support whenever there appears to be a lull in the international space race; it is only aroused to fresh efforts by some external challenge. The drastic cut in NASA's budgetary request for 1970 indicates the basic satisfaction of Congress with the race to the moon. Intervals of reduced support are therefore followed by periods of intense activity, in which frantic attempts are made to compensate for lost time and opportunity. The off-again, on-again nature of the program makes it impossible to provide optimum use of money and manpower, and makes it difficult to engage in the kind of long-range planning essential for the scientific aspects of such a program.

Europeans have often noted the spasmodic nature of American support of science and the lack of any over-all planning, and they note that the waste attendent upon such practices would only be tolerable in a society of plenty. They also point out that even in the United States the method of responding to challenges has serious defects. Not only is there duplication, high cost, and a series of "feudal fiefs which escape the demands of policy," wrote a recent visitor, but research sectors which seem irrelevant to the current challenges are neglected and a serious imbalance occurs. A second unfortunate result, continued the same observer, is that the United States may have "started a certain hardening of the enormous organism." The resources which are committed to current projects are so vast that a change of front would mean serious social and economic upheavals.[24] A massive change of directions because of new possibilities, therefore, is practically out of the question.

While increasing involvement with industry and government has led to both growing affluence and increasing strains within the scientific community, another internal change within science has increased the tendency toward formal organization and also provided potent sources of dissatisfaction as conflicts arise with the older ideology of science. This is the change which has occurred in some areas of research as a result of the development of instruments which can keep many individual research units busy at the same time. The most notable case of this kind has occurred in nuclear physics, where the cyclotron and betatron have become indispensable tools of research. Lee DuBridge, former President of Cal Tech and until recently scientific adviser to President Nixon, offered an early description of the resulting situation. "Several problems," he said, "could be carried on in parallel, and the combined efforts of all groups are needed to keep the machine in operation and to carry on continued improvements." In general, he said, "some of the facilities required for modern work in nuclear physics are so large and so expensive that a large staff is required to operate and make full use of them. . . . I believe it is inevitable that a few great research centers will grow up, and that they will be of greatest importance in the advance of nuclear physics."[25]

As DuBridge noted, the increasing importance of instruments in scientific research has had consequences of the most profound character. Much of the work now being done in science depends absolutely upon the availability of new instruments, which in recent years has depended upon the ready availability of federal funds. One prominent scientist estimated in 1964 that approximately half the funds requested in research proposals were expended in equipment and supplies. New apparatus has led to the creation of whole new fields of research; for example, radio and radar astronomy. The development of a sensitive infrared device capable of detecting a quantum of infrared radiation has made possible new discoveries about the planetary bodies. Space research utilizes recently invented equipment, including launch vehicles, satellites, and instrument packages. Discoveries in nuclear physics are dependent on the accelerators or the detectors; chemistry has become increasingly dependent on instrumentation; biochemistry and molecular biology depend upon such instruments as the analytical ultracentrifuge, radioactive tracers, or the amino acid analyzer. Even the behavioral sciences have felt the impact of new equipment. As the editor of *Science* noted in 1964, some psychology departments were using more electronic equipment than most physics departments used a decade or two before.[26]

The trend toward the increasing dependence on new instruments began during the last three decades of the nineteenth century, when the camera and the spectroscope revolutionized astronomical research. Before that time the traditional work of an observatory had been mapping the sky with exquisite

precision, the preliminary step to reducing all celestial phenomena to the law of gravity. This was the kind of problem-solving characteristic of scientific work after a paradigm had been
accepted; it was assumed that nothing essentially new would
be discovered. It was generally understood that astronomers
sought to say *where* any heavenly body is, and not *what* it is.
The best that one could hope for in that stage of astronomy was
that he might discover a new comet or a new planet which would
be like all the rest, obey the same laws, and have no theoretical
impact on the science. By the end of the century, however, it
was apparent that the days of the old-fashioned descriptive
astronomical observer were gone—the men who occupied themselves with looking through a telescope and making notes were a
vanishing breed. The future of astronomy, many were beginning
to see, lay in applying the methods of physics, especially spectro-
scopy and the photographic plate, to astronomical studies. Stellar
photography and spectrum analysis, both perfected in the
mid-nineteenth century, enabled Langley, Hale and other pioneers
in the New Astronomy of astrophysics to step beyond celestial
surveying and explore the actual composition of the universe. A
photographic plate could record pinpoints of light too faint to be
detected by the human eye. The spectrum of a distant star's light
told its chemical composition with laboratory precision.

Such new tools of research as these stimulated astronomical
investigation, but their full utilization required even more
specially designed instruments such as photographic telescopes,
spectrographs, photometers, and spectroheliographs. The great
telescopes of the 1840's, which had been admirably suited to the
Old Astronomy, gave only mediocre results when adapted to the
New. To finance their new equipment and research the astro-
physicists of the Gilded Age turned with great success to private
philanthropy. The Lick Observatory, endowed as a monument to
himself by an eccentric California millionaire, was the first of
a whole succession of telescopes, each larger than the previous
one. An observatory was a glamorous piece of scientific apparatus; a great dome rising on the skyline would serve as a monument to the donor and a testimony to the cultural achievement
of the community. The thirty-six-inch Lick telescope yielded to
the forty-inch Yerkes, and it in turn to the one-hundred-inch
reflector at Mount Wilson, which Edwin P. Hubble in the 1920's
used to establish the expanding nature of the universe. In 1928,
Cal Tech received a grant of $6,000,000 for construction of a

two-hundred-inch mirror. The Mount Palomar telescope, with which observations were begun in November 1949, has not yet exhausted its potential, but already the radio-telescope is reaching out further into space.

American industrialists, fired by the slogan of bigger and better machines, became the patrons of astronomers to such an extent that by the end of the nineteeeth century they were definitely overstocked with instruments. "We have nearly as many telescopes as exist in all other countries of the world combined," said the Carnegie Institution Advisory Committee in 1902. "The first need of astronomy in this country," the report continued, "is not for more buildings and instruments as for more astronomical workers to use the appliances . . . already provided." Some permanent endowment for salaries and operating costs was provided by the Carnegie Institution; universities provided most of the rest. But the imbalance between men and equipment was not completely adjusted until the coming of large-scale government support after World War II.

At any rate, American preoccupation with the New Astronomy finally made this country the center of the astronomical progress of the world. And since astronomy is definitely not of immediate practical importance, it helped to change the image of American science held by the rest of the world. The main reason for American success has been the willingness of philanthropists to pay the costs of bigger and better telescopes and more magnificent astronomical observatories. Equipped with these colossal instruments, our researchers have been able to gather cosmological data for which no other observatories in the world had been equipped; and with their virtual monopoly of the instruments, they have, until recently, been able to dominate astronomy.

What is true in astronomy is also true in other areas of science. With large and expensive machines, Americans have made notable contributions to such fields as nuclear physics and space research. To foreigners, Americans sometimes seem dependent on their machines. C. H. Waddington, British geneticist, commented after a tour of the United States in 1967 that "quite a number of the American Nobel prize-winners have received their awards for discoveries depending on large machines . . . or the organization of large teams of research associates" rather than for fundamental theoretical breakthroughs.[27]

Problems of all kinds increase as scientists become more

and more dependent upon expensive apparatus for their research. Efficiency is demanded and becomes increasingly important as the investment tied up in equipment increases. Observatories must be built in isolated areas where the number of "seeing hours" is greatest, and the scientists who man them must operate under conditions considered antithetical to their own best interests. They must live and work in a remote, uncivilized area, or, at best, an area offering none of the cultural refinements to which they are accustomed. They must, in most cases, work far from the university atmosphere and the contacts which they feel they should have. Such pressures as these were evident in recent years when the National Radio Astronomy Observatory at Green Bank, West Virginia, was turned into a commuting station manned by a skeleton crew. Planned as a facility to be permanently manned by ninety scientific and technical personnel in 1956, within nine years its headquarters—and all its top scientists—were moved to Charlottesville, Virginia, where more of the amenities of civilization were present.[28] Despite the wishes of scientific statesmen, sites for instruments apparently cannot be successfully chosen on their scientific merits alone. Some compromise with the human factor seems to be demanded.

Connected with the problem of securing maximum efficiency in the use of expensive equipment are the increased organizational problems of a science based on instrumentation. Especially with the more expensive and scarce equipment, there comes a pressing necessity for centralized control and direction. Given the finite number of hours in a day that an instrument can be used, it is often necessary to assign priorities—to decide what is the more worthwhile and what project can be delayed or not entered into at all. For example, in the 1920's one photographic exposure of a tiny piece of the heavens measuring $\frac{1}{10}$ of an inch by $\frac{1}{30}$ of an inch might take up to 75 hours of telescope time; and during this time, of course, the telescope could not be used for other projects. The ideology of science asserts that all unsolved problems are of equal importance, and scientists presently are resisting governmental efforts to have them set priorities,[29] but the ideology is slowly being subverted by the increasing reliance on instruments and the tendency of those instruments to become ever more costly.

There is one further consequence of the dependence on instruments that is seldom noted. It is possible that such dependence, for the most part on commercially produced instruments, has contributed both to the problem-solving orientation and to the stimulation of fad fields in science. If one is bound in his researches by the possibilities built into a piece of equipment, he is almost sure to discover nothing fundamentally new. A new piece of equipment, designed for a particular task, can often produce startling new information—while it is new. Afterwards, work with the machine usually becomes merely a process of further refinement. Thus Michelson's first interferometer led him to a startling new conclusion; further refinements of the instrument and further experiments with it led merely to refinements of his measurements. After the first experiment, the interferometer, in effect, ceased being a tool of basic science and became simply a useful piece of hardware that produced little of theoretical interest.

Whatever the justice of placing the blame on instrumentation, both faddism and limited achievement have been frequently commented upon as characterizing recent science. Or, as one influential scientist put it, science in recent decades has been marked by intense activity and the rapid exhaustion of new, accessible areas of scientific research, which, on the whole, have not been accompanied by marked alterations in the basic structure of most of science. The great activity in high-energy physics, for example, has not altered our concepts of large parts of physics, nor has it much affected our view of phenomena in the range of nucleon-binding energies. Neither has the impact of solid-state physics been great on the theoretical level, although it has led to important practical devices.

Much the same is true of chemistry. The basic principles that guide the chemist have altered only slightly. Progress in gaining detailed comprehension of mechanisms and kinetics of organic chemical reactions has modified, but not revolutionized, the pre-existing structure of knowledge. The noble-gas compounds were a surprising development, which led to frenzied activity for several years in the early 1960's, but discoveries concerning them have not overturned the periodic table of elements, as enthusiasts sometimes predicted they would. Scientists, to put it bluntly, are doing more research, working with more expensive instruments, and publishing more, but they are accomplishing less in some of the classical disciplines than at any time during the last one hundred years.

Accompanying the development of fad fields has been the tendency for the research frontier to exert excessive influence on curricula in the universities. Some universities, for example, which have a strong orientation toward research in nuclear physics tend to emphasize that subject in their graduate teaching. As a result, training in classical physics has suffered and there are recurrent shortages of men so trained. Departments of engineering and branches of geophysics such as meteorology and oceanography complain because their students have little opportunity to obtain the appropriate basic physics training. Industrial and government research organizations report that it is difficult today to find young men who have adequate training in mechanics. As one director of research said, most of the young physicists he sees have trouble proving that a ball will drop.[30]

Another consequence of the narrow specialization in fad fields is early obsolescence. It is more than a witticism when scientists remark that if one has not earned a Nobel prize by the time he is thirty-five, it is too late for him. As the editor of *Science* magazine put it:

> Areas of science which are at the center of the stage at one time are destined to be mined out in a few years. As the mining process nears completion many concern themselves with ever more specialized and trivial aspects. Ultimately they discover that the rest of the world has passed them by, that few others are even slightly interested in what they are doing. They face the need, first of overthrowing deep-seated prejudices and then of acquiring a whole new body of knowledge and techniques. Few succeed. Some turn sour and in effect die intellectually thirty years before they are buried.[31]

A permanent obsolescence is a special danger for those in industry. Teachers have generally been compelled, in order to carry out their teaching function, to retain a certain breadth; it is therefore easier for them to re-tool intellectually than it is for their industrial counterparts. Besides, those with academic tenure can afford to waste two or three years finding their way into a new field. Since the industrial scientist has no such luxury, he has no choice but to continue the problem solving in his now mined-out field. At this point, he has become, in effect, a technician.

Since most scientists have become technicians, it is hardly surprising that much of current activity is devoted to applied research, in which field it has been notably successful. This effort exploits the treasury of fundamental knowledge that has been accumulated over centuries, but it only occasionally—and by accident—produces new knowledge. The fact that little new is being produced is disguised by the tremendous success of the practical effort. The permutations and combinations in which existing konwledge can be assembled and used are innumerable; there is no sign that any area is nearing exhaustion in this respect.

And as science becomes more expensive, there is increasing demand that it be even more practically oriented. Thus in 1966, President Lyndon Johnson called a meeting of National Institutes of Health officials to tell them pointedly that it was now time "to zero in on the targets by trying to get our knowledge fully applied." The billion-dollar-plus agency, he pointed out, was then spending only a small percentage of its budget on clinical research that might lead to new drugs and treatments for human beings.[32] President Johnson was no doubt responding to the Wooldridge Report of the preceding year, which had concluded that the National Institutes of Health research programs had not been well managed and had not produced gratifying results. In a similar vein, Congressman Daddario, head of an influential subcommittee of the House Committee on Science and Astronautics, has called for more attention by the National Science Foundation to the major problems of society: pollution, transportation, and the like. And he periodically recommends changes which will make NSF "more effective in meeting contemporary demands."[33] The Department of Defense, in its celebrated 1966 study, Project Hindsight, also reached conclusions that disturbed the statesmen of science who had propounded the ideology that science pays off best when it is left free to follow its own curiosity. The major theme that emerged from the study of the science and technology embodied in twenty major weapons systems is that the Defense Department's huge investment in basic research has had little direct consequence for advanced weaponry. "A clear understanding of a DOD need," the report unequivocally stated, "motivated 95 per cent of all events."[34]

Because of the increasing pressures toward practicality, a new concept of research has appeared and is threatening to extinguish the older concept of the disinterested search for truth. This is the notion of "mission oriented basic research," or "basic research on a problem." Of course, there is no reason why research on a practical problem must be less basic than research that is of purely scientific interest. One can cite many instances of work on eminently practical problems—Pasteur, the Curies, and many others—which led to theoretical innovations of profound importance. Having a goal in mind is no handicap for a researcher. Yet, when all this is said, it remains true that most purely theoretical problems do not have immediate applicability. And when greater numbers of scientists are concentrating their efforts in the few areas designated "major problems," the audience for research in the interstitial areas of science—areas that may be of the greatest fundamental interest—is likely to be diminished. There is, therefore, a natural snowballing of interest in those areas where research funds are easier to come by and where recognition may more easily be obtained. When the choice of research problems is not dictated by purely scientific criteria, there always exists at least the danger that research will be subordinated to social or political ends, and areas vital to the further progress of the theoretical structure of science may be neglected. This fear appears to be the primary motivation behind much of the outcry against the tendency for an increasing amount of research to be funded by the mission-oriented agencies of government rather than, say, by the National Science Foundation, which to date has the primary mission of furthering science *per se*.

Ironically, the very success of science in twentieth-century America shows signs of threatening its autonomy. "The success of science in a democratic society," observed a recent student, "may be due as much to its being relatively ignored as to the felicitious agreement between the basic values of science and democracy."[35] Certainly it is no longer ignored. A part of the reason for the new attention paid to science is, of course, that it is of immensely greater importance in our lives than at any previous time in human history. The atomic bomb and the hydrogen bomb, for example, have vividly impressed this new importance upon the minds of men. Knowledge that the products of scientific research may very well kill us all forces men to take account of it.

But perhaps equally important—at least in the short run— is the financial aspect, for a big science facility means big money for the area in which it is located. The following analysis of the implications of the 200 BEV accelerator, now being constructed at Weston, Illinois, thirty miles west of Chicago, shows clearly the vast consequences such a facility can have for regional development, and explains why 117 formal proposals from 46 states were received before the site was selected:

> The New accelerator will bring 2,000 of the nation's top scientists and 1,000 supporting technicians, engineers, and chemists, with an annual income of $21 million to the Chicago area.
> It will cost between $300 to $375 million in construction, much of which will go to Chicago area industries.
> A billion dollars in supporting industry will be attracted to the area.
> Industries will be given the advantage of top scientists and technicians.
> It will attract new "science-oriented" industries.
> Added business to manufactures, construction and suppliers can be anticipated.
> The accelerator may have an annual budget of $60 million.
> Retail sales will jump $9 million.
> There will be a need for 90 more retail establishments in the Weston area.
> 1,900 more service industries may be needed.
> A population increase of 60,000 people in the Weston area is foreseen.
> Illinois students will have more opportunities in scientific fields.
> Top science personnel may choose to stay in the Midwest rather than to go to the coasts.

These predictions, gathered from Chicago newspapers in the months before Weston was selected late in 1966, may be slightly overoptimistic, but they probably do not miss the mark by very much.[36] Scientists insist that sites for scientific facilities must be selected strictly in terms of their scientific suitability, but as facilities become more costly and, as a result, more closely associated with regional economic prosperity, there are increasing political pressures for science to compromise some of its demands. When some areas of the country are suffering an economic blight, it seems irresponsible to pour further benefits into the already-wealthy areas. The choice of the Midwest for the site, in fact, may be related to the intensified and broadcast concerns emanating from that area in recent years about equity in the distribution of federal funds for research and development. There has long been a serious imbalance, with the bulk of the research funds going to a few East Coast and West Coast institutions; outcries from the Midwest and other "deprived" areas, have led some scientists to speak ominously of a "Science pork barrel." Glenn Seaborg, then chairman of the Atomic Energy Commission, protested in 1964 that "we must not let our national support of science and technology degenerate to the point where no state—no Congressional district—is complete without a Post Office, a reclamation project and a new science laboratory."[37]

There are now tendencies at work within the science-society relation which suggest that Seaborg's fears may not be unfounded. Basic to the change is the fact that the Cold War is losing its force as an impetus for scientific spending. "Outdoing the Russians" now seems less urgent than the mounting number of domestic problems. With the loss of the Cold War as a rationale, it is beginning to be recognized that the nation will have to come to accept public well-being, rather than national defense, as the principle motive for large-scale support of research and development. This changed rationale will have profound consequences, some of which are already evident. For example, on the public well-being argument, science must compete with other activities for the limited funds available: slum clearance, urban transit, increased social security benefits, and pollution control seem quite as important to many Americans as moon landings and 200 BEV accelerators. Many voices, including some from within the scientific community, have been raised in protest at the imbalance, and the pressures can be expected to increase.

Within science itself, certain changes are already becoming evident. The practice that led to the current domination of research funds by the East and West coasts—that of rewarding existing excellence in the allocation of research contracts—has already come under powerful attack and shows signs of being changed. Under the pressure of the national defense argument, it seemed reasonable to award contracts to those institutions which had a proven capacity to carry out the work most efficiently. When the survival of the society is presumed to be at stake, any other course would be irresponsible. This practice, however, begun under the Office of Scientific Research and Development during World War II, had the effect of making the rich still richer and the poor still poorer, and it had the further effect of bringing a relatively small group of scientists to the high councils of government. But under the public well-being argument, regional economic development and the development of new geographical areas of scientific excellence will have to be given weight. In 1965, a presidential memorandum, generally referred to as the "Share the Wealth Directive," ordered that wide use be made of R&D funds for creating new research capacity. A resolution introduced the following year by Senator Carl T. Curtis (R–Nebraska) requested NSF to recommend changes in existing laws necessary "to provide for a more equitable distribution of [R&D] funds to all qualified institutions of higher learning to avoid the concentration of such activities in any geographical area and to ensure a reservoir of scientific and teaching skills and capacities throughout the several states."

Thus far, the scientific community has met such pleas with cries of anguish and with the insistence that special funds should be allocated for such purposes and none be taken away from existing programs. But the amount of money available for science no longer rises exponentially and cannot be expected to do so again in the future, and the pressures for spreading the wealth are certain to increase. This means that inevitably have-not institutions will be depriving MIT, Harvard, Berkeley, Cal Tech, and the other giants of some of their research funds, and of their influence in the councils of government. The pork barrel is, indeed, on the horizon, but the barrel is now known to have a bottom.

One indicator of the shift is in the 1966 appointments to the National Science Board, top advisory board of NSF. In that year, terms of eight of the twenty-four members expired; their replacements represented IBM, Cornell, the University of Nebraska, Humble Oil, the University of South Carolina, Louisiana State, the University of Minnesota, and Reed College. At about the same time, Samuel M. Nabrit, President of Texas Southern University, was appointed to the five-member Atomic Energy Commission, and Cambridge representation on the eighteen-member Presidential Science Advisory Committee was reduced from a long-standing one-third down to one member.

How far this tendency toward a redistribution of power will go is yet unknown. At any rate, it is clear that as science becomes increasingly politicized, it will increasingly be subjected to shifts in the political winds. And science, in 1971, is far too important not to be politicized in a democratic society. The days when a scientist could plead the purity of his research and claim exemption from the ordinary workings of the social process are gone forever, for that exemption was predicated largely upon his assumed irrelevance to the day-to-day functioning of society. Power, in a democratic society, must bring responsibility, and responsibility means an increasing subjection to political, social, and economic forces. Whether the scientist is working in industry, in government, or in academia the story is more or less the same. He no longer has the choice of being aloof, and his new involvement, while it has brought both prestige and affluence, has also brought unprecedented tensions and problems of an order that the scientist has never before experienced. It is not likely that American science will lose its vigor. What is likely is that it will lose the last vestiges of its autonomy, and that the conditions for the pursuit of science will change even more profoundly in the next decade than they have in the past.

Notes,
Bibliography,
and Index

NOTES

PREFACE

1.   I. Bernard Cohen: "The New World as a Source of Science for Europe," *Congrès International d'Histoire des Sciences*, IX (1959), 96.

CHAPTER I

1.   Margaret T. Hodgen: "Ethnology in 1500: Polydore Vergil's Collection of Customs," *Isis*, LVII (1966), 315.

2.   The best discussion of the reasoning that led Columbus to the New World is in Edmundo O'Gorman: *The Invention of America* (Bloomington, Ind., 1961), Chap. 1. For the general geographical background, see George T. Kimble: *Geography in the Middle Ages* (London, 1938).

3.   José de Acosta: *Historia natural y moral de las Indias . . . ahora fielmente reimpresa de la primera edición* (1st edn., Seville, 1590; Madrid, 1894). Quotations are from the English translation of Edward Grimstone of 1604, Clements R. Markham ed., London, 1880.

4.   O'Gorman: *Invention of America*, p. 58.

5.   Thomas Kuhn: *The Copernican Revolution: Planetary Astronomy in the Development of Western Thought* (Cambridge, Mass., 1957), pp. 146–7.

6.   George Alsop to "My Father," January 17 (1659?) in C. C. Hall, ed.: *Original Narratives of Early American History*, 19 vols. (New York, 1906–17), XI, 378.

7.   Gonzalo Fernández de Oviedo y Valdés: *Natural History of the West Indies*, trans. Sterling A. Stoudemire (Chapel Hill, N.C., 1959), pp. 29–30, 24–5, 109.

8.   Cited in J. D. Bernal: *Science in History* (London, 1954), pp. 277–8.

9.   Cited in Kimble, *Geography in the Middle Ages*, p. 241.

10.   Cited in Lynn Thorndike: *A History of Magic and Experimental Science* (New York, 1941), VI, 275–7.

11.   Richard Eden: *The Decades of the Newe World . . . Wrytten in the Latine Tounge by Peter Martyr of Angleria* (London, 1555), p. 230.

12.   Ibid., pp. 215–16.

13.   Cited in F. L. V. Baumer, ed.: *Main Currents of Western Thought* (New York, 1952), p. 251.

14.   Thorndike: *History of Magic and Experimental Science*, VI, 295–6.

15.   Ibid., p. 272.

16.   Elsa G. Allen: "History of American Ornithology Before Audubon," American Philosophical Society *Transactions*, XLI (1951), Pt. 3.

17.   W. S. C. Copeman: *Doctors and Disease in Tudor Times* (London, 1960), p. 142.

18.   Thorndike: *History of Magic and Experimental Science*, VI, 332–3.

19.   Daniel Gookin, in his "Historical Collections of the Indians in New England," Massachusetts Historical Society *Collections*, II (1793), 142–69, written in 1674, summarizes all the currently held theories concerning the origin of the Indians.

20.   Cited in Louis B. Wright, ed.: *The Elizabethans' America* (Cambridge, Mass., 1965), p. 215.

21.   Custis to Collinson, May 28, 1737, December 5, 1737, July 18, 1738, and December 26, 1738, in E. G. Swem, ed.: "Brothers of the Spade; Correspondence of Peter Collinson, of London, and of John Custis, of Williamsburg, Virginia, 1734–1746," American Antiquarian Society *Proceedings*, LVIII (1948), 60–1, 64, 69, 60–73.

Chapter II

1.   Cartier's journals of his voyages to Canada are in James P. Baxter: *A Memoir of Jacques Cartier* (New York, 1906).

2.   Germán Somolino d'Ardois: *Vida y obra de Francisco Hernández, precidida de España y Nueva España en la época de Felipe II por José Miranda* (Mexico, 1960).

3.   Daymond Turner: "Gonzalo Fernández de Oviedo's *Historia General y Natural*—First American Encyclopedia," *Journal of Inter-American Studies*, VI (1964), 267.

4.   Howard F. Cline: "The *Relaciones Geográficas* of the Spanish Indies, 1577–1586," *Hispanic American Historical Review*, XLIV (1964), 341.

5.   Luis Nicolan d'Oliver: *Fray Bernardino de Sahagún, 1499–1590* (Mexico, 1952).

6.   Higginson's letter was published in Massachusetts Historical Society *Collections*, I (1792); quote, p. 117.

7.   Winthrop Tilley: *The Literature of Natural and Physical Science in the American Colonies from the Beginning to 1765* (Ph.D. dissertation, Brown Univ., 1933), p. 23.

8.   Joseph and Nesta Ewan: "John Banister and His Natural History of Virginia, 1679–1692," *International Congress of the History of Science*, X (1962), 927.

9.   Sarah P. Stetson: "Traffic in Seeds and Plants from England's Colonies in North America," *Agricultural History*, XXIII (1949), 45–56.

10.   Raymond P. Stearns: "The Royal Society of London, Retailer in Experimental Philosophy, 1660–1800," *Dargan Historical Essays* (Albuquerque, N.M., 1952), 39, 49–50.

11.   Robert C. Black, III: *The Younger John Winthrop* (New York, 1966), p. 244.

12.   Fulmer Mood: "John Winthrop Jr. on Indian Corn," *New England Quarterly*, 1937, p. 122.

13.   Eva G. R. Taylor, ed.: *The Original Writings and Correspondence of the Two Richard Hakluyts* (London, 1935), I, 193–5.

14.   Walter Artelt: "The 'Theatrum rerum naturalium Brasiliae' of 1660 of the former Preussische Staatsbibliothek," *International Congress of the History of Science*, X (1962), 925.

15.   William J. Robbins: "French Botanists and the Flora of the Northeastern United States; J. G. Milbert and Élias Durand," American *Philosophical Society Proceedings*, CI (1957), 362–7. William J. Robbins and Mary Christine Howson: "André Michaux's New Jersey Garden and Pierre Paul Saunier, Journeyman Gardener," American *Philosophical Society Proceedings*, CII (1958), 351–70.

16.   Michael Kraus: *The Eighteenth-Century Origins of the Atlantic Civilization* (Ithaca, N.Y., 1949), p. 164.

17.   Martii Kerkkonen: *Peter Kalm's North American Journey: Its Ideological Background and Results* (Helsinki, 1959), p. 71.

18.   Léon Lortie: "La Trame scientifique de l'histoire du Canada," in G. F. G. Stanley, ed.: *Pioneers of Canadian Science* (Toronto, 1966), p. 6. See also Jacques Rousseau: *Jacques Cartier et la "Grosse Maladie"* (Montreal, 1953).

19.   John Morgan: *A Discourse upon the Institution of Medical Schools in America* (Philadelphia, 1765), pp. 52–3.

Chapter III

1.   The best account of the international natural history circle is in E. St. John Brooks: *Sir Hans Sloane* (London, 1954), *passim*.

2.   Robert C. Black, III: *The Younger John Winthrop* (New York, 1966).

3.   Edmund Berkeley and Dorothy Smith Berkeley: *The Reverend John Clayton: A Parson with a Scientific Mind* (Charlottesville, Va., 1965), pp. 96–7.

4.   Joseph and Nesta Ewan: "John Banister and his Natural History of Virginia, 1679–1692," *International Congress of the History of Science*, X (1962), 927–9.

5.   Winthrop Tilley: "The Literature of Natural and Physical Science in the American Colonies from the Beginning to 1765" (Ph.D. dissertation, Brown Univ. 1933), p. 25.

6.   Michael Kraus: *The Eighteenth-Century Origins of the Atlantic Civilization* (Ithaca, N.Y., 1949), p. 164.

7.   J. E. Smith: *A Selection of the Correspondence of Linnaeus and Other Naturalists from the Original Manuscripts*, 2 vols. (London, 1821), I, 34.

8.   Arthur Vallée: *Michel Sarrazin* (Quebec, 1927).

9.   Edmund Berkeley and Dorothy Smith Berkeley: *John Clayton, Pioneer of American Botany* (Chapel Hill, N.C., 1963), p. 84.

10.   Marie Luise Gotheim: *A History of Garden Art*, ed. Walter P. Wright, trans. Mrs. Archer Hind, 2 vols. (London, 1928), I, 445.

11.   *Gentleman's Magazine*, 1756, pp. 278–9.

12.   E. G. Swem, ed.: "Brothers of the Spade: Correspondence of Peter Collinson, of London, and of John Custis, of Williamsburg, Virginia, 1734–1746," *American Antiquarian Society Proceedings*, LVIII (1948), 40.

13.   Raymond P. Stearns: "James Petiver, Promoter of Natural Science, *c.* 1663–1718," *American Antiquarian Society Proceedings*, LX (1952), 262–3, 291.

14.   Ibid., pp. 265–6.

15.   William Darlington: *Memorials of John Bartram and Humphrey Marshall, with Notices of Their Botanical Contemporaries* (Philadelphia, 1849).

16.   Kraus: *Eighteenth-Century Origins*, p. 164.

17.   Berkeley: *The Reverend John Clayton*, xxxiv.

18.   Ibid., pp. 3–5.

19.   Massachusetts Historical Society *Proceedings,* Ser. 1, Vol. XVI, p. 211.

20.   Fothergill to Bartram, October 29, 1768, in the Bartram Papers, Historical Society of Pennsylvania.

21.   François Jean Chastellux: *Travels in North America, in the Years 1780, 1781, and 1782,* trans. with introduction and notes by Howard C. Rice, Jr. (Chapel Hill, N.C., 1963).

22.   Smith: *Correspondence,* I, 316.

23.   Berkeley: *The Reverend John Clayton,* p. 59.

24.   Black: *The Younger John Winthrop,* pp. 309–10.

25.   Conway Zirkle: "Early Records of Plant Hybrids," *Journal of Heredity,* 1932, p. 446.

26.   Frederic R. Kellogg: "A Man and A Method: John Winthrop, Professor of Natural and Experimental Philosophy at Harvard, 1738–1779" (Honors paper, Harvard, 1964).

27.   Daniel Boorstin, in *The Americans: The Colonial Experience* (New York, 1958), pp. 251–9, makes approximately the same point, but he seems inclined to give more credit to the naïveté. The best source for Franklin's electrical experiments—on which this account is largely based, is I. Bernard Cohen: *Franklin and Newton: An Inquiry into Speculative Newtonian Experimental Science and Franklin's Work in Electricity as An Example Thereof,* American Philosophical Society *Memoirs,* XLIII (1956).

28.   Brooke Hindle: "Cadwallader Colden's Extension of the Newtonian Principles," *William and Mary Quarterly,* XIII (October 1956), 459–75, is an excellent account of Colden's effort.

29.   Kenneth B. Murdock: *Increase Mather* (Cambridge, Mass., 1925), pp. 147–8.

30.   *The Letters and Papers of Cadwallader Colden,* in New York Historical Society *Collections,* L (1917), 50; LI (1918), 272.

31.   Brooke Hindle: *The Pursuit of Science in Revolutionary America, 1735–1789* (Chapel Hill, N.C., 1956), pp. 5, 53.

32.   Darlington: *Memorials,* pp. 407, 409, 410, 411.

33.   Swem: "Brothers of the Spade," p. 160.

Chapter IV

1.   Phyllis Allen: "Science in English Universities in the Seventeenth Century," *Journal of the History of Ideas,* X (1949), 219.

2.   For the Hartlib correspondence, see Massachusetts Historical Society *Proceedings,* LXXII, 36–67.

3. Mel Gorman: "Gassendi in America," *Isis*, LV (December 1964), 409–17.

4. All almanacs referred to in this section are at the Massachusetts Historical Society.

5. Francisco Guerra: "Medical Almanacs of the American Colonial Period," *Journal of the History of Medicine and Allied Sciences*, XVI (1961), 234–55. See also his "Medical Colonization of the New World," *Medical History*, VII (1963), 147–54.

6. Samuel Eliot Morison: "Astronomy at Colonial Harvard," *New England Quarterly*, VII (1934), 217.

7. Chester E. Jorgenson: "The New Science in the Almanacs of Ames and Franklin," *New England Quarterly*, VIII (1935), 551–61.

8. The following account of Lee is based on that of Theodore Hornberger: "Samuel Lee (1625–1691), A Clerical Channel for the Flow of New Ideas to Seventeenth-Century New England," *Osiris*, I (1936), 341–55.

9. Otho T. Beall, Jr., and Richard H. Shryock: *Cotton Mather, first significant figure in American medicine* (Baltimore, 1954).

10. *Early Piety Exemplified* (London, 1689), reprinted in Mather's *Magnalia Christi Americana* (Hartford, Conn., 1855), II, 158.

11. Quoted in Hornberger: "Samuel Lee," p. 352.

12. The best source for the philosophical aspects of seventeenth-century science is still E. A. Burtt: *The Metaphysical Foundations of Modern Physical Science*, Anchor edn. (Garden City, N.Y., 1955).

13. Massachusetts Historical Society *Collections*, Ser. 4, Vol. VIII (1868), p. 63.

14. Hancock's commonplace book is at the Houghton Library, Harvard University.

15. Quoted in Theodore Hornberger: "The Date, the Source, and the Significance of Cotton Mather's Interest in Science," *American Literature*, VI (1935), 418–19.

16. *Brontologia Sacra: The Voice of the Glorious God in the Thunder* (London, 1695), p. 5.

17. Cotton Mather: *The Wonderful Works of God Commemorated* (Boston, 1690).

18. Hornberger: "Cotton Mather's Interest in Science," pp. 417–19.

19. Houghton Library, Harvard.

20. The manuscript of *Biblia Americana* is at the Massachusetts Historical Society. For an analysis of one section of it, see Theodore Hornberger: "Cotton Mather's Annotations on the First Chapter of Genesis," *University of Texas Studies in English*, 1938, pp. 112–22.

21. Thomas Rodney to his son, 1808; cited in Simon Gratz: "Thomas Rodney," *Pennsylvania Magazine of History and Biography*, XLV (1921), 61–3.

22.   Winthrop's Commonplace Book is in the Harvard Archives.

23.   Robie's Commonplace Book is at the Massachusetts Historical Society.

24.   John Cotton: *Way of the Congregational Churches Cleared* (London, 1648), p. 42. Samuel Nowell: *Abraham in Arms* (Boston, 1678), p. 11. John Richardson: *The Necessity of A Well Experienced Souldiery* (Cambridge, 1679), p. 6.

25.   Perry Miller: *The New England Mind, The Seventeenth Century* (Cambridge, Mass., 1954), p. 200.

26.   Charles Chauncey: *The Earth Delivered from the Curse* (Boston, 1756), p. 11.

27.   Morton's *Compendium Physicae* has been edited by Theodore Hornberger and published in the Colonial Society of Massachusetts *Publications*, XXXIII.

28.   Theodore Hornberger: "Samuel Johnson of Yale and King's College: A Note on the Relation of Science and Religion in Provincial America," *New England Quarterly*, VIII (1935), 378–97.

29.   The point is made in Clarence L. Ver Steeg: *The Formative Years, 1607–1763* (New York, 1964), p. 208.

30.   Quoted in Morison: "Astronomy at Colonial Harvard," p. 3.

31.   Carl Bridenbaugh, ed.: *Gentleman's Progress: The Itinerarium of Dr. Alexander Hamilton, 1744* (Chapel Hill, N.C., 1948), p. 110.

32.   The best general account of the relation of Puritanism to the New Science is in Miller: *The New England Mind: The Seventeenth Century.*

33.   L. L. Tucker: "President Thomas Clap of Yale College: Another Founding Father of American Science," *Isis*, LII (1961), 55–77.

34.   *The Mystery of Israel's Salvation* (London, 1669), pp. 28–9.

35.   Michael G. Hall: "The Introduction of Modern Science into 17th Century New England: Increase Mather," *Proceedings of the Tenth International Congress of the History of Science* (Paris, 1964), pp. 261–4.

36.   David Rittenhouse: "An Oration Delivered February 24, 1775, before the American Philosophical Society" (Philadelphia, 1775), pp. 14–15.

37.   Commonplace Book, about 1711, n.p., Massachusetts Historical Society.

38.   Garden to Linnaeus, March 15, 1755, in J. E. Smith: *A Selection of the Correspondence of Linnaeus and Other Naturalists from the Original Manuscripts*, 2 vols. (London, 1821), I, 284.

39.   Colden to Linnaeus, February 9, 1748/9, ibid., pp. 452–5.

40.   Linnaeus, as quoted in Philip C. Ritterbush: *Overtures to Biology: The Speculations of Eighteenth-Century Naturalists* (New Haven, Conn., 1964), p. 113.

41.   Garden to Linnaeus, June 3, 1763, in Smith: *Correspondence*, I, 314. Garden to Colden, in New-York Historical Society *Collections*, LIV, 32–3; refers to again, August 14, 1756, in ibid., p. 90.

Chapter V

1.   John Dove: *Strictures on Agriculture* (London, 1770), Preface.

2.   Martti Kerkkonen: *Peter Kalm's North American Journey: Its Ideological Background and Results* (Helsinki, 1959), p. 56.

3.   Edwin T. Martin: *Thomas Jefferson: Scientist* (New York, 1952), pp. 44–5.

4.   Benjamin Franklin: *Complete Works*, John Bigelow ed., 10 vols. (New York, 1887–8).

5.   Brooke Hindle: *David Rittenhouse* (Princeton, 1964) is an excellent study of Rittenhouse.

6.   Brooke Hindle: *The Pursuit of Science in Revolutionary America* (Chapel Hill, N.C., 1956), pp. 107, 108.

7.   Daniel J. Boorstin: *The Lost World of Thomas Jefferson* (New York, 1948), p. 216.

8.   Martin: *Thomas Jefferson*, p. 35.

9.   The agreement is in the Pemberton Papers, XXII, p. 71, at the Historical Society of Pennsylvania.

10.   Martin: *Thomas Jefferson*, p. 35.

11.   Lyman Carrier: *The Beginnings of Agriculture in America* (New York, 1923), pp. 185–6.

12.   Martin: *Thomas Jefferson*, pp. 9–10.

13.   Francis Home: *Principles of Agriculture* (London, 1757), pp. 201–7.

14.   Carrier: *Agriculture in America*, pp. 200–1.

15.   Francis Moore: *A Voyage to Georgia* (London, 1744), pp. 53–5. See also James W. Holland: "The Beginning of Public Agricultural Experimentation in America: The Trustees' Garden in Georgia," *Agricultural History*, XII (1938), 294.

16.   E. St. John Brooks: *Sir Hans Sloane* (London, 1954), p. 203.

17.   J. E. Smith: *A Selection of the Correspondence of Linnaeus and Other Naturalists from the Original Manuscripts*, 2 vols. (London, 1821), I, 443–4, 385–7, 402–19.

18.   Martin: *Thomas Jefferson*, pp. 105–6.

19.   Victor S. Clark: *History of Manufactures in the United States, 1607–1860* (Washington, 1916) gives numerous examples of colonial legislative efforts to promote manufactures.

20.   E. N. Hartley: *Ironworks on the Saugus: The Lynn and Brain-tree Ventures of the Company of Undertakers of the Ironworks in New England* (Norman, Okla., 1957) is a good account of Winthrop's ventures. On trans-Atlantic cross-purposes, see Bernard Bailyn: *The New England Merchants in the Seventeenth Century* (Cambridge, Mass., 1955), pp. 66–9.

21.   Robert C. Black, III: *The Younger John Winthrop* (New York, 1966), pp. 125 6. See also George H. Haynes: "The Tale of Tantiusques," American Antiquarian Society *Proceedings*, new ser., XIV, 471–97.

22.   Benjamin Franklin, London, to Humphrey Marshall, April 22, 1771, in the Gratz Collection, Historical Society of Pennsylvania.

23.   Smith: *Correspondence*, I, 443.

24.   C. E. Hatch, Jr., and T. G. Gregory: "The First American Blast Furnace, 1619–1622: The Birth of a Mighty Industry at Falling Creek in Virginia," *Virginia Magazine of History and Biography*, LXX (1962), 259–96.

25.   Bartram to Garden, March 14, 1756, in William Darlington: *Memorials of John Bartram and Humphrey Marshall, with Notices of Their Botanical Contempories* (Philadelphia, 1849), p. 393.

26.   Quoted in Samuel A. Mitchell: "Astronomy During the Early Years of the American Philosophical Society," American Philosophical Society *Proceedings*, LXXXVI (1942), 16–17.

27.   Whitfield J. Bell: "The Scientific Environment of Philadelphia, 1775–1790," American Philosophical Society *Proceedings*, XCII (1948), 13.

28.   Clark: *History of Manufactures*, p. 217.

CHAPTER VI

1.   For a suggestive account of institutional beginnings, see e.g., George B. Goode: *The Origin of the National Scientific and Educational Institutions of the U.S.A.* (New York, 1890).

2.   Perry Miller: *The Life of the Mind in America* (New York, 1965), p. 274.

3.   Ibid., p. 272.

4.   Brooke Hindle: *The Pursuit of Science in Revolutionary America, 1735–1789* (Chapel Hill, N.C., 1956), pp. 250–1.

5.   Ibid., p. 251.

6.   William M. Smallwood and Mabel S. C. Smallwood: *Natural History and the American Mind* (New York, 1941), p. 286.

7.   Dirk J. Struik: *The Origins of American Science (New England)* (New York, 1957), p. 203.

8.   Ibid., p. 265.

9.   Smallwood: *Natural History*, p. 242.

10.   Christine C. Robbins: "David Hosack's Herbarium and Its Linnaean Specimens," American Philosophical Society *Proceedings*, CIV (1960), 293–313.

11.   John C. Greene: "American Science Comes of Age, 1780–1820," *Journal of American History*, LV (1968), 31–2.

12.   Ibid., p. 32.

13.   For Mitchill's comment, see *The Medical Repository*, VI (1802), 434. For Silliman, see George P. Fisher: *Life of Benjamin Silliman* (New York, 1866), I, 159.

14.   Whitfield J. Bell: "The Scientific Environment of Philadelphia, 1775–1790," American Philosophical Society *Proceedings*, XCII (1948), 13.

15.   Ibid., p. 12.

16.   See, e.g., Richard H. Gaines: "Richmond's First Academy Projected by M. Quesuay de Blaurepaire in 1786," *Collections of the Virginia Historical Society*, II (1891), 168.

17.   John C. Greene: "Science and the Public in the Age of Jefferson," *Isis*, XLIX (1958), 35.

18.   Harris to LeConte, February 18, 1830, in the LeConte Papers, Academy of Natural Sciences of Philadelphia.

19.   The fate of the Lewis and Clark specimens is detailed in Paul R. Cutright: *Lewis and Clark: Pioneering Naturalists* (Urbana, Ill., 1969), Chap. 22.

20.   "A Short Account of the Present State of the College, Academy and Charitable School of Philadelphia," *Columbian Magazine*, V (1790), 275.

21.   Anon.: "A Charge Which Ought to be Delivered to the Graduates in the Arts, in all the Colleges in the United States," *Columbian Magazine*, V (1790), 78–9.

22.   "Observations on botany as applicable to rural economics," in William Darlington: *Memorials of John Bartram and Humphrey Marshall, with Notices of their Botanical Contemporaries* (Philadelphia, 1849), pp. 582–5.

23.   Donald J. D'Elia: "Dr. Benjamin Rush and the American Medical Revolution," American Philosophical Society *Proceedings*, CX (1966), 227–34. For an example of Rush's interest in native remedies, see his "Natural History of Medicine among the Indians," in *Medical Inquiries and Observations Upon the Diseases of the Mind* (Philadelphia, 1812), I, 151.

24. Hindle: *The Pursuit of Science*, p. 238.

25. Bell: "Scientific Environment of Philadelphia," p. 13.

26. D'Elia: "Dr. Benjamin Rush," p. 230.

27. [F. C. Gray]: "American Forest Trees," *North American Review*, XLIV (1837), 361.

28. "American Medicine," *Philadelphia Journal of the Medical and Physical Sciences*, IX (1824), 407, 408.

29. "Dr. Webster on St. Michael and the Azores," *North American Review*, XIV (1822), 50.

30. "Review of Major Long's Second Expedition," *North American Review*, XXI (1825), 178, 179.

31. "Redfield's and Espy's Theories," *New York Review*, VII (1840), 300.

32. "Astronomy," *American Quarterly Review*, III (1828), 319.

33. C. O. Paullin: "Early Movements for A National Observatory, 1802–1842," *Records of the Columbia Historical Society*, XXV (1923), 42–3. Richard Rathbun: "The Columbian Institute for the Promotion of Arts and Sciences," U.S. National Museum *Bulletin*, No. 101 (Washington, 1917), p. 64.

34. "National Institute," *American Whig Review*, II (1845), 238.

35. AAAS *Proceedings*, II (1849), 86.

36. *American Journal of Science*, XVI (1829), note, p. 225.

37. Ethel M. McAlister: *Amos Eaton* (Philadelphia, 1941), p. 296.

38. Dana to Haldeman, March 14, 1848, in the Haldeman Papers, Academy of Natural Sciences of Philadelphia.

39. Struik: *Origins of American Science*, p. 333.

40. Quoted in Bell: "Scientific Environment of Philadelphia," pp. 13–14.

41. Ibid., p. 14.

42. Olive M. Gambrill: "John Beale Bordley and the Early Years of the Philadelphia Agricultural Society," *Pennsylvania Magazine of History and Biography*, LXVI (1942), 439.

43. The preceding material on societies is summarized from my *American Science in the Age of Jackson* (New York, 1968), Chap. 1.

44. H. C. Bolton: "Early American Chemical Societies," *Popular Science Monthly*, LI (1897), 821–6.

45. The preceding material on scientific journals is summarized from my *American Science in the Age of Jackson*, Chap. 1.

46. Greene: "American Science Comes of Age," p. 41.

Chapter VII

1. Brooke Hindle: *The Pursuit of Science in Revolutionary America,* 1735–89 (Chapel Hill, N.C., 1956), p. 358.

2. Ibid., p. 357.

3. Ibid., p. 384.

4. Ibid., p. 223.

5. Donald Fleming: "American Science and the World Scientific Community, *Journal of World History*, Vol. VIII, No. 4 (1965), pp. 669–70.

6. Anon.: "A Charge which Ought to be delivered to the Graduates in the Arts, in all the Colleges in the United States," *Columbian Magazine*, V (1790), 78–9.

7. Whitfield J. Bell: "Some Aspects of the Social History of Pennsylvania," *Pennsylvania Magazine of History and Biography*, LXII (1938), 294.

8. Ibid., p. 294.

9. Edwin T. Martin: *Thomas Jefferson: Scientist* (New York, 1952), p. 218.

10. A. Hunter Dupree: *Science in the Federal Government: A History of Policies and Activities to 1940* (Cambridge, Mass., 1957), p. 21.

11. *The Medical Repository*, II (1815), 259–60.

12. John C. Greene: "Science and the Public in the Age of Jefferson," *Isis*, XLIX (1958), 19–21.

13. Dirk J. Struik: *The Origins of American Science (New England)* (New York, 1957), p. 208.

14. Francis Wayland: *Thoughts on the Present Collegiate System in the United States* (Boston, 1842), p. 154.

15. George Boutwell, as quoted in Samuel Eliot Morison: *Three Centuries of Harvard, 1636–1936* (Cambridge, Mass., 1936), p. 287.

16. Struik: *Origins of American Science*, p. 216.

17. Quoted in Edward Lurie: "Science in American Thought," *Journal of World History*, Vol. VIII, No. 4 (1965), p. 656.

18. Francis Hopkinson, as quoted in Hindle: *Pursuit of Science*, p. 269.

19. Struik: *Origins of American Science*, p. 216.

20. Samuel Tyler: "The Influence of the Baconian Philosophy," *Princeton Review*, XV (1843), 483.

21. D. March: "Physical Science and the Useful Arts in Their Relation to Christian Civilization," *New Englander*, IX (1851), 492.

22. Greene: "Science and the Public," p. 23.

23. G. P. Marsh, Constantinople, to S. F. Baird, August 23, 1850, in William H. Dall: *Spencer Fullerton Baird* (Philadelphia and London, 1915), p. 215.

24. Leonard D. White: *The Jeffersonians: A Study in Administrative History, 1801–1829* (New York, 1951), pp. 247–8.

25. Alexander Dallas Bache: "Replies to a Circular in relation to the occurrence of an unusual Meteoric Display . . .," *American Journal of Science*, XXVIII (1835), 309.

26. *American Journal of Science*, XXXIII (1838), 135.

27. American Philosophical Society *Transactions*, new ser., II (1825), 438.

28. American Association for the Advancement of Science *Proceedings*, I (1849), 273.

29. Dall: *Baird*, p. 191.

30. Dewey is quoted in A. Hunter Dupree: *Asa Gray, 1810–1888* (Cambridge, Mass., 1959), p. 102. For Jefferson, see his letter to Dr. John Manners, February 22, 1814, quoted in full in Edmund H. Fulling: "Thomas Jefferson, His Interest in Plant Life as Revealed in His Writings, II," *Bulletin of the Torrey Botanical Club*, Vol. LXXII, No. 3 (1945), p. 250.

31. American Association for the Advancement of Science *Proceedings*, IV (1851), xli–xlv.

32. Joseph Henry to Dallas Bache, April 16, 1844, August 9, 1838, in the Henry Papers, Smithsonian Institution.

33. Alexander Dallas Bache: "Presidential Address," AAAS *Proceedings*, VI (1851), xliv, xlvi–xlvii.

34. Daniel Vaughn to David A. Wells, February 18, 1857, in the John Warner Papers, American Philosophical Society.

35. AAAS *Proceedings*, IX (1855), 284.

36. Benjamin Silliman, Jr., to Dallas Bache, August 17, 1860, in the Rhees Papers, Huntington Library.

37. John Warner to J. P. Lesley, June 26, 1858, in the Warner Papers, American Philosophical Society; Peirce to John A. LeConte, May 23, 1858, in the LeConte Papers, American Philosophical Society.

38. Oliver Wendell Holmes: "The Positions and Prospects of the Medical Student," in *Currents and Counter Currents in Medical Science* (Boston, 1861), pp. 316–18, 319.

39. These generalizations are based on material in my article, "The Process of Professionalization in American Science: The Emergent Period," *Isis*, LVIII (1967), 157–60.

40.   A. Hunter Dupree: "The Founding of the National Academy of Sciences—A Reinterpretation," *American Philosophical Society Proceedings*, CI (October 1957), 434–40.

CHAPTER VIII

1.   For a good account of the expanding role of government during that period, see the early chapters of A. Hunter Dupree: *Science in the Federal Government: A History of Policies and Activities to 1940* (Cambridge, Mass., 1957).

2.   AAAS *Proceedings*, I (1849), 98.

3.   Nathan Reingold: *Science in Nineteenth-Century America: A Documentary History* (New York, 1964), p. 60. I owe the entire conception of this chapter to Reingold's brief introduction.

4.   The story is elaborated in Dupree: *Science in the Federal Government*, p. 26.

5.   Daniel Boorstin: *The Lost World of Thomas Jefferson* (New York, 1948), pp. 24, 36–40.

6.   George H. Daniels: *American Science in the Age of Jackson* (New York, 1968), p. 26.

7.   Dupree: *Science in the Federal Government*, pp. 99–100.

8.   George R. Taylor: *The Transportation Revolution, 1815–1860* (New York, 1951), p. 94.

9.   William Goetzmann: *Army Exploration in the American West, 1803–1863* (New Haven, Conn., 1959), pp. 201–3.

10.   Ibid., p. 320.

11.   Susan D. McKelvey: *Botanical Exploration of the Trans-Mississippi West, 1790–1850* (Jamaica Plains, Mass., Arnold Arboretum of Harvard Univ., 1955).

12.   Goetzmann: *Army Exploration*, pp. 326–31.

13.   Ibid., pp. 307–8.

14.   Quoted in Goetzmann: *Army Exploration*, p. 343.

15.   Edsel K. Rintala: *Douglas Houghton [1809–45], Michigan's Pioneer Geologist* (Detroit, 1954), pp. 78–82.

16.   Dupree: *Science in the Federal Government*, pp. 56–61.

17.   Ibid., pp. 52–3.

18.   Ibid., pp. 100–4.

19.   All of the state legislative acts relating to geological surveys are printed in George P. Merrill: *Contributions to the History of State*

Geological and Natural History Surveys (Washington, 1920). Unless otherwise indicated, all details on the state surveys are from this volume.

20.   Walter B. Hendrickson: "Nineteenth-Century State Geological Surveys: Early Government Support of Science," *Isis*, LII (September 1961), p. 367.

21.   *Niles Weekly Register*, XLVI (1834), 386.

22.   John D. Wright: "Robert Peter and the First Kentucky Geological Survey," *Kentucky Historical Society Register*, LII (1954), 202, 205.

23.   M. V. Philips to James Hall, 1847, quoted in John M. Clarke: *James Hall of Albany, Geologist and Paleontologist, 1811–1898* (Albany, N.Y., 1923), pp. 175–6.

24.   Ibid., pp. 138, 183–4.

25.   James Hall: "The New York Geological Survey," *Popular Science Monthly*, XXII (1883), 824.

26.   Clarke: *James Hall*, pp. 356–9, 292–3.

27.   Wright: "Robert Peter," p. 206.

28.   Ibid., p. 211.

29.   Clarke: *James Hall*, pp. 79, 112–13.

30.   Ibid., pp. 331–2.

31.   Gerald D. Nash: "The Conflict Between Pure and Applied Science in Nineteenth-Century Public Policy: The California State Geological Survey, 1860–1874," *Isis*, LV (June 1963), 224–5.

32.   Ibid., p. 223.

33.   George H. Daniels: "The Process of Professionalization in American Science: The Emergent Period, 1840–1860," *Isis*, LVIII (1967), 157–8.

34.   Clarke: *James Hall*, p. 529.

CHAPTER IX

1.   Henry D. Rogers: *Address Delivered at the Meeting of the Association of American Geologists and Naturalists, Held in Washington, May, 1844* (New York, 1844), p. 44.

2.   James E. DeKay: *Anniversary Address on the Progress of Natural Sciences in the United States: Delivered before the Lyceum of Natural History of New York, February, 1826* (New York, 1826), p. 70.

3. John M. Clarke: *James Hall of Albany, Geologist and Paleontologist, 1811–1898* (Albany, N.Y., 1923), pp. 38–9.

4. James D. Dana: "Address," AAAS *Proceedings*, IX (1855), 10.

5. Benjamin Silliman: "Address before the Association of American Geologists and Naturalists, Assembled at Boston, April 24, 1842," *American Journal of Science*, XLIII (1842), 248.

6. Quoted in Charles C. Gillispie: *Genesis and Geology: A Study in the Relations of Scientific Thought, Natural Theology and Social Opinion in Great Britain, 1790–1850* (Cambridge, Mass., 1951), p. 100.

7. Ibid., p. 101.

8. George P. Merrill: *The First One Hundred Years of American Geology* (New Haven, 1924), pp. 294–5.

9. Dana: "Address," pp. 10–11.

10. Leonard G. Wilson: "The Emergence of Geology as a Science in the United States," *Journal of World History*, X (1967), 416.

11. Dana: "Address," p. 10.

12. Edward Hitchcock: *Elementary Geology*, 8th edn. (New York, 1847), p. 279.

13. Quoted in Merrill: *American Geology*, p. 16.

14. W. Phillips: "Broughan on Natural Theology," *North American Review*, XLII (1836), 473.

15. J. J. Dana: "The Religion of Geology," *Bibliotheca Sacra*, X (1853), 508–9.

16. J. Cummings: "Divine Agency in Material Phenomena," *Methodist Quarterly Review*, III (1851), 13.

17. Francis Bowen: "Review of Chalmers Theology," *North American Review*, LIV (1842), 359.

18. Anon.: "Utility of Physical Sciences," *The Catholic Expositor and Literary Magazine*, II (1842), 253.

19. Hitchcock: *Elementary Geology* (1847), pp. 284–5.

20. Anon.: "The Religion of Geology," *Bibliotheca Sacra*, LXV (1860), 57.

21. Edward Hitchcock: *The Religion of Geology and Its Connected Sciences* (Boston, 1851), pp. 151–2.

22. Mark Hopkins: "Argument from Nature, for the Divine Existence," *American Quarterly Observer*, II (1834), 306.

23. Ibid., p. 309.

24. Edward Hitchcock: *Elementary Geology* (New York, 1862), pp. 377–80; Hitchcock: *The Religion of Geology and Its Connected Sciences* (Boston, 1860), pp. 146–76, 339–41. For a further discus-

sion, see John A. DeJong: "American Attitudes Toward Evolution Before Darwin" (Ph.D. thesis, Univ. of Iowa, 1962), pp. 131–5.

25. Hitchcock: *Religion of Geology*, p. 521.

26. Ibid., Lecture 6.

27. George P. Fisher: *Life of Benjamin Silliman*, 2 vols. (New York, 1866), II, 136.

28. Hitchcock: *Religion of Geology*, p. 104.

29. George Featherstonehaugh: "Introduction," *Monthly American Journal of Geology and Natural Science*, I (1831), 9.

30. Conrad Wright: "The Religion of Geology," *New England Quarterly*, XIV (1941), 339.

31. Edward Hitchcock: *Reminiscences of Amherst College* (Northampton, Mass., 1863), 291.

32. G. W. M.: "Geology—Its Facts, and Its Inferences," *Universalist Quarterly and General Review*, II (1845), 5–21.

33. A. G.: "A Theory of Creation," *North American Review*, LX (1845), 448.

34. E. P. Barrows: "The Mosaic Six Days and Geology," *Bibliotheca Sacra*, XIV (January 1857), 61.

35. Anon.: "Hayden's Geological Essays," *North American Review*, XII (1821), 135.

36. Anon.: "Geology," *American Quarterly Review*, VI (1829), 103.

CHAPTER X

1. Francis Darwin, ed.: *The Life and Letters of Charles Darwin* (New York, 1897), I, 437–8.

2. For initial reactions, see the introduction to my *Darwinism Comes to America* (Waltham, Mass., 1968).

3. Asa Gray: "Darwin and His Reviewers," *Atlantic Monthly*, (October 1860), p. 407.

4. The best account of the background of evolutionary theory is John C. Greene: *The Death of Adam: Evolution and Its Impact on Western Thought* (Ames, Iowa, 1959).

5. Robin G. Collingwood: *The Idea of Nature* (Oxford, Eng., 1945), p. 34.

6. George P. Merrill: *The First One Hundred Years of American Geology* (New Haven, 1924), pp. 12–15.

7.   Thomas Jefferson: *Notes on the State of Virginia,* ed. with an Introduction and Notes by William Pedey (Chapel Hill, N.C., 1955), p. 19.

8.   For an excellent discussion, see John A. DeJong: *American Attitudes Toward Evolution Before Darwin* (Ph.D. dissertation, Univ. of Iowa, 1962), pp. 82–4.

9.   For representative American reviews of *Vestiges,* see, for example, Tayler Lewis: *American Review,* I (May 1845), 527; Francis Bowen: *North American Review,* LX (April 1846), 295; James D. Dana: *American Journal of Science,* 2nd Ser., Vol. I (March 1846), pp. 250–4; Asa Gray: *North American Review,* LXII (April 1846), 500–1; Albert B. Dod: *Biblical Repertory and Princeton Review,* XVII (October 1845), 533; J. D. Whelpley: *American Review,* III (April 1846), 395; and unidentified authors in *American Methodist Quarterly,* VI (April 1846), 295; *New Englander,* IV (January 1846), 113–27; *Christian Examiner,* XL (1846), 333–49; and *Littell's Living Age,* VI (September 20, 1845), 564.

10.   *Christian Examiner,* XL (1846), 233–4.

11.   S. S. Haldeman: "Enumeration of the Recent Freshwater Mollusca which are Common to North America, with Observations on Species and their Distribution," *Boston Journal of Natural History,* IV (January 1844), 479. Rafinesque, Haldeman and other early American evolutionists are discussed in DeJong: *American Attitudes Toward Evolution.*

12.   Ibid., p. 477.

13.   Ibid.

14.   The following account is from my *Darwinism Comes to America,* pp. xiii–xv.

15.   Quoted in Edward J. Pfeifer: *The Reception of Darwinism in the United States, 1859–1880* (Ph.D. dissertation, Brown Univ., 1957), p. 36.

16.   J. Lawrence Smith: "Presidential Address," AAAS *Proceedings,* XXII (1873), 11.

17.   D. R. Goodwin: "Darwin on the Origin of Species," *American Theological Review,* II (1860), 330–44.

18.   Loren Eisley: *Darwin's Century* (London, 1958) is the best source for the scientific arguments against Darwinism.

19.   Pfeifer: *Reception of Darwinism,* pp. 87–9.

20.   James McCosh: "Religious Aspects of the Doctrine of Development," in P. Schaff and S. Prince, eds.: *History, Essays, Orations, and other Documents of the Sixth General Conference of the Evangelical Alliance* (New York, 1874), p. 270.

Chapter XI

1.   Emma Brace: *The Life of Charles Loring Brace, Chiefly Told in His Own Letters* (New York, 1894), p. 302.

2.   Quoted in George H. Daniels, ed.: *Darwinism Comes to America* (Waltham, Mass., 1968), p. 113.

3.   Quoted in Harvey Wish: *Society and Thought in America* (New York, 1952), II, 327.

4.   John Fiske: "The Destiny of Man Viewed in the Light of His Origin," in *Miscellaneous Writings* (Boston, 1884), IX, 19.

5.   John Fiske: "Evolution and Religion," in *Miscellaneous Writings*, VII, 271.

6.   John Fiske: "The Idea of God as Affected by Modern Knowledge," in *Miscellaneous Writings*, VII, 184.

7.   Woodrow Wilson, *Constitutional Government in the United States* (New York, 1908).

8.   David Starr Jordan: "Science and the Colleges," *Popular Science Monthly*, XLII (1893), 733.

9.   Wayne Dennis: "The Historical Beginnings of Child Psychology," *Psychological Bulletin*, XLVI (1949), 225.

10.   John C. Greene: *Darwinism and the Modern World View* (New York, 1963), pp. 80–4, has a good discussion of Spencer.

11.   Quoted in Greene: Darwinism, pp. 84–5.

12.   Brooks Adams: *The Law of Civilization and Decay* (New York, 1896).

13.   Josiah Strong: *Our Country: Its Possible Future and Its Present Crisis*, new edn., Jurgen Herbst (Cambridge, Mass., 1963), p. 214.

14.   William Z. Ripley: "The European Population of the United States," *Annual Report of the Board of Regents of the Smithsonian Institution, 1909* (Washington, 1910), p. 606.

15.   Cited in Milton Berman: *John Fiske, The Evolution of a Popularizer* (Cambridge, Mass., 1961), p. 245.

16.   Edward A. Ross: *The Old World in the New: The Significance of Past and Present Immigration to the American People* (New York, 1914), p. 97.

17.   Quoted in Samuel P. Hays: *The Response to Industrialism, 1885–1914* (Chicago, 1957), p. 166.

18.   Simons's approach to history is found in *Social Forces in American History* (New York, 1913); Adams's best statement is in *The Law*

of Civilization and Decay (New York, 1897); Morgan's is in *Ancient Society* (New York, 1877).

19.   An excellent account of the new point of view on which the following section is largely based is Robert G. McCloskey: *American Conservatism in the Age of Enterprise, 1865–1910* (New York, 1964).

20.   William Graham Sumner: *The Challenge of Facts and Other Essays* (New Haven, 1914), p. 90; Sumner: *Earth-Hunger and Other Essays* (New Haven, 1913), pp. 351–2; Sumner: *Challenge of Facts,* p. 27.

21.   Henry Adams: *The Education of Henry Adams* (Boston and New York), p. 225.

22.   Robert Wiebe: *The Search for Order, 1877–1920* (New York, 1967), p. 136.

23.   Andrew Carnegie: *Autobiography of Andrew Carnegie* (Boston, 1920), p. 327.

24.   John D. Rockefeller, as quoted in Richard Hofstadter: *Social Darwinism in American Thought* (Boston, 1955), p. 45.

25.   As quoted in Edward C. Kirkland: *Dream and Thought in the Business Community, 1860–1890* (Ithaca, N.Y., 1956), p. 125.

26.   Ibid., p. 130.

27.   Generalizations are drawn from ibid., *passim.*

28.   The point is made by Charles Rosenberg: "Science and American Social Thought," in David D. Van Tassel and Michael G. Hall: *Science and Society in the United States* (Homewood, Ill., 1966), p. 148.

Chapter XII

1.   A. Hunter Dupree: *Science in the Federal Government: A History of Activities and Policies to 1940* (Cambridge, Mass., 1957), pp. 132–3.

2.   Ibid., p. 137.

3.   Ibid., p. 135.

4.   Ibid., pp. 182, 155–6.

5.   F. W. Clarke: "The Chemical Work of the United States Geological Survey," *Science,* XXX (August 1909), 161–71.

6.   David Starr Jordan: "Science and the Colleges," *Popular Science Monthly,* XLII (1893), 721.

7.   Dupree: *Science in the Federal Government,* p. 151.

8.   *Science,* new ser., II (August 2, 1895), 115.

9.   A. Hunter Dupree: "The Founding of the National Academy of Sciences—A Reinterpretation," American Philosophical Society *Proceedings*, CI (October 1957), 434–40.

10.   Benjamin Apthorp Gould: "Presidential Address," AAAS *Proceedings*, XVIII (1869), 30.

11.   Henry A. Rowland: "A Plea for Pure Science," AAAS *Proceedings*, XXXII (1883), 106.

12.   *Popular Science Monthly*, X (1876–7), 86.

13.   J. Lawrence Smith: "Presidential Address," AAAS *Proceedings*, XXII (1873), 18.

14.   Edward Beardsley: *The Rise of the American Chemistry Profession, 1850–1900* (Gainesville, Fla., 1964), p. 33.

15.   E. A. Atkinson: "The Application of Science to the Production and Consumption of Food," AAAS *Proceedings*, XXXIV (1885), 425.

16.   Robert H. Wiebe: *The Search for Order, 1877–1920* (New York, 1967), p. 129.

17.   Samuel Rezneck: "The Emergence of a Scientific Community in New York a Century Ago," *New York History*, XLIII (July 1962), 212.

18.   For quotations, see Charles Rosenberg: "Science and American Social Thought," in David D. Van Tassel and Michael G. Hall: *Science and Society in the United States* (Homewood, Ill., 1966), pp. 154–7.

19.   David Starr Jordan: *Popular Science Monthly*, XLII (1893), 42.

20.   Charles S. Minot, as cited by W. A. Setchell, in "The Baltimore Meeting of the American Society of Naturalists," *Science*, new ser., I (1895), 40.

21.   Rowland: "A Plea for Pure Science," p. 106.

22.   Frank W. Clarke: "American Colleges versus American Science," AAAS *Proceedings*, XVII (1878), 138.

23.   Nicholas Murray Butler: Introduction to Friedrich Paulsen: *The German Universities, Their Character and Historical Development* (New York, 1895), p. xxiii.

24.   C. Hart Merriam: "Biology in our Colleges," *Science*, XXI (1893), 352.

25.   F. H. Getman: *The Life of Ira Remsen* (Easton, Pa., 1840), p. 122.

26.   Gould: "Presidential Address," p. 30.

27.   For references, see George H. Daniels: "The Pure Science Ideal and Democratic Culture," *Science*, CLVI (1967), 1–7.

28.   Alexander Winchell: "Science and the State," *The Forum*, I (1886), 2.

29.   Pinchot Papers, Library of Congress.

30.   S. A. Miller: "Letter," in *Science*, XXI (1893), 67.

31.   Joint Commission to Consider the Present Organization of the Signal Service, Geological Survey, Coast and Geodetic Survey, and the Hydrographic Office of the Navy Department, With a View To Secure Greater Efficiency and Economy of Administration of the Public Service in Said Bureau, *Testimony*, 17 December, 1885; 49th Cong., 1st Sess., Senate Misc. Document 82 (Ser. 2345), p. 693. Hereafter referred to as Allison Commission: *Testimony*.

32.   L. S., "Letter," in *Science*, VII (1886), 12.

33.   "The Consolidation of the Government Scientific Work," *Science*, V (1885), 336.

34.   Carroll Pursell: "Science and Government Agencies," in David D. Van Tassel and Michael G. Hall, eds.: *Science and Society in the United States* (Homewood, Ill., 1966), p. 230.

35.   *Science*, VII (1886), 427.

36.   Simon Newcomb to Walcott Gibbs, December 21, 1887, in the Gibbs Papers, Franklin Institute, Philadelphia.

37.   A. F. Harvey: "Letter," in *Science*, V (1885), 394; editorial comment, p. 393.

38.   W. J. Beal: "Agriculture: Its Needs and Opportunities," AAAS *Proceedings*, XXXII (1883), 280.

39.   B. A. Gould to H. W. Bellows, December 3, 1868, in the Bellows Papers, Massachusetts Historical Society.

40.   Allison Commission: *Testimony*, p. 645.

41.   *Popular Science Monthly*, XXIX (1886), 415; XXVII (1885), 846.

42.   Ouida: "Some Fallacies of Science," *North American Review*, CXLII (1886), 142.

43.   Miller: "Letter," p. 67.

CHAPTER XIII

1.   Samuel P. Hays: *Conservation and the Gospel of Efficiency* (Cambridge, 1959), pp. 207–8.

2.   Nathan Reingold: *Science in Nineteenth-Century America, A Documentary History* (New York, 1964).

3.   In sketching in the general ideological framework, I have profited from reading an unpublished manuscript by John C. Burnham of Ohio State University.

4.   P. G. Nutting: "Organized Knowledge and National Welfare," *Science*, XLVI (1917), 251.

5.   The generalizations on World War I research are based on R. M. Yerkes, ed.: *The New World of Science: Its Development During the War* (New York, 1920); and Grosvenor B. Clarkson: *Industrial America in the World War: The Strategy Behind the Line* (New York, 1923).

6.   G. W. Herrick: "Some Obligations and Opportunities of Scientists in the Up-Building of Peace," *Science*, LII (1920), 99.

7.   G. E. Hale: "Cooperation in Research," *Science*, L (1919), 146–50.

8.   Mark H. Haller: *Eugenics: Hereditarian Attitudes in American Thought* (New Brunswick, N.J., 1963).

9.   Quoted in John C. Burnham: "The New Psychology: From Narcissism to Social Control," in John Brahman, ed.: *Change and Continuity in Modern America* (Columbus, Ohio, 1968), p. 356.

10.   Ibid., p. 358.

11.   W J McGee: "Fifty Years of American Science," *Atlantic Monthly*, LXXXII (1898), 320.

12.   Ira Remsen: "Scientific Investigation and Progress," *Popular Science Monthly*, LXIV (1904), 300; F. H. Getman: *The Life of Ira Remsen* (Easton, Pa., 1940), pp. 61–7.

13.   Thomas C. Chamberlin: "The Function of Scientific Study in A True Education," *Elementary School Teacher*, III (1903), 343.

14.   Simon Newcomb: "The Relation of Scientific Method to Social Progress," *Bulletin of the Philosophical Society of Washington*, IV (1881), 40.

15.   Nutting: "Organized Knowledge," p. 251.

16.   For an excellent brief account of McGee's conservation ideas, see Hays: *Conservation and the Gospel of Efficiency*, pp. 102–9; quotations, p. 124.

17.   E. A. Ross: *Social Control: A Survey of the Foundations of Order* (New York, 1901).

18.   Richard T. Ely: *Social Aspects of Christianity, and Other Essays* (New York, 1889), p. 25.

19.   As quoted in Burnham: "The New Psychology," p. 360.

20.   Abram Lipsky: *Man the Puppet: The Art of Controlling Minds* (New York, 1925), p. 11.

21.   Theodore Roosevelt, as quoted in Hays, *Conservation and the Gospel of Efficiency*, pp. 267–8.

22.   Quoted in Hays: *Conservation and the Gospel of Efficiency*, p. 66.

23.   A. Hunter Dupree: *Science in the Federal Government: A History of Policies and Activities to 1940* (Cambridge, Mass., 1957), p. 239.

24.   Hays: *Conservation and the Gospel of Efficiency*, pp. 129, 123.

25.   Charles Macdonald: "Presidential Address," American Society of Civil Engineers *Proceedings*, XXXIV (August 1908), 34, 254.

26.   Quoted in Dupree: *Science in the Federal Government*, p. 269.

27.   Sir William Roberts: "Science and Modern Civilization," *Nature*, LVI (1897), 623.

28.   Remsen: "Scientific Investigation and Progress," p. 301.

29.   "Legislation and Science," *Popular Science Monthly*, LXIX (1906), 190–1.

30.   Hays: *Conservation and the Gospel of Efficiency*, p. 12.

31.   Samuel Haber: *Efficiency and Uplift: Scientific Management in the Progressive Era, 1890–1920* (Chicago, 1964), p. 48.

32.   See quotations in Hays: *Conservation and the Gospel of Efficiency*, pp. 124–5.

33.   *Science*, IV (1884), 428.

34.   Elizabeth A. Osborne, ed.: *From the Letter Files of S. W. Johnson* (New Haven, Conn., 1913), p. 207.

35.   Oscar E. Anderson, Jr.: *The Health of A Nation: Harvey W. Wiley and the Fight for Pure Food* (Chicago, 1958), p. 127.

36.   Haber: *Efficiency and Uplift*, pp. 61, 53.

37.   Roy Lubove: *The Progressives and the Slums* (Pittsburgh, 1962), p. 239.

38.   Haber: *Efficiency and Uplift*, pp. x, xi, 24.

39.   E. L. Thorndike: "The Nature, Purposes, and General Methods of Measurements of Educational Products," *Seventeenth Yearbook of the National Society for the Study of Education*, (Bloomington, Ill., 1918), Pt. 2, p. 16.

40.   Laurence A. Cremin: *The Transformation of the School* (New York, 1964), pp. 185–9. This section is based largely on Cremin's book.

41.   H. Edwin Mitchell: "Some Social Demands on the Course of Study in Arithmetic," *National Society for the Study of Education Yearbook, 1918*, pp. 7–8.

42.   Haber: *Efficiency and Uplift*, p. 26.

43.   Dupree: *Science in the Federal Government*, p. 288.

44.   Ibid., p. 250.

CHAPTER XIV

1.   Bernard Jaffe: *Men of Science in America* (New York, 1944), p. 372. For a recent account of the series of experiments, see Loyd S. Swenson, Jr.: "The Michelson-Morley-Miller Experiments Before and After 1905," *Journal for the History of Astronomy*, I (1970), 56–78.

2.   William Gilman: *Science: U.S.A.* (New York, 1965), p. 7.

3.   Kendall Birr: "Science in American Industry," in David D. Van Tassel and Michael G. Hall, eds.: *Science and Society in the United States* (Homewood, Ill., 1966), p. 68.

4.   Robert A. Millikin: *Autobiography* (New York, 1950).

5.   U.S. Bureau of the Census: *Statistical Abstract of the United States* (Washington, 1969), pp. 523–36.

6.   Walter Hirsch: "Knowledge for What?", *Bulletin of the Atomic Scientists*, XXI (1965), p. 28.

7.   Spencer Klaw: "The Industrial Labyrinth," *Science and Technology*, No. 86 (February 1969), p. 38.

8.   Ibid., p. 36.

9.   Bernard Barber: *Science and the Social Order* (Glencoe, Ill., 1952), p. 166.

10.   A. Hunter Dupree: *Science in the Federal Government: A History of Policies and Activities to 1940* (Cambridge, Mass., 1957), p. 168.

11.   Donald F. Hornig: "United States Science Policy: Its Health and Future Direction," *Science*, CLXIII (1969), 527–8.

12.   For a perceptive analysis, see D. S. Greenberg: "News and Comment," *Science*, CXLVIII (1965), 1304–5. See also Alan T. Waterman: "Federal Support of Science," *Science*, CLIII (1966), 1359–61.

13.   Dupree: *Science in the Federal Government*, Chap. 18.

14.   *Science*, CLXVI (1969), 583.

15.   Barber: *Science and the Social Order*, pp. 122–3.

16.   Howard S. Miller: "Science and Private Agencies," in Van Tassel and Hall: *Science and Society in the United States*, pp. 218–19.

17.   Millikin: *Autobiography*, pp. 180–4, 212–15.

18.   Ibid., p. 212.

19.   Barber: *Science and the Social Order*, pp. 126–7.

20.   Ibid., p. 128.

21.   Loyd Swenson: *This New Ocean: A History of Project Mercury* (Washington, D.C., 1966), p. 508.

22.   *Science,* CLXIII (1969), 649–54.

23.   Frederick Seitz: "Science and the Space Program," *Science,* CLII (1966), 1720.

24.   Théo Lefèvre, as cited in Philip M. Boffey: "American Science Policy: OECD Publishes a Massive Critique," *Science,* CLIX (1968), 177.

25.   Barber: *Science and the Social Order,* pp. 128–9.

26.   Philip H. Abelson, in *Science,* CXLIX (1964), 219.

27.   C. H. Waddington, quoted in *Science,* CLIX (1968), 177.

28.   *Science,* CL (1965), 722–4.

29.   For a typical statement, see, e.g., Richard D. Alexander's letter in *Science,* CLVII (1967), 135–6.

30.   Philip H. Abelson: "Bigotry in Science," *Science,* CXLIX (1964), 371.

31.   Ibid., p. 371.

32.   *Science,* CLIII (1966), 150.

33.   Emilio Q. Daddario: "A Revised Charter for the Science Foundation," *Science,* CLII (1966), 42.

34.   *Science,* CLIV (1966), 873.

35.   Norman Storer: "The Future of American Science," *Science,* CXXXIX (1963), 465.

36.   The newspaper accounts were gathered and summarized for me by my student Ronald Marose.

37.   Glenn Seaborg, in *Science,* CXLIX (1964), 457. See also Philip H. Abelson: "The Research and Development Pork Barrel," *Science,* CIL (1965), 11.

BIBLIOGRAPHY

Research for this book was primarily based on published sources, although manuscript sources were consulted at the Huntington Library, the Houghton Library (Harvard), the Massachusetts Historical Society, the Pennsylvania Historical Society, the American Philosophical Society, the Franklin Institute, the Academy of Natural Sciences of Philadelphia, and the Library of Congress. In the following bibliography I have listed, by periods, the books and articles that were helpful to me in preparing this volume. Among primary sources, I have listed only the major books. The dates are, of course, approximate; many of the subjects overlap, but I have listed each work only once, in the period where its major emphasis falls.

GENERAL

BATES, RALPH S.: *Scientific Societies in the United States*. New York: John Wiley; 1945.

BERNAL, J. D.: *Science in History*. London: Watts; 1954.

BURTT, EDWIN A.: *The Metaphysical Foundations of Modern Physical Science*. Garden City, N.Y.: Doubleday, Anchor Books; 1955.

CLARK, VICTOR S.: *History of Manufactures in the United States, 1607–1860*. Washington, D.C.: Carnegie Institution; 1916.

DUPREE, A. HUNTER: *Science in the Federal Government: A History of Policies and Activities to 1940*. Cambridge, Mass.: Harvard University Press; 1957.

EVANS, HERBERT M. (ed.): *Men and Movements in the History of Science*. Seattle: University of Washington Press; 1959.

JAFFE, BERNARD: *Men of Science in America*. New York: Simon & Schuster; 1958.

KLOSE, NELSON: *America's Crop Heritage*. Ames, Iowa: Iowa State University Press; 1950.

LURIE, EDWARD: "Science in American Thought," *Journal of World History*, Vol. VIII, No. 4 (1965), pp. 638–65.

SHRYOCK, RICHARD H.: *Medicine and Society in America, 1660–1860*. New York: New York University Press; 1960.

SMALLWOOD, WILLIAM M., and SMALLWOOD, MABEL S. C.: *Natural History and the American Mind*. New York: Columbia University Press; 1941.

TORY, H. M. (ed.): *A History of Science in Canada*. Toronto: Ryerson Press; 1939.

TRUE, A. C.: *A History of Agricultural Experimentation and Research in the United States, 1607–1925*. Miscellaneous Publication 251, U.S. Department of Agriculture. Washington, D.C.: U.S. Government Printing Office; 1937.

VAN TASSEL, DAVID D., and HALL, MICHAEL G. (eds.): *Science and Society in the United States*. Homewood, Ill.: Dorsey Press; 1966.

VERDOORN, FRANZ: *Plants and Plant Science in Latin America*. Waltham, Mass.: Chronica Botanica; 1945.

1500–1749

ACOSTA, JOSÉ DE: *Historia natural y moral de las Indias . . . ahora fielmente reimpresa de la primera edición*. Seville, 1590. Madrid, 1894.

————: *The Natural & Moral History of the Indies, by Father Joseph de Acosta*. Reprinted from the English Translated Edition of Edward Grimstone, 1604, Clements R. Markham, ed. London, 1880.

ALLEN, ELSA G.: "History of American Ornithology Before Audubon," American Philosophical Society, *Transactions*, Vol. XLI, Pt. III (1951), 387–591.

ALLEN, PHYLLIS: "Science in English Universities in the Seventeenth Century," *Journal of the History of Ideas*, 10 (1949), 219.

American Philosophical Society: *Early History of Science in America*. Philadelphia: American Philosophical Society; 1942.

ARTELD, WALTER: "The 'Theatrum rerum naturalium Brasiliae' of 1660 of the former Preussische Staatsbibliothek," *Congrès International d'Histoire des Sciences*, Vol. X, Pt. II (1962), 925–6.

ASHBURN, P. M.: *The Ranks of Death: A Medical History of the Conquest of America*. New York: Coward-McCann; 1947.

ASTUTO, PHILIP LOUIS: "Scientific Expeditions and Colonial Hispanic America," in *Thought Patterns*, VI. Brooklyn, N.Y.: St. John's University Press; 1959.

BAILYN, BERNARD: *The New England Merchants in the Seventeenth Century*. Cambridge, Mass.: Harvard University Press; 1955.

BAKELESS, JOHN: *The Eyes of Discovery*. Philadelphia: Lippincott; 1950.

BARTON, BENJAMIN S.: "Some Account of Mr. John Banister, the Naturalist," *Philadelphia Medical and Physical Journal*, Vol. II, Pt. II (1806), 134–9.

————: Memorandums of the Life and Writings of Mr. John Clayton, the Celebrated Botanist of Virginia," *Philadelphia Medical and Physical Journal*, Vol. II, Pt. I (1806), 139–45.

————: "Memorandums of the Lives and Literary Labours of Mr. William Vernon and Dr. David Krieg," *Philadelphia Medical and Physical Journal*, Vol. II, Pt. II (1806), 139–43.

BELL, WHITFIELD J.: "Medical Practice in Colonial America," *Bulletin of the History of Medicine*, XXXI (September–October 1957), 442–53.

BERKELEY, EDMUND, and BERKELEY, DOROTHY SMITH: *John Clayton, Pioneer of American Botany*. Chapel Hill, N.C.: University of North Carolina Press; 1963.

————: *The Reverend John Clayton: A Parson with a Scientific Mind*. Charlottesville, Va.: University of Virginia; 1965.

BLACK, ROBERT C., III: *The Younger John Winthrop*. New York: Columbia University Press; 1966.

BLANTON, WYNDHAM B.: *Medicine in Virginia in the Seventeenth Century*. Richmond, Va.: William Byrd Press; 1930.

————: *Medicine in Virginia in the Eighteenth Century*. Richmond, Va.: Garrett and Massie; 1931.

BREBNER, J. B.: *The Explorers of North America, 1492–1806*. Garden City, N.Y.: Doubleday; 1955.

BRIDENBAUGH, CARL (ed.): *Gentleman's Progress: The Itinerarium of Dr. Alexander Hamilton, 1744*. Chapel Hill, N.C.: University of North Carolina Press; 1948.

BRIEGER, HENRY F.: "Botanical Prospecting for Ore Deposits, Used in Peru before 1600: Alvaro Alonso Barba," *El Serrano* (Lima), XII (1961), 7–9.

BROOKS, E. ST. JOHN: *Sir Hans Sloane*. London: Batchworth Press; 1954.

BROWNE, SIR THOMAS: *Pseudodoxia Epidemica*. 2nd. edn. London; 1650.

BULLOCK, CHARLES J.: "Life and Writings of William Douglass," *Economic Studies*, Vol. II, No. 5 (1897).

CARRIER, LYMAN: "Dr. John Mitchell, Naturalist, Cartographer, and Historian," American Historical Association, *Annual Report for 1918*, I, 207.

————: *The Beginnings of Agriculture in America*. New York: McGraw-Hill; 1923.

CATESBY, MARK: *Natural History of Carolina, Florida and the Bahamas*. 2 vols. London, 1732–43.

CAWLEY, ROBERT R.: *Unpathed Waters: Studies in the Influence of the Voyagers on Elizabethan Literature*. Princeton, N.J.: Princeton University Press; 1940.

CLINE, HOWARD F.: "The *Relaciones Geográficas* of the Spanish Indies, 1577–1586," *Hispanic American Historical Review*, XLIV (1964), 341.

COHEN, J. BERNARD: "The Compendium Physicae of Charles Morton," *Isis*, XXXIII (1942), 657–67.

COLDEN, CADWALLADER: *The Colden Letter Books*. 12 vols. New-York Historical Society, *Collections*, Vols. ix–x (1877–8).

————: *The Letters and Papers of Cadwallader Colden.* 9 vols. New-York Historical Society, *Collections*, vols. 50–6, 67–8 (1917–23, 1934–5).

COPEMAN, W. S. C.: *Doctors and Disease in Tudor Times.* London: Dawson's; 1960.

DUTTON, RALPH: *The English Garden.* 2nd edn., rev. London and New York: Batsford; 1950.

EDEN, RICHARD: *The Decades of the Newe Worlde or West India, conteynyng the navigations and conquestes of the Spanyardes . . . Wrytten in the Latine Tounge by Peter Martyr of Angleria, and translated into Englysshe by Richard Eden.* London; 1555.

EWAN, JOSEPH, and EWAN, NESTA: "John Banister and His Natural History of Virginia, 1679–1692," Proceedings of the Tenth International Congress of the History of Science, Ithaca (N.Y.) meeting, Pt. 2 (1962), pp. 927–9.

FISHER, M. S.: *Robert Boyle, Devout Naturalist: A Study in Science and Religion in the Seventeenth Century.* Philadelphia: Oshiver Studio Press; 1945.

GANONG, W. F.: *Crucial Maps in the Early Cartography and Place-Nomenclature of the Atlantic Coast of Canada.* Toronto: University of Toronto Press; 1964.

————: *The Description and Natural History of the Coasts of North America (Acadia) by Nicolas Denys.* Toronto: The Champlain Society; 1908.

GARCIA, RODOLFO: "Historia das explorações científicos," *Dicionário Histórico y Geográfico Brasileiro* (Rio de Janeiro), I (1922), 856–910.

GITIN, LEWIS L.: "Cadwallader Colden, as Scientist and Philosopher," *New York History*, XVI (1935), 169.

GORMAN, MEL: "Gassendi in America," *Isis*, LV (1964), 409–17.

GOTHEIN, MARIE LUISE: *A History of Garden Art.* 2 vols. Walter P. Wright ed. Mrs. Archer Hind trans. London: J. M. Dent; 1928.

GUERRA, FRANCISCO: "Medical Almanacs of the American Colonial Period," *Journal of the History of Medicine*, XVI (July 1961), 234–55.

————: "Medical Colonization of the New World," *Medical History*, VII (1963), 147–54.

HALL, MICHAEL G.: "The Introduction of Modern Science into 17th Century New England: Increase Mather." Proceedings of the Tenth International Congress of the History of Science, Ithaca (N.Y.) Meeting (Paris), Pt. 1 (1962), 261–4.

HANKE, LEWIS: "The Dawn of Conscience in America: Spanish Experiments and Experiences with Indians in the New World," American Philosophical Society, *Proceedings*, CVII (1963), 83–92.

————: *Bartolomé de Las Casas: An interpretation of his life and writings.* The Hague: M. Nijhoff; 1951.

HARTLEY, E. N.: *Ironworks on the Saugus: The Lynn and Braintree Venture of the Company of Undertakers of Ironworks in New England.* Norman, Okla.: University of Oklahoma Press; 1957.

HATCH, C. E., JR., and GREGORY, T. G.: "The First American Blast Furnace, 1619–1622: The Birth of a Mighty Industry at Falling Creek in Virginia," *Virginia Magazine of History and Biography,* LXX (1962), 259–96.

HAYNES, GEORGE H.: "The Tale of Tantiusques," *American Antiquarian Society Proceedings,* new ser. XIV, 471–97.

HINDLE, BROOKE: "Cadwallader Colden's Extension of the Newtonian Principles," *William & Mary Quarterly,* Vol. XIII, No. 4 (October 1956), 459–75.

HODGEN, MARGARET T.: "Ethnology in 1500: Polydore Vergil's Collection of Customs," *Isis,* LVII (1966), 315–24.

HOLLAND, JAMES W.: "The Beginning of Public Agricultural Experimentation in America: The Trustees' Garden in Georgia," *Agricultural History,* XII (1938), 294.

HORNBERGER, THEODORE: "Acosta's Historia Natural y Moral de las Indias: A Guide to the Source and the Growth of the American Scientific Tradition," *University of Texas Studies in English* (1939), 3–43.

——: "Samuel Johnson of Yale and King's College: A Note on the Relation of Science and Religion in Provincial America," *New England Quarterly,* VIII (1935), 378–97.

——: "The Date, the Source, and the Significance of Cotton Mather's Interest in Science," *American Literature,* VI (1935), 413–20.

——: "Samuel Lee (1625–1691), a clerical channel for the flow of new ideas to seventeenth-century New England," *Osiris,* I (1936), 341–55.

——: "The Effects of the New Science upon the Thought of Jonathan Edwards," *American Literature,* IX (1937), 196–207.

——: "Puritanism and Science: The Relationship Revealed in the Writings of John Cotton," *New England Quarterly,* X (1937), 503–15.

——: "Cotton Mather's Annotations on the First Chapter of Genesis," *University of Texas Studies in English* (1938), 112–27.

——: "Notes on the Christian Philosopher," in Thomas J. Holmes: *Cotton Mather: A Bibliography,* I, pp. 133–8. Cambridge, Mass.: Harvard University Press; 1940.

——: "The Scientific Ideas of John Mitchell," *Huntington Library Quarterly,* X (1947), 277–96.

JONES, C. K.: "The Transmission and Diffusion of Culture in the Spanish American Colonies," in A. C. Wilgus: *Colonial Hispanic America.* New York: Russell & Russell; 1963.

JORGENSON, CHESTER E.: "The New Science in the Almanacs of Ames and Franklin," *New England Quarterly,* VIII (1935), 551–61.

KELLOGG, FREDERIC R.: "A Man and a Method: John Winthrop, Professor of Natural and Experimental Philosophy at Harvard, 1738–1779." Unpublished Honors Paper, Harvard College, 1964.

KILGOUR, FREDERICK G.: "The Rise of Scientific Activity in Colonial New England," *Yale Journal of Biology and Medicine,* XXII (1949), 123–38.

———: "Thomas Robie (1689–1729), Colonial Scientist and Physician," *Isis,* XXX (1939), 473–90.

KIMBLE, GEORGE H. T.: *Geography in the Middle Ages.* London: Methuen & Co.; 1938.

KITTREDGE, GEORGE L.: "Cotton Mather's Scientific Communications to the Royal Society," *American Antiquarian Society Proceedings,* new ser. XXVI (1916), 18–57.

KUHN, THOMAS: *The Copernican Revolution: Planetary Astronomy in the Development of Western Thought.* Cambridge, Mass.: Harvard University Press; 1957.

LAFITAN, JOSEPH FRANÇOIS: *Moeurs des sauvages amériquains, comparées aux moeurs des premiers temps.* Paris; 1724.

LANNING, JOHN TATE: *Academic Culture in the Spanish Colonies.* London and New York: Oxford University Press; 1940.

LEONARD, IRVING A.: *Don Carlos de Sigüenza y Góngora, A Mexican Savant of the Seventeenth Century.* Berkeley, Calif.: University of California Press; 1929.

MENDELSOHN, EVERETT: "John Lining and His Contribution to Early American Science," *Isis,* LI (1960), 278–92.

MILLER, PERRY: *The New England Mind, The Seventeenth Century.* Cambridge, Mass.: Harvard University Press; 1954.

MOOD, FULMER: "John Winthrop, Jr. on Indian Corn," *New England Quarterly,* X (1937), 120.

MORISON, SAMUEL E.: "Astronomy at Colonial Harvard," *New England Quarterly,* VII (1934), 3–24.

———: *Harvard College in the Seventeenth Century.* 2 vols. Cambridge, Mass.: Harvard University Press; 1936.

MURDOCK, KENNETH B.: *Increase Mather.* Cambridge, Mass.: Harvard University Press; 1925.

O'GORMAN, EDMUNDO: *The Invention of America.* Bloomington, Ind.: Indiana University Press; 1961.

ORNSTEIN, MARTHA: *The Role of Scientific Societies in the Seventeenth Century.* Chicago: University of Chicago Press; 1928.

OVIEDO Y VALDÉS, GONZALO FERNÁNDEZ DE: *Natural History of the West Indies,* Sterling A. Stoudemire trans. Chapel Hill, N.C.: University of North Carolina Press; 1959.

ROUSSEAU, JACQUES: "La Botanique canadienne à l'époque de Jacques Cartier," *Annales de l'ACFAS,* III (1950), 151.

———: "Michel Sarrazin, J. F. Gaulthier et l'étude pré-Linnéenne de la flor canadienne," in *Les Botanistes français en Amérique du nord avant 1850, Collogues internationaux du Centre na-*

tional de la recherche scientifique (Paris), LXIII (1957), 149.

SCHNEER, CECIL: "The Rise of Historical Geology in the Seventeenth Century," *Isis*, XLV (1954), 256–68.

SOMOLINO D' ARDOIS, GERMÁN: *Vida y obra de Francisco Hernández. Precidida de España y Nueva España en la Época de Felipe II, por José Miranda.* 1st edn. Mexico City, D.F.: Universidad Nacional de México; 1960.

STEARNS, RAYMOND P.: "Colonial Fellows of the Royal Society of London, 1661–1788," *Osiris*, VIII (1948), 73–121.

———: "James Petiver, Promoter of Natural Science, c. 1663–1718," *American Antiquarian Society Proceedings*, new ser. LX (1952), 243–365.

———: "The Royal Society of London: Retailer in Experimental Philosophy, 1660–1800," in *Dargan Historical Essays*. Albuquerque, N.M.: University of New Mexico Press; 1952. Pp. 39–54.

STETSON, SARAH P.: "Traffic in Seeds and Plants from England's Colonies in North America," *Agricultural History*, XXIII (1949), 45–56.

SWEM, E. G. (ed.): "Brothers of the Spade: Correspondence of Peter Collinson, of London, and of John Custis, of Williamsburg, Virginia, 1734–1746," *American Antiquarian Society Proceedings*, LVIII (1948), 17–190.

TAYLOR, EVA G. R. (ed.): *The Original Writings and Correspondence of the Two Richard Hakluyts.* 2 vols. London: Hakluyt Society; 1935.

THORNDIKE, LYNN: *A History of Magic and Experimental Science.* Vol. VI. New York: Macmillan; 1941.

TILLEY, WINTHROP: "The Literature of Natural and Physical Science in the American Colonies from the Beginning to 1765." Unpublished Ph.D. dissertation, Brown University, 1933.

TOLLES, FREDERICK B.: *James Logan and the Culture of Provincial America.* Boston: Little, Brown; 1957.

TURNER, DAYMOND: "Gonzalo Fernández de Oviedo's Historia General y natural—First American Encyclopedia," *Journal of Latin-American Studies*, VI (1964), 267.

VALLÉE, ARTHUR: *Michel Sarrazin.* Quebec: Impr. de L. A. Proulx; 1927.

VER STEEG, CLARENCE L.: *The Formative Years, 1607–1763.* New York: Hill & Wang; 1964.

WHITE, G. W.: "Early American Geology," *Science Monthly*, LXXVI (March 1953), 134–41.

WILSON, IRIS HIGBIE: "Scientists in New Spain: The Eighteenth Century Expeditions," *Journal of the West*, I (1962), 24–44.

WRIGHT, LOUIS B. (ed.): *The Elizabethans' America.* Cambridge, Mass.: Harvard University Press; 1965.

———: *Middle-Class Culture in Elizabethan England.* Chapel Hill, N.C.: University of North Carolina Press; 1935.

1750–1845

ABBOT, FREDERICK K.: "The Role of the Civil Engineer in Internal Improvements: The Baldwins, Father and Son, 1776–1838." Unpublished Ph.D. dissertation, Columbia University, 1952.

ADAMS, ALEXANDER B.: *John James Audubon, A Biography.* New York: Putnam's; 1966.

AGRICOLA: "Essays on Agriculture," *American Magazine or Monthly Chronicle* (February 1758), 234–6; (May 1758), 382–7; (July 1758), 486–93; (August 1758), 541–6; (October 1758), 621–3.

American Magazine, 1769.

American Magazine, 1 vol. 1786–8.

American Museum, 12 vols. 1787–92.

BALCH, EDWIN S.: "Arctic Expeditions Sent from the American Colonies," *Pennsylvania Magazine of History and Biography,* XXXI (1907), 420.

BARTRAM, WILLIAM: *Travels through North & South Carolina, Georgia, East and West Florida.* Philadelphia; 1791.

BAUSMAN, O. R., and MUNROE, J. A.: "James Tilton's Notes on the Agriculture of Delaware in 1788," *Agricultural History,* XX (1946), 176–87.

BELL, WHITFIELD J., JR.: *Early American Science: Needs and Opportunities for Study.* Williamsburg, Va.: Institute of Early American History and Culture; 1955.

————: *John Morgan, Continental Doctor.* Philadelphia: University of Pennsylvania Press; 1965.

————: "Philadelphia Medical Students in Europe, 1750–1800," *Pennsylvania Magazine of History and Biography,* LXVII (1943), 1–29.

————: "Science and Humanity in Philadelphia, 1775–1790." Unpublished Ph.D. dissertation, University of Pennsylvania, 1947.

————: "The Scientific Environment of Philadelphia, 1775–1790," American Philosophical Society, *Proceedings,* XCII (1948), 6–14.

————: "Some Aspects of the Social History of Pennsylvania," *Pennsylvania Magazine of History and Biography,* LXII (1938), 281–308.

BERMAN, ALEX: "Social Roots of the 19th Century Botanico-Medical Movement in the United States" in *Actes du VIII Congrès International d'Histoire des Sciences.* Florence and Milan, II (1956), 561–5.

BERRY, ROBERT ELTON: *Yankee Stargazer: The Life of Nathaniel Bowditch.* New York: McGraw-Hall; 1941.

BIDWELL, PERCY W.: "The Agricultural Revolution in New England," *American Historical Review,* XXVI (July 1921), 683–702.

BIGELOW, JOHN (ed.): *The Complete Works of Benjamin Franklin.* 10 vols. New York; 1887–8.

BOLTON, H. C.: "Early American Chemical Societies," *Popular Science Monthly*, V (1897), 819–26.

BOORSTIN, DANIEL J.: *The Lost World of Thomas Jefferson*. New York: Holt; 1948.

BRASCH, FREDERICK E.: "Newton's First Critical Disciple in the American Colonies, John Winthrop," in *Sir Isaac Newton, 1727–1927*. Baltimore: Williams & Wilkins; 1928.

———: "The Newtonian Epoch in the American Colonies," *American Antiquarian Society Proceedings*, new ser. XLIX (1939), 314–32.

BRODERICK, FRANCIS L.: "Pulpit, Physics, and Politics: The Curriculum of the College of New Jersey, 1746–1794," *William & Mary Quarterly*, 3rd ser. VI (1949), 57.

BROWN, MARION E.: "Adam Kuhn: Eighteenth Century Physician and Teacher," *Journal of the History of Medicine*, V (1950), 177.

BUTLER, JUNE RAINSFORD: "America—A Hunting Ground for Eighteenth Century Naturalists with Special Reference to Their Publications about Trees," Bibliographical Society of America, *Papers*, XXXII (1938), 1–16.

BUTTERFIELD, LYMAN H.: "Benjamin Rush as a Promoter of Useful Knowledge," American Philosophical Society, *Proceedings*, XCII (1948), 26–36.

CHAPIN, SEYMOUR L.: "Les Associés libres de l'Académie royale des Sciences," *Revue d'Histoire des sciences et de leurs applications*, Vol. XVIII, No. 1 (1965), p. 7.

CHASTELLUX, FRANÇOIS JEAN: *Travels in North America, in the Years 1780, 1781, and 1782*. Translated by and with introductions and notes by HOWARD C. RICE, JR. Chapel Hill, N.C.: University of North Carolina; 1963.

[CLINTON, DeWITT], HIBERNICUS: *Letters on the Natural History and Internal Resources of the State of New York*. New York; 1822.

COHEN, I. BERNARD: "How Practical Was Benjamin Franklin's Science?" *Pennsylvania Magazine of History and Biography*, LXIX (1945), 284–93.

———: "Science and the Revolution," *Technology Review*, Vol. XLVII, No. 6 (1945), 367–80.

———: "Science in America: The Nineteenth Century," in Arthur M. Schlesinger, Jr. and Morton White (eds.) *Paths of American Thought*. Boston: Houghton Mifflin; 1963.

———: "Some Reflections on the State of Science in America During the Nineteenth Century," National Academy of Sciences, *Proceedings*, XIV (May 1959).

Columbian Magazine, 5 vols. 1786–90.

CRUICKSHANK, HELEN G.: *John and William Bartram's America*. New York: Devin-Adair; 1957.

CUTLER, WILLIAM P., and CUTLER, JULIA P.: *Life, Journals, and Correspondence of Rev. Manasseh Cutler*. 2 vols. Cincinnati; 1888.

CUTRIGHT, PAUL L.: *Lewis and Clark: Pioneering Naturalists.* Urbana Ill.: University of Illinois; 1969.

DANIELS, GEORGE H.: *American Science in the Age of Jackson.* New York: Columbia University Press; 1968.

DARLINGTON, WILLIAM: *Memorials of John Bartram and Humphrey Marshall, with Notices of Their Botanical Contemporaries.* Philadelphia; 1849.

D'ELIA, DONALD J.: "Dr. Benjamin Rush and the American Medical Revolution," American Philosophical Society, *Proceedings,* CX (1966), 227–34.

DEKAY, JAMES E.: Anniversary Address on the Progress of the Natural Sciences in the United States: Delivered before the Lyceum of Natural History of New York, February, 1826. New York; 1826.

DENNY, MARGARET: "Linnaeus and His Disciple in Carolina: Alexander Garden," *Isis,* XXXVIII (1948), 161.

[DEXTER, SAMUEL]: *The Progress of Science.* N.p., 1780.

DOVE, JOHN: *Strictures on Agriculture.* London; 1770.

EDELSTEIN, SIDNEY M.: "The Chemical Revolution in America from the Pages of the 'Medical Repository,'" *Chymia* (1959), 155–79.

FISHER, GEORGE P.: *Life of Benjamin Silliman.* 2 vols. New York; 1866.

GAMBRILL, OLIVE M.: "John Beale Bordley and the Early Years of the Philadelphia Agricultural Society," *Pennsylvania Magazine of History and Biography,* LXVI (1942), 410–39.

GILLISPIE, CHARLES C.: *Genesis and Geology: A Study in the Relations of Scientific Thought, Natural Theology, and Social Opinion in Great Britain, 1790–1850.* Cambridge, Mass.: Harvard University Press; 1951.

GOETZMAN, WILLIAM: *Army Exploration in the American West, 1803–1863.* New Haven: Yale University Press; 1959.

GOODE, GEORGE B.: *The Origin of the National Scientific and Educational Institutions of the U.S.A.* Reprint from papers of the American Historical Association. New York; 1890.

GOODRICH, CARTER: *Canals and American Economic Development.* New York: Columbia University Press; 1961.

———: "Public Spirit and American Improvements," American Philosophical Society, *Proceedings,* XCII (1948), 305–9.

GREENE, JOHN C.: "American Science Comes of Age," *Journal of American History,* LV (1968), 22–41.

———: "Science and the Public in the Age of Jefferson," *Isis,* XLIX (1958), 13–25.

HALL, COURTNEY R.: *A Scientist of the Early Republic: Samuel Latham Mitchill, 1764–1831.* New York: Columbia University Press; 1934.

HASKELL, D. C.: *The United States Exploring Expedition, 1838–1842, and Its Publications, 1844–1874: A Bibliography.* New York: New York Public Library; 1942.

HAYS, DONALD R.: "Douglas Houghton, Chemist," *Michigan History*, L (1966), 341–8.

HENDRICKSON, WALTER B.: "Nineteenth-Century State Geological Surveys: Early Government Support of Science," *Isis*, LII (1961), 357–71.

HINDLE, BROOKE: *David Rittenhouse*. Princeton, N.J.: Princeton University Press; 1964.

———: *The Pursuit of Science in Revolutionary America, 1735–1789.* Chapel Hill, N.C.: University of North Carolina Press; 1956.

JOHNSON, AMANDUS: *The Journal and Biography of Nicholas Collin, 1746–1831.* Philadelphia: Swedish Colonial Foundation; 1936.

KERKKONEN, MARTTI: *Peter Kalm's North American Journey: Its Ideological Background and Results.* Helsinki: Finnish Historical Society; 1959.

KRAUS, MICHAEL: *The Eighteenth Century Origins of the Atlantic Civilization.* Ithaca, N.Y.: Cornell University Press; 1949.

———: "Scientific Relations Between Europe and America in the Eighteenth Century," *Science Monthly*, LV (September 1942), 259–72.

LEMAY, J. A. L.: "Franklin and Kinnersley," *Isis*, LII (1961), 575–81.

MCALLISTER, ETHEL M.: *Amos Eaton*. Philadelphia: University of Pennsylvania Press; 1941.

MCKELVEY, SUSAN D.: *Botanical Exploration of the Trans Mississippi West*. Jamaica Plains, Mass.: Arnold Arboretum; 1955.

MARTIN, EDWIN T.: *Thomas Jefferson: Scientist*. New York: H. Schuman; 1952.

MERRILL, G. P.: *The First One Hundred Years of American Geology*. New Haven: Yale University Press; 1924.

MERZ, JOHN T.: *A History of European Thought in the Nineteenth Century*, I. Edinburgh; 1896.

MILLER, PERRY: *The Life of the Mind in America*. New York: Harcourt, Brace; 1965.

MILLER, SAMUEL: *Brief Retrospect of the Eighteenth Century*. New York; 1803.

MITCHELL, SAMUEL A.: "Astronomy During the Early Years of the American Philosophical Society," American Philosophical Society, *Proceedings*, LXXXVI (1942), 16.

NETTELS, CURTIS D.: *The Emergence of a National Economy, 1775–1815.* New York: Holt; 1962.

NORTH, S. M. D., and NORTH, R.: *Simeon North*. Concord, N.H.: Rumford Press; 1913.

Pennsylvania Magazine. 1775, 1776.

PLATE, ROBERT: *Alexander Wilson, Wanderer in the Wilderness*. New York: McKay; 1966.

RINTALA, EDSEL, K.: *Douglas Houghton* [1809–45], *Michigan's Pioneer Geologist*. Detroit: Wayne State University Press; 1954.

RITTERBUSH, PHILIP C.: *Overtures to Biology: The Speculations of Eighteenth Century Naturalists*. New Haven: Yale University Press; 1964.

ROBBINS, CHRISTINE CHAPMAN: "David Hosack's Herbarium and Its Linnaean Specimens," American Philosophical Society, *Proceedings*, CIV (1960), 293–313.

ROBBINS, WILLIAM J.: "French Botanists and the Flora of the Northeastern United States: J. G. Milbert and Elias Durant," American Philosophical Society, *Proceedings*, CI (1957), 362–7.

ROBBINS, WILLIAM J., and HOUSON, MARY CHRISTINE: "André Michaux's New Jersey Garden and Pierre Paul Saunier, Journeyman Gardener," American Philosophical Society, *Proceedings*, CII (1958), 351–70.

ROGERS, H. D.: *Address Delivered at the Meeting of the Association of American Geologists and Naturalists, Held in Washington, May, 1844*. New York and London; 1844.

RUFUS, WILL CARL.: "Astronomical Observations in the United States Prior to 1848," *Science Monthly*, XIX (1924), 120–39.

SIMPSON, GEORGE G.: "The Discovery of Fossil Vertebrates in North America," *Journal of Paleontology*, 17 (1943), 26–38.

SMITH, C. EARLE, JR.: "Henry Muhlenberg—Botanical Pioneer," American Philosophical Society, *Proceedings*, CVI (1962), 443–60.

SMITH, J. E.: *A Selection of the Correspondence of Linnaeus and Other Naturalists from the Original Manuscripts*. 2 vols. London; 1821.

STANNARD, JERRY: "Early American Botany and Its Sources," *Bibliography of Natural History* (1966), pp. 73–102.

STRUIK, DIRK J.: *The Origins of American Science (New England)*. New York: Cameron Associates; 1957.

TAYLOR, GEORGE R.: *The Transportation Revolution, 1815–1860*. New York: Rinehart; 1951.

TUCKER, LEONARD: "President Thomas Clap of Yale College: Another 'Founding Father' of American Science," *Isis*, LII (1961), 55–77.

WEINER, CHARLES: "Joseph Henry and the Relations Between Teaching and Research," *American Journal of Physics*, 34 (1966), 1093–1100.

WHITE, GEORGE W.: "Lewis Evans' Contributions to Early American Geology, 1743–1755," Illinois Academy of Science, *Transactions*, 44 (1951), 152–8.

———: "Lewis Evans' Early American Notice of Isostasy," *Science*, 114 (1951), 302–3.

WHITE, LEONARD D.: *The Jeffersonians: A Study in Administrative History, 1801–1829*. New York: Macmillan; 1959.

WOOLF, HARRY: *The Transits of Venus: A Study of Eighteenth Century Science*. Princeton, N.J.: Princeton University Press; 1959.

Worcester Magazine, 3 vols. 1786–8.

1846–1900

AGASSIZ, G. R.: *Letters and Recollections of Alexander Agassiz*. Boston: Houghton Mifflin; 1913.

ANDERSON, OSCAR E., JR.: *The Health of a Nation: Harvey W. Wiley and the Fight for Pure Food*. Chicago: University of Chicago Press; 1958.

BAILEY, S. I.: *The History and Work of Harvard Observatory 1839–1927*. New York: McGraw-Hill; 1931.

BATES, R. S.: *The American Academy of Arts and Sciences 1780–1940*. Boston: The American Academy; 1940.

BEARDSLEY, EDWARD H.: *The Rise of the American Chemistry Profession, 1850–1900*. University of Florida Monographs, Social Science, No. 23 (Summer 1964).

BERKELMAN, ROBERT: "Clarence King: Scientific Pioneer," *American Quarterly*, 5 (1953), 301–24.

BERMAN, MILTON: *John Fiske: The Evolution of a Popularizer*. Cambridge, Mass.: Harvard University Press; 1961.

BRACE, EMMA: *The Life of Charles Loring Brace*. New York, 1894.

CHITTENDEN, R. H.: *History of the Sheffield Scientific School*. 2 vols. New Haven: Yale University Press; 1928.

CLARKE, JOHN M.: *James Hall of Albany, Geologist and Paleontologist, 1811–1898*. Albany: John M. Clarke; 1923.

COHEN, I. B.: "Science and the Civil War," *Technology Review*, Vol. 48, No. 3 (January 1946), p. 167.

COULSON, THOMAS: *Joseph Henry: His Life and Work*. Princeton, N.J.: Princeton University Press; 1950.

DALL, WILLIAM HEALEY: *Spencer Fullerton Baird: A Biography, Including Selections from his Correspondence with Audubon, Agassiz, Dana, and Others*. Philadelphia and London: Lippincott; 1915.

DANA, E. S. (ed.): *A Century of Science in America*. New Haven: Yale University Press; 1918.

DANIELS, GEORGE H.: *Darwinism Comes to America*. Waltham, Mass.: Blaisdell; 1968.

———: "The Process of Professionalization in American Science: The Emergent Period, 1840–1860," *Isis*, LVIII (1967), 151–66.

———: "The Pure Science Ideal and Democratic Culture," *Science*, CLVI (1967), 1699–1705.

DEJONG, JOHN A.: *American Attitudes Toward Evolution Before Darwin*. Unpublished Ph.D. thesis, University of Iowa, 1962.

DUPREE, A. HUNTER: "The Founding of the National Academy of Sciences, A Reinterpretation," *American Philosophical Society, Proceedings*, CI (1957), 434–40.

————: *Asa Gray, 1810–1888*. Cambridge, Mass.: Harvard University Press, 1959.

EDENSTEIN, WILLIAM: "The Early Reception of the Doctrine of Evolution in the U.S." *Annals of Science*, 4 (1939), 306–18.

EISLEY, LOREN: *Darwin's Century: Evolution and the Men Who Discovered It*. Garden City, N.Y.: Doubleday; 1958.

ELLEGARD, ALVAR: "The Darwinian Revolution: A Review Article," *Lychnos* (1960–1), 55–85.

FISKE, JOHN: *Miscellaneous Writings*. 12 vols. Boston, 1884.

FLEMING, DONALD: "American Science and the World Scientific Community," *Journal of World History*, Vol. VIII, No. 4 (1965), pp. 666–78.

GETMAN, F. H.: *The Life of Ira Remsen*. Easton, Pa.: Journal of Chemical Education; 1940.

GLICKSBERG, CHARLES L.: "William Cullen Bryant and Nineteenth Century Science," *New England Quarterly*, XXIII (1950), 91–6.

GOODE, G. BROWN: *The Smithsonian Institution, 1846–1896: The History of the First Half Century*. Washington, D.C.; 1897.

GRAY, J. L.: *Letters of Asa Gray*. 2 vols. Boston, 1893.

GREEN, JOHN C.: *The Death of Adam: Evolution and Its Impact on Western Thought*. Ames, Iowa: Iowa State University Press; 1959.

HAAR, CHARLES M.: "E. L. Youmans: A Chapter in the Diffusion of Science in America," *Journal of the History of Ideas*, 9 (1948), 193–213.

HABER, L. F.: *The Chemical Industry During the Nineteenth Century*. Oxford: Clarendon Press; 1958.

HOFSTADTER, RICHARD: *Social Darwinism in American Thought*, rev. edn. Boston: Beacon Press; 1955.

HOLMES, OLIVER WENDELL: *Currents and Counter Currents in Medical Science, with other Addresses and Essays*. Boston; 1861.

————: *Writings*. 14 vols. Boston; 1897–8.

JONCICH, GERALDINE: "Scientists and the Schools of the Nineteenth Century: The Case of American Physicists," *American Quarterly*, 18 (1966), 667–85.

KREUTER, KENT K.: *The Literary Response to Science, Technology and Industrialism: Studies in the Thought of Hawthorne, Melville, Whitman and Twain*. Unpublished Ph.D. thesis, University of Wisconsin, 1963.

LEVERETTE, WILLIAM EDWARD, JR.: *Science and Values: A Study of Edward L. Youmans' Popular Science Monthly, 1872–1887*. Unpublished Ph.D. thesis, Vanderbilt University, 1963.

LOWENBERG, BERT J.: "The Reaction of American Scientists to Darwinism," *American Historical Review*, 38 (1932–3), 687–701.

————: "The Controversy over Evolution in New England, 1859–1893," *New England Quarterly*, VIII (1935), 232–57.

LURIE, EDWARD: *Louis Agassiz: A Life in Science*. Chicago: University of Chicago Press; 1960.

MANNING, THOMAS G.: *Government in Science: The U.S. Geological Survey, 1867–1897.* Lexington: University of Kentucky Press; 1967.

McCLOSKEY, ROBERT G.: *American Conservatism in the Age of Enterprise, 1865–1910.* New York: Harper & Row, Torchbooks; 1964.

MERRILL, G. P.: *Contributions to the History of State Geological and Natural History Surveys.* Bulletin 109. Washington, D.C.: U.S. National Museum; 1920.

MILLER, HOWARD S.: "A Bounty for Research: The Philanthropic Support of Scientific Investigation in America, 1838–1902." Unpublished Ph.D. thesis, University of Wisconsin, 1964.

NASH, GERALD D.: "The Conflict Between Pure and Applied Science in Nineteenth Century Public Policy: The California State Geological Survey, 1860–1874," *Isis,* LV (1963), 217–28.

NEWCOMB, SIMON: *The Reminiscences of an Astronomer.* Boston: Houghton Mifflin; 1903.

PASSER, HAROLD C.: *The Electrical Manufacturers, 1875–1900.* Cambridge, Mass.: Harvard University Press, 1953.

PENNIMAN, T. K.: *A Hundred Years of Anthropology.* London: Duckworth; 1935.

PFEIFER, EDWARD J.: "The Reception of Darwinism in the United States, 1859–1880." Unpublished Ph.D. thesis, Brown University, 1957.

REINGOLD, NATHAN: "Science in the Civil War; the Permanent Commission of the Navy Department," *Isis,* XLIX (1958), 307–18.

———: *Science in Nineteenth Century America, A Documentary History.* New York: Hill & Wang; 1964.

REZNECK, SAMUEL: "The Emergence of a Scientific Community in New York a Century Ago," *New York History,* XLIII (1962), 211–38.

RICE, WILLIAM NORTH: "Scientific Thought in the Nineteenth Century," *Transactions of the Connecticut Academy of Arts and Sciences,* XI (1901–3, Centennial Volume), 12–23.

ROBERTS, WINDSOR HALL: "The Reaction of the American Protestant Churches to the Darwinian Philosophy, 1860–1900." Unpublished Ph.D. dissertation, University of Chicago, 1936.

RODGERS, ANDREW D., III: *American Botany, 1873–1892: Decades of Transition.* Princeton, N.J.: Princeton University Press; 1944.

ROGERS, EMMA (ed.): *Life and Letters of William Barton Rogers.* 2 vols. Boston; 1896.

ROSS, EARLE D.: *Democracy's College: The Land-Grant Movement in the Formative Years.* Ames, Iowa: Iowa State University Press; 1942.

SHALER, N. S.: *Autobiography.* Boston: Houghton Mifflin Co.; 1909.

SHRYOCK, RICHARD H.: "American Indifference to Basic Science During the Nineteenth Century," *Archives Internationales d'Histoire des Sciences.* Vol. 5 (1948), pp. 50–65.

SILLIMAN, BENJAMIN, JR., and GOODRICH, C. R.: *Science and Mechanism, Illustrated by Examples in the New York Exhibition, 1853–54.* New York; 1854.

——: *The World of Science, Art and Industry, Illustrated from Examples in the New York Exhibition, 1853–54.* New York; 1854.

STARR, RICHARD J.: *The Beginnings of Graduate Education in America.* Chicago: University of Chicago Press; 1953.

THOMPSON, ROBERT L.: *Wiring a Continent: A History of the Telegraph Industry in the United States, 1832–1866.* Princeton, N.J.: Princeton University Press; 1947.

United States Congress: *A Memorial of Joseph Henry.* Washington, D.C.; 1880.

WHEELER, JOHN A., and BAILEY, HERBERT S., JR.: "Joseph Henry (1797–1878), Architect of Organized Science," *American Scientist,* 34 (1946), 619–32.

WIEBE, ROBERT H.: *The Search for Order, 1877–1920.* New York: Hill & Wang; 1967.

WILSON, LEONARD G.: "The Emergence of Geology as a Science in the United States," *Journal of World History,* Vol. X (1967), 416–37.

WRIGHT, CONRAD: "The Religion of Geology," *New England Quarterly,* XIV (1941), 335–58.

WRIGHT, JOHN DEAN: "Robert Peter and the First Kentucky Geological Survey," *Kentucky Historical Society Register,* 52 (July 1954), 201–12.

1901–1970

ALPEROVITZ, GAR: *Atomic Diplomacy: Hiroshima and Potsdam, The Use of the Atomic Bomb and the American Confrontation with Soviet Power.* New York: Random House, Vintage Books; 1967.

ASHBY, SIR ERIC: *Technology and the Academics: An Essay on Universities and the Scientific Revolution.* New York: St. Martin's Press; 1963.

BAILEY, JOSEPH C.: *Seaman A. Knapp.* New York: Columbia University Press; 1945.

BARKER, BERNARD: *Science and the Social Order.* Glencoe, Ill.: Free Press; 1952.

BARTLETT, H. R.: "The Development of Industrial Research in the United States," in *Research—A National Resource,* II. Washington, D.C.: U.S. Government Printing Office; 1941.

BIRR, KENDALL: *Pioneering in Industrial Research, The Story of the General Electric Research Laboratory.* Washington, D.C.: Public Affairs Press; 1957.

Bulletin of the Atomic Scientists.

BURNHAM, JOHN C.: "The New Psychology: From Narcissism to Social Control," in *Change and Continuity in Recent American History.* Athens: Ohio University Press; 1969.

———: "Psychiatry, Psychology and the Progressive Movement," *American Quarterly,* XII (1960), 457–65.

BUSH, VANNEVAR: *Science, The Endless Frontier.* Washington, D.C.: U. S. Government Printing Office; 1945.

CLARKSON, GROSVENOR B.: *Industrial America in the World War: The Strategy Behind the Line, 1917–1918.* Boston: Houghton Mifflin; 1923.

CLEVELAND, FREDERIC N.: *Science and State Government.* Chapel Hill, N.C.: University of North Carolina Press; 1959.

COX, DONALD W.: *America's New Policy Makers: The Scientists' Rise to Power.* Philadelphia: Chilton Books; 1964.

CREMIN, LAWRENCE A.: *The Transformation of the School.* New York: Random House, Vintage Books; 1964.

DAVIS, W. M.: "The Progress of Geography in the United States," *Annals of the Association of American Geographers,* 14 (1924), 159–215.

GILMAN, WILLIAM: *Science: U.S.A.* New York: Viking; 1965.

GOODENOUGH, FLORENCE L.: *Mental Testing: Its History, Principles and Applications.* New York: Rinehart; 1949.

GREEN, JOHN C.: *Darwin and the Modern World View.* New York: New American Library, Mentor Books; 1963.

HABER, SAMUEL: *Efficiency and Uplift: Scientific Management in the Progressive Era, 1890–1920.* Chicago: University of Chicago Press; 1964.

HASKINS, CARYL P.: *The Scientific Revolution and World Politics.* New York: Harper & Row; 1964.

HAYS, SAMUEL P.: *Conservation and the Gospel of Efficiency.* Cambridge, Mass.: Harvard University Press; 1959.

———: *The Response to Industrialism, 1885–1914.* Chicago: University of Chicago Press; 1957.

HEYWOOD, CHARLES WILLIAM: *Scientists and Society in the U.S., 1900–1940: Changing Concepts of Social Responsibility.* Unpublished Ph.D. thesis, University of Pennsylvania, 1954.

HORNIG, DONALD F.: "United States Science Policy: Its Health and Future Direction," *Science,* 163 (1969), 523–8.

KARL, BARRY D.: *Executive Reorganization and Reform in the New Deal.* Cambridge, Mass.: Harvard University Press; 1963.

KLAW, SPENCER: *The New Brahmins.* New York: Morrow; 1968.

LAPP, RALPH E.: *The New Priesthood: The Scientific Elite and the Uses of Power.* New York: Harper & Row; 1965.

LASSWELL, HAROLD D.: "The Political Science of Science," *American Political Science Review,* Vol. L, No. 4 (December 1956), pp. 961–79.

LUBOVE, ROY: *The Progressives and the Slums*. Pittsburgh: University of Pittsburgh Press; 1962.

MARCSON, SIMON: *The Scientist in American Industry; Some Organizational Determinants in Manpower Utilization*. Princeton, N.J.: Princeton University Press; 1960.

MILLER, JOHN A.: *Yankee Scientist: William David Coolidge*. Schenectady, N.Y.: Mohawk Development Service; 1963.

MILLIKIN, ROBERT A.: *The Autobiography of Robert A. Millikin*. New York: Prentice-Hall; 1950.

ORLANS, HAROLD: "Federal Expenditures and the Quality of Education," *Science*, 142 (1963), 1625–9.

PRICE, DON K.: *Government and Science*. New York: Oxford University Press; 1962.

PRICE, DON K., *et al.*: "Current Trends in Science Policy in the United States," *Impact of Science on Society*, Vol. X, No. 3 (1960), pp. 187–213.

RABINOWITCH, EUGENE I.: *The Dawn of a New Age; Reflections on Science and Human Affairs*. Chicago: University of Chicago Press; 1963.

REMSEN, IRA: "The Development of Chemical Research in America," *Journal of the American Chemical Society*, XXXVII (1915), 1.

ROSHOLT, ROBERT L.: *An Administrative History of NASA, 1958–1963*. Washington, D.C.: U.S. Government Printing Office; 1966.

SEITZ, FREDERICK: "Science and the Space Program," *Science*, 152 (1966), 1720.

STEWART, IRWIN: *Organizing Scientific Research for War: The Administrative History of the Office of Scientific Research and Development*. Boston: Little, Brown; 1948.

STOCKING, GEORGE W., JR.: "Lamarckianism in American Social Science, 1890–1915," *Journal of the History of Ideas*, 23 (1962), 239–56.

SWENSON, LLOYD S., *et al.*: *This New Ocean: A History of Project Mercury*. Washington, D.C.: U.S. Government Printing Office; 1966.

WEAVER, WARREN: *Scene of Change: A Lifetime in American Science*. New York: Scribner's; 1970.

WRIGHT, HELEN: *Explorer of the Universe: A Biography of George Ellery Hale*. New York: Dutton; 1966.

INDEX

About the Author

George Daniels is presently Professor of the History of Science and Director of the Center for Interdisciplinary Study of Science and Technology at Northwestern University, Evanston, Illinois. Born in Tennessee in 1935, Professor Daniels took his A.B. at East Tennessee State (1959); his M.A. at the University of Iowa (1961); and his Ph.D. at the same university in 1963. He taught at U.C.L.A. before coming to Northwestern in 1964.

A NOTE ON THE TYPE

The text of this book was set in a typeface called Primer, designed by Rudolph Ruzicka for the Mergenthaler Linotype Company and first made available in 1949. Primer, a modified modern face based on Century broadface, has the virtue of great legibility and was designed especially for today's methods of composition and printing.

Primer is Ruzicka's third typeface. In 1940 he designed Fairfield, and in 1947 Fairfield Medium, both for the Mergenthaler Linotype Company.

Ruzicka was born in Bohemia in 1883 and came to the United States at the age of eleven. He attended public schools in Chicago and later the Chicago Art Institute. During his long career he has been a wood engraver, etcher, cartographer, and book designer. For many years he was associated with Daniel Berkeley Updike and produced the annual keepsakes for The Merrymount Press from 1911 until 1941.

This book was composed, printed and bound by Haddon Craftsmen, Inc., Scranton, Pa.

AL